BT COMMUNICATIONS TECHNOLOGY SERIES 9

Mobile and Wireless Communications: Key Technologies and Future Applications

Other volumes in this series:

Mobile and Wireless Communications: Key Technologies and Future Applications

Edited by
Peter Smyth

The Institution of Electrical Engineers

Published by: The Institution of Electrical Engineers, London,
United Kingdom

© 2004: British Telecommunications plc

The Institution of Electrical Engineers,
Michael Faraday House,
Six Hills Way, Stevenage,
Herts. SG1 2AY, United Kingdom

www.iee.org

British Library Cataloguing in Publication Data

A catalogue record for this product is available from the British Library

ISBN 0 86341 368 4

Typeset in the UK by Bowne Global Solutions Ltd, Ipswich, Suffolk
Printed in the UK by T J International, Padstow, Cornwall

CONTENTS

PREFACE

There are few technologies that have had a more profound effect on people's lives than mobile communications. As recently as twenty years ago no one had a mobile phone, while today 1.4 billion men, women and children depend on them. This now exceeds the number of landline users, where it took the preceding one hundred years to reach the 1 billion mark. The ability to make mobile voice calls turns out to be the answer to a deeply felt need across different cultures who simply want to communicate.

Just as the range of modes of Internet usage has expanded over its relatively short life — from e-mail to file transfer, Web browsing, and peer-to-peer — mobile communications has widened from voice to text messaging, to multimedia messaging, to games and to location services. Many of these changes, and the rate of adoption, were not predicted by 'the experts'.

Mobile communications today has reached its third generation (3G). While the first two focused on voice communications, 3G is aimed at bringing together the mobile and Internet industries, giving birth to the concept of 'true mobility' — where the most desirable attributes of fixed line, mobile and data services are linked together into a services mesh, making it far simpler to communicate than in today's separate worlds of wireline, mobile and data, all on different devices, with different features and different experiences.

Although there is great scepticism about 3G at the moment, it will nevertheless provide a mobile communications infrastructure with wide area coverage for voice and data, up to and beyond peak speeds of 384 kbit/s. This, coupled with new types of mobile terminal and access to the Internet, will lead to surprising applications of this new technology over the next decade, equally not predicted by 'the experts'.

While mobile is aimed at ubiquitous coverage, WLANs will provide an increased access speed in the order of several Mbit/s with lower costs than cellular in homes, offices and hot-spots such as those provided by BT Openzone. This technology, currently used for data communications, will move on in the future to support voice, video communications, and mobility across a cluster of WLAN cells: the ubiquitous broadband wireless infrastructure for homes, businesses and public places — true mobility.

Ultimately, users of these new forms of mobile application will decide which network technology is best for them and where they use it. The future is about bringing together these two access technologies, along with other innovations, in areas such as presence, messaging, risk management and content repurposing, into a single terminal and/or a single service — a new form of seamless access.

To understand the future of mobility requires an understanding of a bewildering number of choices, ranging from the different wireline and wireless access technologies, to how these can be confederated in a core, all-IP network with mobility management, quality of service, and the associated billing systems.

Of course, all of this depends on creating economic value. To do this, operators may use the PC industry model for telecommunications in order to simulate a large number of innovative mobile applications. They would use Web Services to access, via APIs, their own intelligent network. On this network, application developers could, for example, find the terminal location, and then use new forms of directory service to locate users' networks, and identify which protocols to use for multimedia communications: effectively an operating system for telecommunications.

I am delighted to recommend the contents of this book, which I am sure will chart the reader through the bewildering number of mobile and wireless communications technologies and choices on the way to reaching a world where telecommunications is truly unplugged, whether it be on a fixed, cellular or WLAN network.

Matt Bross
Chief Technology Officer, BT Group

INTRODUCTION

In his preface, Matt Bross explained that behind the emergence of 3G and WLAN access technologies there is the convergence between the worlds of mobile and the Internet, two of the fastest growing sectors of communications today. This presents not only great opportunities for the future of mobile and wireless telecommunications, but also great challenges in understanding the relative position of different technologies in this future. The aim of this book is to review the contribution of different wireless access technologies to that future, considering, in particular, how these different technologies can be confederated in a core, all-IP network. This edition also reviews the opportunities of opening up access to telecommunications systems, via application programming interfaces (APIs), for third party application developers — mobile Web Services. All of this depends on the economic value created by mobile applications. These are reviewed with an emphasis on the user interface as well as opportunities for video-streaming. Throughout, this book considers the economic and regulatory issues associated with wireless communications, with the final chapter reviewing the history and potential future of mobility from a user perspective.

The first chapter, by Mark Birchler et al, is a joint publication between BT Exact and Motorola that positions the roles of 3G and WLANs in mobility by using scenario modelling of future applications and services with detailed techno-economic modelling of various networks to support these. The chapter also identifies new opportunities in the local loop for photonic networks using radio-on-fibre techniques. This technology is described in detail in the following chapter by Pete Smyth.

In the next chapter, Louise Burness et al chart how WLANs will evolve over the next few years, from today, where they are primarily used for 'best effort' support of data, to the eventual support of a mixture of real-time services and broadband data with quality of service mechanisms.

'It works whenever you work, seamlessly connecting all of your mobile devices. Creating unprecedented productivity ...' is how Bluetooth is described by the Bluetooth Special Interest Group. Steve Buttery and Andy Sago, in their chapter on future applications of Bluetooth, position this technology in terms of other wireless access technologies and the opportunities it brings.

The radio spectrum is a scarce resource and efficient use is of paramount importance for economic value. Ultra-wideband technology has attracted the attention of many people in this regard. It offers the potential for low-cost, hundreds of Mbit/s, transmission systems as well as low power and positioning applications. Regulators are also interested in its potential, since it may avoid the partition problems of spectrum allocation for specific uses for decades to come. Xuanye Gu and Larry Taylor review this exciting technology in Chapter 5.

Traditionally, wireless networks have been deployed with fixed radio access points connected to a backhaul network. It is the responsibility of this radio access point to manage the communications between it and various mobile terminals. An alternative approach is 'opportunity driven multi-access' where individual mobile terminals act together as relay stations for onward transmission to the radio access points, effectively extending the range of coverage. These can be described as either *ad hoc* or symbiotic networks, and they provide exciting opportunities in many applications. Richard Gedge sets the scene with his chapter on *ad hoc* networks, Sverrir Olafsson then looks at their scalability, capacity and local connectivity, while Alan Readhead and Soukhim Trill review this technology for mobility. Clearly this approach may lead to a security issue and this is reviewed by Ben Strulo et al in Chapter 9.

So far we have discussed terrestrial systems in the delivery of mobile multimedia. Michael Fitch, in Chapter 10, reviews the use of satellites for multimedia communications.

The chapter from Dave Wisely and Enric Mitjana is a joint publication between BT Exact and Siemens that describes systems beyond 3G, using IP to combine 3G, wireless LANs, and other access technologies, to offer users much greater flexibility and choice for mobile communications — a future vision of 'oneness'.

Systems beyond 3G may require different economic models from the conventional value chain of the mobile industry. Gabriele Corliano and Kashaf Khan present this argument in the next chapter, on economic tussles in the public mobile access market.

The next part of this edition contains two chapters on the theme of enabling operators to open access to their intelligent network for the deployment of third party applications — providing an operating system for telecommunications. The first, by Martin Yates, does this from the perspective of mobile applications, while Richard Stretch reviews the Parlay Web Services standard.

The next chapter, by Johnny Dixon, deals with radio spectrum regulation and frequency allocation for present and future wireless systems.

Julie Harmer, in her chapter on mobile applications, reviews the new application opportunities for mobile operators, while Simon Ringland and Frank Scahill consider the mobile user interface on small form factor terminals and the future of the wireless user interface — multimodality. Matt Walker et al review BT Exact's video-streaming technology for mobile terminals in Chapter 18.

As Matt Bross has pointed out in his foreword, the ability to make mobile voice calls turns out to answer a deeply felt need across different cultures. Hazel Lacohée et al, in the final chapter, review the social history of mobile and present a futurologist's view of the next decade of mobile communications.

I would like to thank all the authors and reviewers for their valuable contributions to this book, which I hope will provide you with an insight into some of the key technologies and future applications for mobile and wireless communications.

Pete Smyth
Venture Leader for Mobility and Wireless Systems, BT Exact
peter.p.smyth@bt.com

CONTRIBUTORS

B Baker, Wireless Access Technologies, Motorola Labs, USA

M Birchler, Wireless Access Technologies, Motorola Labs, USA

L Burness, Mobility Research, BT Exact, Adastral Park

S Buttery, Market Development, BT Group, Adastral Park

G Corliano, Communications Control, BT Exact, Adastral Park

J S Dixon, Spectrum Management, BT Group, Adastral Park

J Farr, Peer-to-Peer Motivational Research, BT Exact, Adastral Park

M Fitch, Satellite Network Development, BT Exact, Adastral Park

R Gedge, Seamless Wireless Networks, BT Exact, Adastral Park

X Gu, Networks Research, BT Exact, Adastral Park

J A Harmer, Portfolio Management, BT Exact, Adastral Park

D Higgins, Product Development, BT Global Services, Felixstowe

T Jebb, Video Codec Development, BT Exact, Adastral Park

K Khan, Communications Control, BT Exact, Adastral Park

H Lacohée, Regional Manager, BT Retail

G Martinez, Composite Radio Development, Motorola Labs, France

E Mitjana, Mobile Internet, Siemens, Germany

M Nilsson, Mobile Video Systems, BT Exact, Adastral Park

S Olafsson, Networks Research, BT Exact, Adastral Park

I Pearson, Futurologist, BT Exact, Adastral Park

A Readhead, Business Systems, BT Exact, Adastral Park

S P A Ringland, Mobile Internet, BT Exact, Adastral Park

A Sago, Wireless Solutions Design, BT Exact, Adastral Park

F J Scahill, Mobile Internet, BT Exact, Adastral Park

A Smith, Software Research, BT Exact, Adastral Park

P P Smyth, Mobility and Wireless Systems, BT Exact, Adastral Park

R Stretch, Network Intelligence, BT Exact, Adastral Park

B Strulo, Software Research, BT Exact, Adastral Park

L Taylor, WLAN Consultant

P Thorpe, Multimedia Design, BT Exact, Adastral Park

S Trill, Business Finance, BT Exact, Adastral Park

R Turnbull, Multimedia Streaming, BT Exact, Adastral Park

N Wakeford, Sociology and Social Methodology, University of Surrey

M D Walker, Multimedia Streaming, BT Exact, Adastral Park

D Wisely, Seamless Mobility Research, BT Exact, Adastral Park

M J Yates, Research Programme Manager, BT Exact, Adastral Park

1

FUTURE OF MOBILE AND WIRELESS COMMUNICATIONS

M Birchler, P P Smyth, G Martinez and M Baker

1.1   Introduction

One of the primary challenges in addressing a broad issue such as 'the future of wireless communications' is the distributed nature of relevant information. As the wireless communications industry spans many players, including equipment vendors, operators, content providers and backhaul network providers, to name just a few, to have any hope of generating an accurate assessment, the relevant information from these players must be identified, integrated and converted into a coherent framework. The likelihood of any single player having access to a critical mass of this information is exceedingly low. Thus, the need to develop collaborative relationships across industry players is paramount.

1.2   Enabling Fruitful Cross-Company Collaboration

In our case, combining the intellectual capital of BT Exact and Motorola Labs provided an exceptional opportunity to create the critical mass required for such an endeavour. However, there remained the significant issue of just how we could effectively organise and execute on such a joint effort.

The vehicle selected upon which to base our collaboration is called the Strategic Landscape Process. This process provided two essential elements for a successful joint effort:

- a well-defined framework for identification of the problem space;

- an evaluation tool that provides an integrated business/technology value modelling capability.

The key dimensions of the landscape framework are the potential network types and dominant applications. The framework's focus is to generate connectivity between technological solutions and the market value that they could create. The existence of this common framework allowed for a focused, productive discussion on the specific problem that we could address together.

The availability of a common business model also played a key role in enabling a productive collaboration. Discussions on work partitioning and definition were made significantly more efficient through use of this common vehicle. In addition, we were able to work independently with the knowledge that the resulting information could be easily conveyed and validated by the other party. Finally, this common model supported the generation of consensus results.

1.3 Techno-Economic Comparisons

1.3.1 Background

The level of uncertainty associated with predicting the composition of next-generation wireless systems has rarely, if ever, been higher than it is now. The primary reason for this situation is uncertainty surrounding the services that will drive future wireless usage growth. In addition, the exponential growth of the Internet has challenged the prevailing understanding of network organisation and ownership. And finally, the increasing use of portable computing, organising and communications devices is forcing a reappraisal of the form factors and capabilities of future wireless subscriber units.

Our work has sought to respond to this situation by the creation of a techno-economic framework that supports the identification and evaluation of potential future scenarios. We have proposed both a fundamental scenario space and the set of key strategic issues that must be addressed in order to evaluate future system viability within this space. Our goal has not been to predict the future. Rather, we have sought to envision the set of potential futures and wireless systems that may come into being. By so doing, we have enabled evaluation of the fit between future conditions and systems. The desired end result has been to understand which system configurations provide the greatest value for which potential futures.

This effort has been based on the principle that only those who envision, measure and understand the characteristics of numerous potential futures can win in the actual future. Failure to successfully execute on this strategic level will inexorably lead to a continuous string of surprises and missed opportunities.

The key requirements for effective strategy generation that have driven this process development are:

- front-end connectivity between business and technology development processes;
- 'big picture' issues identified and addressed;
- clear, common framework for information generation and evaluation;
- enable 'apples-to-apples' comparison of systems and potential future scenarios;

- create a common language between technologists and strategic marketers;
- integration into corporate strategy process.

The following section provides a description of the process utilised to fulfil the above described strategy requirements.

1.3.2   The Strategic Landscape Process

The Strategic Landscape Process has been used to create a context within which the evolution of wireless solutions can be effectively investigated. The high level of uncertainty and complexity introduced by the emergence of new solutions, such as wireless LANs (WLANs), necessitated the creation of this novel investigative approach. Our primary goal has been to generate connectivity between technological solutions and the market value that they could create. The 'landscape' defined to support this goal is shown in Fig 1.1.

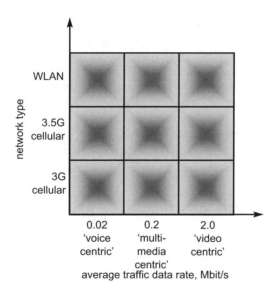

Fig 1.1   Strategic landscape investigation framework.

The key dimensions of this landscape are the potential network types and dominant applications. Three application scenarios were considered:

- a voice-centric world, which focuses on typical cellular usage today for voice and low-speed messaging/data, with an average traffic data rate of 20 kbit/s per user;
- a multimedia-centric world, where there is now also significant demand for mobile applications such as high-speed file transfer, higher quality for voice, music and video-streaming services as well as mobile videoconferencing to

terminals with small screens, with an average traffic data rate of 200 kbit/s per user;

- a video-centric world, which supports high-definition video on mobile terminals with larger displays, as well as all the other applications of the other scenarios with even greater download speeds, with an average traffic data rate of 2 Mbit/s per user (see Chapter 16).

At any one time, depending on the number of users per cell and the distribution of application types, instantaneous data rates may be higher than the average traffic data rate. Web browsing is a typical example that for the user would offer apparently significantly higher peak data rates than the long-time average.

In addition to the assumption of an average traffic rate, a unique tariff model was developed for each scenario. Assuming the same traffic models[1], the monthly average revenue per user (ARPU) spend is $29 for the voice-centric scenario, $45 for the multimedia-centric scenario and $106 for the video-centric scenario. The tariff for multimedia services also includes the voice services, and likewise the tariff for video also includes the multimedia and voice services.

It is even possible to see these scenarios as an evolution from voice services today to high-speed multimedia in the future in which case the techno-economic modelling will help to establish a viable technology roadmap. Motorola Labs [1] has collaborated with BT Exact to generate an integrated business/technology model focused on predicting operator value for each network type across the dominant applications. This 'landscape' model includes all significant capital and operations system costs. Technical performance information such as capacity and range are integrated into the cost model through their impact on the number of required sites.

A high-level block diagram of the 'landscape' business/technology model is shown in Fig 1.2.

1.3.3   Techno-Economic Studies

The following material represents the results of exploratory research rather than a statement of either BT's or Motorola's technical or business position.

1.3.3.1   WLANs and Cellular Networks

One specific study conducted jointly by BT Exact and Motorola Labs assumed use of cellular 'micro-cells' and WLAN 'pico-cells' to provide blanket coverage for a dense urban deployment scenario (e.g. London's financial district). For cellular, both the standard UMTS system (3G, Release '99) and the enhanced 3G technology

---

[1] The model assumes 3% busy hour utilisation, resulting in 2.16, 21.6 and 216 Mbit/s per busy hour delivered to the user for voice, multimedia and video, respectively.

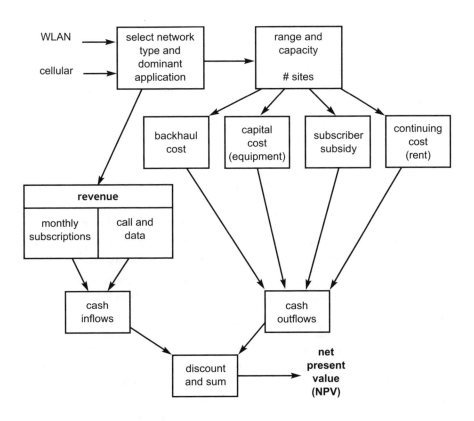

Fig 1.2 'Landscape' business model block diagram.

being developed in Motorola Labs (3.5G) were modelled. For WLAN, the equivalent of IEEE802.11a was modelled (see Chapter 3 for more details).

- Technical analysis

    The technical performance parameter that most strongly influences business results is capacity. Motorola Labs conducted system level simulations for all three network types using common propagation models and coverage criteria. In addition, range and equipment cost data was generated to support the business modelling.

- Business modelling

    The business results for a study of this type are targeted on understanding the relative value proposition across different technologies rather than predicting absolute value in a totally realistic sense. For example, we did not attempt to model all of the competitive factors that will influence the absolute value of a specific deployment.

The key conclusions of this study are:

- WLANs could be a cost-effective complement to 3G systems in urban blanket deployments for multimedia applications;

- access network costs represent a key controlling variable for business value that currently is highly volatile and uncertain;

- cellular is clearly more cost effective than WLAN in less densely populated areas or for voice-only applications.

The key implications of this work are:

- WLANs could potentially be widely deployed in urban areas as multimedia and/ or video applications become popular;

- cellular and WLAN operators could maximise value through co-operation;

- significant additional evolution of WLAN technology will be required to meet requirements for high mobility multimedia and video applications;

- it is also possible that radio spectrum will need to be allocated to operators for such use in order to maintain quality of services, as discussed in Chapter 16.

It should be noted that studies of this type are not designed to predict that specific business results will occur. Rather, they seek to predict the potential operator value if the supporting applications and networks were created and accepted by the market-place. This distinction points to the critical need to develop both the supporting applications and the new systems required to fuel the next generation of wireless business growth. In summary, the Strategic Landscape Process has proved to be both an effective vehicle for collaboration and a powerful tool to identify the market value of technological solutions.

1.3.3.2   Central Processing Utilising RF Over Fibre

Motorola Labs and BT Exact built on the success of the Strategic Landscape Process by co-operating to evaluate other opportunities within the cellular industry. Our second effort was a techno-economic evaluation of the 'central processing using remote radio frequency' concept for designing cellular networks in dense urban areas, using optical micro-cells. This is covered in more detail in Chapter 2.

The Central Processing Concept and Potential Advantages

The central processing concept is made possible by the extended deployment of fibre-optics in metropolitan areas, and by recent developments in optical technology, which allow the conversion of RF analogue signals to light and their transport via fibre-optics, with very low loss and distortion. This technology, called 'RF-on-fibre' (RoF), allows the concentration of all the base-station processing in a central office, connected via fibre-optics to simple low-power RF repeaters at micro-cell sites. Figure 1.3 shows how a typical cellular system is deployed. All the base-station transceiver system (BTS) processing equipment and the RF head are located at the cell site. However, there are several disadvantages to this conventional deployment:

- no inter-BTS communication;
- expensive ongoing site costs (OPEX);
- difficult to adapt to new air interfaces;
- expensive capital equipment.

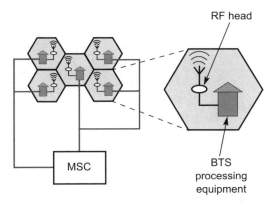

Fig 1.3   Typical cellular system.

With a central processing deployment (see Fig 1.4), the RF head remains at the cell site, but much of the BTS processing equipment can be moved to the central office.

This architecture opens new possibilities in the design and implementation of cellular networks. It has been the subject of intense discussions in the 3GPP standards body [2]. The following is a list of key advantages of a central processing oriented wireless system as applied to 3G and beyond technologies:

- trunking gains of 10-30%;
- decrease in redundant equipment;
- decreased operational expenses due to less site hardware;
- fewer site service calls;
- inter-BTS communications easily facilitated;
- allows new revenue generating services;
- utilise software defined radio (SDR) technology in central processing hardware architecture for maximum flexibility.

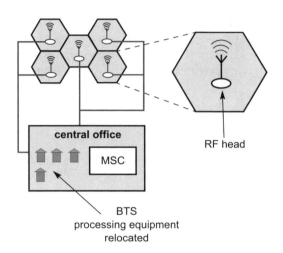

Fig 1.4   Central processing system.

Business Modelling Results

Motorola and BT Exact worked together to assess the potential business impact of the optical micro-cell system. We started with the strategic landscape net present value (NPV) model for a traditional cellular system and modified a number of parameters in order to estimate NPV for an optical micro-cell system. Motorola Labs simulated the propagation environment and estimated range and capacity for the optical micro-cells. BT Exact provided future estimates for access costs and advised on the optical implementation and Motorola provided cost estimates for future cellular equipment.

Figure 1.5 shows our results (displayed as relative value with respect to a 3G baseline). In Fig 1.5, we compare a traditional cellular deployment with a central processing cellular deployment using RF on fibre. Our results show that optical micro-cells have approximately the same value as traditional micro-cells for a voice data model. However, for a multimedia data model, optical micro-cells could have a significant advantage over conventional micro-cell systems for both 3G and 3.5G. (This is because many more base-stations are needed to supply the multimedia capacity, and so the lower cost of the optical micro-cell results in a lower system NPV. For voice centric, relatively few base-stations are needed and their costs are a smaller percentage of the network deployment.) Results are not included for video centric as neither 3G nor 3.5G cellular has sufficient capacity to support a high-definition video service at 2 Mbit/s.

We believe these results are very encouraging because they include the high cost of fibre-optic backhaul for the optical micro-cell case. The deployment of fibre-optic backhaul in most cities, combined with the technology of transporting analogue RF on fibre, has the potential of providing an alternative to the current

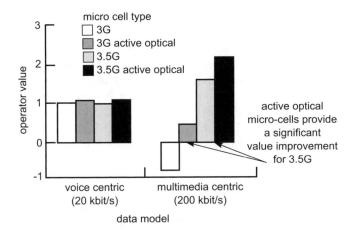

Fig 1.5   Value for optical and traditional micro-cell deployments.

cellular network architecture, at higher flexibility and lower cost. Centralised processing, combined with low-cost remote RF units, opens a new realm of design and implementation possibilities, which have still to be researched and analysed. The 3GPP standards organisation recognised this potential, and initiated feasibility studies with the purpose of standardising the interface. Given the positive value shown, we plan to perform further investigations of this concept, to fully qualify and quantify its advantages and implications for future cellular network deployment.

1.4   Road Map to the Future

1.4.1   2.5G to 3G

Third generation (3G) will clearly build successfully on network operators' investments in second generation (2G) infrastructure equipment and systems, by reusing the same resources, investments and experience. For example, GSM network operators are firstly deploying new data applications on their 2.5G general packet radio service (GPRS) environment while deploying 3G network infrastructure.

Once dual-mode handsets become plentiful, probably around the first quarter of 2004, the successful applications will be available on 3G, which has greater capacity for data applications. Whereas 2.5G GPRS networks will offer typically up to 40 kbit/s, 3G will be able to offer up to 384 kbit/s peak data rates. However, most customers will not buy bytes, bit rates or technology — only interesting and affordable mobile applications and content. Therefore, the success factors of 2G need to be maintained and built upon, going forward to 3G.

1.4.2   3G to WLANs

3G is a cellular technology which supports both voice and data applications successfully up to 384 kbit/s, depending on the cell loading and application types. It is designed from the outset to support high-velocity handover between cells for voice services; and, like all cellular technologies, it is particularly suited to rural areas where the base-stations can cover large geographical areas or in urban areas where there is a high demand requiring smaller cells. The strategic landscaping process has shown that the blanket deployment of WLAN technologies is attractive only for urban areas and, in particular, when there is a high   demand for multimedia applications.

Furthermore, WLANs also have the potential to support both technically and economically high-definition mobile video-based applications.

However, the current generation of WLANs does not support mobile handover between cells. Nevertheless, many of the multimedia and video applications may only require low-speed mobility, less than 10 km per hour, because people will typically use these in stationary situations or at walking speeds. Internet technologies such as mobile IP (see Chapter 11) may provide the mechanism for mobility management in such situations.

Further work is also required on WLAN's media access control to support a mixture of real-time and non-real-time services, so that, as the system reaches its maximum capacity, then non-real-time traffic, such as FTP, is dropped before real-time services such as voice and videoconferencing.

Once 3G has shown that there is a high demand for multimedia applications in urban environments, then WLAN technologies could be potentially an economic upgrade to the 3G network to support even greater capacity demand economically. The sites used for 3G can be shared for WLANs utilising the same access back-connection. WLANs would then have the further advantage of being able to support video applications that may arise later as well.

1.4.3   Access Networks

During our landscape investigations, we found that developing a cost-effective means of transporting voice and data information from the wireless access points to the operator's switching centre was critical. Most of today's cellular transport is copper, but the data rates for future wireless systems will require fibre transport. Sites may require bit rates of up to 10 Mbit/s. In this section, we describe current trends within the fibre industry and possible scenarios for access and backhaul transport of future wireless systems.

1.4.3.1   US Long- and Short-Haul Fibre Deployment and Pricing Trends

Our goal for investigating fibre price and deployment trends is to estimate when fibre will be available for future wireless systems and how much it will cost.

Currently, there is an over-capacity of long-haul fibre between many of the major cities of the world and an over capacity of metropolitan fibre within many of those same cities. This over capacity came about because carriers used overly optimistic estimates of Internet traffic and wireless growth rates to justify fibre build-out in the late 1990s. During 1995 and 1996, Internet traffic grew at an astounding 1000% per year. From 1997 through to 2001 the rate dropped to 100% per year. However, business analysts and even the US Commerce Department continued quoting a 1000% Internet traffic growth rate per year. Going forward, the expected growth rate for 2003 through 2004 is anticipated to be 40% per year [3, 4]. The subsequent collapse of the fibre market has caused many fibre construction companies and carriers to go bankrupt. The remaining carriers' plans for fibre build-out will be based on more reasonable traffic growth rates.

A recent article [5] quotes a severe decline in long-haul fibre pricing over the past two years. During this period, the lease cost for a 2 Mbit/s link from London to New York declined from $22 000 to $5000. As the first wave of carriers emerges from bankruptcy, analysts predict a further decline to $2000. The $2000 figure is seen as the minimum price that a no-debt carrier can provide with a small profit margin. It is anticipated that this additional erosion in fibre prices could cause a second wave of bankruptcies, this time within the primary carriers. Long-haul fibre prices may decline after this second wave, but should recover to a point that offers some profit for carriers.

UBS Warburg gives their optimistic estimate that global carrier spending will drop 24% in 2002 and 4% in 2003 [6]. Barring the invention of a killer application, we should not anticipate additional build-out of the fibre network until at least 2004. The opportunities to utilise the current installed fibre base need to be investigated and documented. This fibre base may be all that is available for future wireless systems for the next decade.

Regarding short-haul or access fibre pricing, we have had significant difficulty getting current information. We suspect that the metropolitan areas that have over-capacity of fibre will follow the same pattern of price reduction due to competition and bankruptcy as has been seen in the long-haul industry. Morgan Stanley issued a report in December of 2001 that provides pricing trend information on local (copper) tail and point-to-point circuits for a number of European countries [7]. Average US pricing is also provided as a reference. Table 1.1 shows the US pricing trends from 1996 to 2000 for local tail circuits.

It should be noticed that prior to the telecommunications collapse, the price seemed to be stabilising. Indeed, during the 1995 to 2000 period, FCC data shows that average prices for local circuits dropped only 2.6%. Notice also that the local 2000 DS1 prices are not significantly lower than our estimated 2004 price for long haul from London to New York. We anticipate a significant decline in short-haul fibre and copper pricing as metropolitan fibre companies, such as Metromedia, emerge from bankruptcy and begin competing with the established carriers.

Table 1.1   Annual cost for US tail circuit pricing.
[Source: Morgan Stanley [7]]

	1996	1997	1998	1999	2000
DS1 (1.5 Mbit/s)	$1764	$1836	$1728	$1746	$1782
DS3 (44.7 Mbit/s)	$13 410	$14 483	$15 019	$16 092	$17 165

1.4.3.2   Opportunities for Wireless Systems — Dark Fibre

As part of the BT/Motorola collaboration we focused on fibre solutions for future wireless systems. Many operators in the USA have deployed dark fibre. This is fibre that has been put in the ground but not yet connected to a functioning network. Dark fibre allows the network designer flexibility to meet the air interface requirements of cellular systems. It can be used to transport either analogue or digital data.

Dark fibre is being deployed by a variety of companies and government bodies, including public utilities, water companies, municipalities, city traffic departments, railways, CLECs, cable companies and telcos. Extensive dark fibre has been employed in and around the major urban areas in the USA and Europe. The prices shown in Table 1.2 are based on a variety of sources and estimates [7-9]. They may vary significantly from market to market, and should be interpreted as only indicative of pricing. For a wireless system, we assume that two fibres are needed per site; however, there are also optical solutions that allow the use of a single fibre. The number of fibres only has an impact, however, on the monthly lease fees. Non-recurring costs are independent of the number of fibre strands per site. These prices do not include DWDM equipment costs for multiple wavelengths on a single fibre strand.

Table 1.2   Dark installation and connection fees.

	Urban	Suburban
Connection fee, building	$10 000	$10 000
Connection fee, aerial	$8000	$8000
Fibre lease/strand/year/Km	$6000	$1200
Fibre installation/Km, buried	$100 000	$50 000
Fibre installation/Km, aerial	$35 000	$25 000

Based on these prices, we calculated the fibre back-haul recurring and non-recurring cost for a typical 100 site network. We assumed 50 urban sites and 50 suburban sites with 1 km and 3 km cell radii respectively. For the urban sites, we assumed 100% fibre availability. For the suburban sites, we assumed that 50% of the fibre would be new construction and that half of this new construction could be done aerially. Costs based on these assumptions are shown in Table 1.3.

Table 1.3   Average dark fibre deployment costs for urban and suburban sites.

	Urban	Suburban
Average recurring per site per year	$9174	$1800
Average non-recurring per site	$19 125	$152 340

We see that urban areas have much lower recurring and non-recurring costs than suburban areas, since we assumed much more fibre being available in urban areas than in suburban areas. For each suburban site, an assessment has to be made of the cost efficiency to add it to the fibre network, or stay with copper. As fibre becomes more available, more sites would be added.

1.4.3.3   *Future Directions — Business Modelling of Back-Haul Costs*

Current fluctuations in costs associated with long- and short-haul fibre networks present significant challenges for implementing future wireless systems. However, the over-deployment of fibre in the 1990s, and the consequent available 'dark fibre', may represent an opportunity for establishing proprietary networks. Further analysis will be required for refining wireless business models based on dark fibre availability.

1.5   Future for DAB and DVB as Mobile Networks

Earlier we explained the challenges in addressing the broad issue of the future of wireless communications. Historically, this discussion about mobile communications would mainly have involved network operators and the equipment infrastructure providers; now it has widened to include the media industry as well — not only for their content but also for the distribution using new digital broadcast networks.

Digital audio broadcasting (DAB) and terrestrial digital video broadcasting (DVB-T) are two technologies that have been developing since the late 1980s for supporting the analogue-to-digital transition of radio and television respectively. Because both technologies are based on OFDM modulations in particular, they demonstrate interesting capabilities in terms of bit-rate-associated mobility [10].

From a deployment perspective, broadcast cells are significantly larger than mobile network cells. Therefore, broadcast networks enable the provision of a cost- and spectrum-efficient radio bearer for delivering point-to-multipoint services.

Association with a mobile network leads to a two-way system with asymmetrical spectrum pattern, which is good for matching the UMTS Forum prediction of 10:1 traffic asymmetry for future multimedia services.

Hybrid networks combining mobile cellular (UMTS/GPRS) and broadcast (DAB, DVB-T) technologies (see Fig 1.6) have recently gained in interest for the following reasons:

- their characteristics, once combined, can provide overall cost-effective infrastructure for delivering multimedia-rich services to mobile users;

- the commercial lack of success of DAB and still limited success of DVB-T has been favouring regulatory moves to enable experimenting with new services and business models that will take advantage of already deployed broadcast infrastructures.

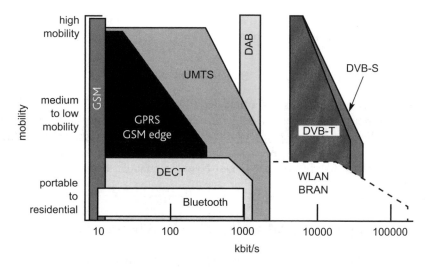

Fig 1.6   Bit rate/mobility mapping of wireless technologies.

As a result, several broadcast-oriented international bodies are starting to evaluate the need for, and attempting open standards development in the area of, mobile/broadcast convergence.

1.5.1   Services

Typical services that benefit from mobile/broadcast synergy are those that exhibit a portion of content that is common to many users while requiring interaction.

While existing networks can support services in a more or less satisfactory manner (depending on provided quality of service, cost to be supported), the combination of networks can improve their attractiveness.

The most promising approach seems to be the combination of a cyclic transmission of data over a broadcast system, associated with terminal storage capabilities (see Table 1.4).

Table 1.4   Typical service performance in uniform or hybrid environments [8].

	Typical services	Mobile	Broadcast	Mobile + broadcast
Entertainment	TV, radio programmes	-	++	++
	Audio, video-on-demand	0	-	0
	Games, interactive TV	+/0	-	++
General information	News, weather, financial information	+	+	++
	Travel, traffic, maps	+	+	++
	Commercial information	+	+	++
Personalised information	Web browsing, file transfer	+	-	++
	Individual traffic information, navigation	+	-	++
	Emergency, location-based services	++	-	++
Business and commerce	Remote access, mobile office	++	-	++
	E-mail, voice, unified messaging	++	-	++
	eCommerce, eBanking	++	-	++

Besides typical interaction to servers, the mobile network can be used to interact with network and service entities that permanently optimise their data cycles in order to decrease access times.

1.5.2   Regulatory and Business Constraints

Regulation related to broadcast networks tends to evolve in the sense that end users may be authenticated for the content they want to consume, but still under the constraint of delivering the same content to multiple users simultaneously. Consequently, broadcast networks appear well adapted to the radio-efficient delivery of (IP) multicast while cellular networks are still required for the one-to-one connections.

Cellular networks are built under the telecommunications regulations that give operators provision for authenticating and measuring the amount of traffic subscribers consume. Typically cellular systems are well suited for one-to-one communications.

Under this framework, a possible business case seems to be the one where a network services provider has agreed capacity under a service level agreement with both cellular and broadcast network operators, with the object of building delivery services to exploit the availability of a radio broadcast bearer with large geographical coverage for delivering multicast traffic [11] (see Fig 1.7).

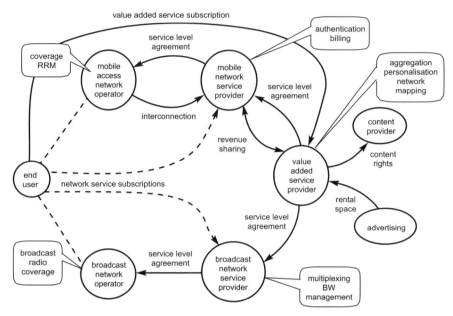

Fig 1.7   Business relationships.

The network services provider role is to optimise the capacity available through the service level agreements based on the QoS requests made by value-added service providers to a network services provider.

1.5.3   Architectural Framework

An architectural framework is being developed in the UMTS *ad hoc* group, part of the DVB consortium. It builds on the intrinsic transport mechanisms of the underlying networks but adds a signalling plane established between the various logical domains (see Fig 1.8).

The service provisioning sub-system (operated by the value-added service provider) provides service descriptions to the delivery sub-system and more precisely to the service access entity. These descriptions include, in particular, QoS information that the delivery sub-system uses to optimise the available bandwidth.

The service access entity is responsible for facilitating to mobile terminals the discovery of services. The network control entity's main role is to configure the networks in order to ensure that services can properly go through decided networks with the required QoS parameters. The configuration can be decided according to the network load levels, the services to be admitted in the different networks, and the terminals that must connect to the service. The network access entity enables the establishment of a user context in the delivery sub-system, in order to enable optimisation of the service descriptions to be presented to users (by means of user profile exchange), or to know about the terminal transmission capabilities in order for the delivery sub-system to decide about the service that the mobile terminal can potentially consume.

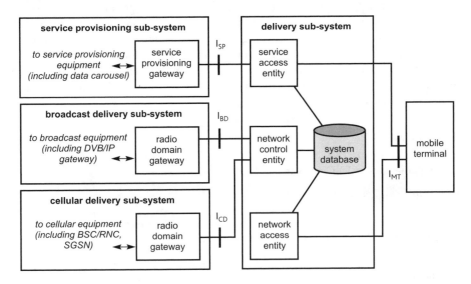

Fig 1.8   DVB-UMTS architectural framework.

1.6   Summary

This chapter has taken an unusual approach in that it has attempted to predict the future roles of mobile and wireless communications by sharing information between a network operator, BT, and an equipment vendor, Motorola, and by using techno-economic modelling techniques applied to 'connected' scenarios. Our goal has not been to predict the future as such but to envision a set of possible futures and the wireless systems that may come into being. Firstly, we considered a blanket network deployment 3G, 3.5G and WLANs against three application scenarios covering voice, multimedia and video. This has identified the opportunities for 3G and 3.5G in supporting voice and lower average speed multimedia applications while WLANs offer a cost-effective complement to 3G in the urban environment for higher average speed multimedia and video applications. Cellular is clearly more cost effective than WLANs in less densely populated areas. It has highlighted the opportunities for cellular and WLANs operators to maximise value through co-operation. We found that access network costs represent a key controlling variable for business value which lead on to collaborative work in centralising processing and utilising RF over fibre networks to lower-cost base-stations. Such co-operation between fixed and cellular operators would significantly increase the commercial value for both.

Terrestrial broadcast technologies are gaining momentum because of their ability to support mobility and therefore integrate into the mobile landscape. Existing technologies such as DAB and DVB-T can already be considered and broadcasters seem rather open towards creating synergies with mobile operators. This situation could clearly create new business opportunities not only to broadcasters, but also to mobile operators who could then leverage on the complementary aspect to terrestrial broadcast networks, i.e. interaction, user authentication and location.

It should be noted that the material presented in this chapter represents the results of exploratory research rather than a statement of either BT's or Motorola's technical or business position.

References

1   Motorola Labs — http://www.motorola.com/labs

2   3GPP — http://www.3gpp.org/

3   Dreazen, Y.: '*Telecom carriers were driven by wildly optimistic data on Internet's growth rate*', Wall Street Journal (September 2002).

4   Berman, D.: '*Innovation outpaced the marketplace*', Wall Street Journal (September 2002)

5   '*Telecom restructuring could reproduce glut*', Wall Street Journal (August 2002).

6   Fuller, M.: '*Wall Street eyes 2004 for possible industry recovery*', Lightwave (November 2002).

7   Morgan Stanley Report on Technology: '*Telecom Equipment*', Wireline (December 2001).

8   '*Benchmark Prices for Dark Fibre for Schools and Universities*', Canet-3-News (June 1999) — http://www.canarie.ca

9   Merryll Lynch: '*Report on passive optical networks*', (May 2001) — http://www.mlhsbc.com/

10 DVB-UMTS Group: '*The Convergence of Broadcast and Telecommunications Platforms*', (August 2001).

11 Martinez, G.: '*Regulatory and business environment of systems beyond 3G*', RMN/MLW 2002 Workshop (May 2002).

2

OPTICAL RADIO — A REVIEW OF A RADICAL NEW TECHNOLOGY FOR WIRELESS ACCESS INFRASTRUCTURE

P P Smyth

2.1   Introduction

Wireless communication has experienced enormous growth during the past decade, primarily driven by mobile cellular systems for voice communications, but more recently by the use of wireless local area networks (WLANs) for corporate enterprises. The coming decade is not expected to be any different. Therefore, the use of wireless data is expected to continue increasing dramatically and it is anticipated that this will be driven by the desire for wireless access to the Internet via 2.5G/3G and WLANs in homes, offices and public hot-spots. The combination of an increased number of users and the trend towards multimedia services means that the capacity of our wireless networks will probably need to be increased by a factor of between 10 and 100 during the next decade.

For the future wireless world one thing is certain — to provide high-capacity wireless communications when there is only a limited amount of spectrum will require the use of a large number of short-range systems. This presents two cost-reduction challenges that will need to be addressed if the future demands for bandwidth are to be met. The first one is the cost of base-stations and the second is the cost of their backhaul access network.

In cellular communications, micro-cells[1] are already being deployed in densely populated urban areas. However, the reason why their use has not become more widespread is the very high cost of deployment. Each radio antenna unit is expensive (even for those specially developed for small cell size) and they contain

---

[1] These occur where the base-station antenna is placed below roof height to provide coverage along city streets within a radio range of a few hundred metres at most.

huge amounts of electronic processing. Furthermore, different radio systems need their own special type of antenna unit.

Although relatively low-cost WLAN radio access points, having a typical operating range of between 20 to 50 m, are being deployed within buildings (for example office blocks, shopping centres and airports), they nevertheless still require dedicated high-bandwidth backhaul access networks, which has associated cost implications.

For either 3G cellular base-stations, or WLAN access points supporting capacities of greater than 2 Mbit/s, fibre backhaul networks are currently deployed. These are based on digital systems.

Using analogue optical networks for delivering radio signals offers one means of addressing the identified cost challenges. The advantages of using such systems for delivering radio signals from a central location to many remote antenna sites have long been recognised [1]. By using single mode optical fibres with very low losses, typically 0.2 dB/km, it is possible to provide a transmission infrastructure that is capable of transmitting the whole of the usable radio frequency (RF) spectrum (DC to 300 GHz). The other advantage of this approach is that all of the signal processing functions normally carried out in the radio access point can now be performed centrally with only the RF signals being transported over the optical network. The remote antenna sites then only require opto-electronic conversion, filtering and linear amplification. Such systems are known as either optical radio or radio-on-fibre.

Of course, the holy grail of optical radio now is to ensure that these simple radio access points are designed to be transparent to the frequency of operation and the underlying protocols used. This will provide solutions that meet the needs of today as well as those of future wireless systems and without any expensive upgrades being required. Microwave Photonics [2] is actively involved with commercialising opportunities offered by electro-absorption modulator (EAM) technology.

The aim of this chapter is to briefly review optical radio and to introduce a new semiconductor device, the electro-absorption modulator, which has great potential for producing low-cost radio access points for WLANs and cellular infrastructure.

2.2   Analogue Optical Fibre Networks

Traditional analogue optical fibre links used for feeding remote antennas have used a combination of a semiconductor laser, as a transmitter, and a photodiode, as a receiver, for each direction of the bidirectional transmission path. Much work has been carried out aimed at reducing the overall insertion loss between the semiconductor laser input and the photodiode output and maximising the system dynamic range that can be achieved.

For mass-market applications, such as use in an enterprise WLAN network, the remote terminal equipment costs and electrical power consumption are important considerations. A packaged semiconductor laser and associated control circuitry for high-frequency operation is costly and consumes a significant amount of power. Using an optical intensity modulator in a loop-back configuration for the return path would remove the need for a remote transmit laser which would reduce power consumption and simplify bias and wavelength control. However, even greater simplification can be achieved if both transmit and receive functions at the remote terminal are performed by a single low-power discrete optoelectronic device. The electro-absorption modulator is such a device.

2.2.1   EAM Radio Transceiver

Typical EAMs are made of indium phosphide (InP) for operation in 1.3 mm and 1.5 mm wavelength fibre windows. InP is a member of the III-V family of semiconductors. III-V compounds have a cubic lattice-like structure with atoms in each corner with an energy bandgap, which makes it opaque for light energy that is higher than the bandgap and transparent for energy levels that are lower. This direct bandgap, which allows electrons to transfer directly between energy levels, supports optical gain, as is required for lasers, and also very high absorption, as is required for functions such as optical modulation and fast photodetection.

Materials such as silicon, which do not have direct bandgaps, are unsuitable for these operations.

Westbrook and Moodie [3] reported the first use of an EAM as a simultaneous transmitter/receiver in a bi-directional analogue optical fibre link. Prior to this, EAMs had been used extensively for the generation of short pulses and OTDM multiplexers/demultiplexers in digital long-haul optical communications systems.

The basic operating principle for the EAM is illustrated in Fig 2.1. Light from the input optical fibre is coupled into an optical waveguide in the EAM. Applying an electric field to the electrical contacts on the optical waveguide can modulate the intensity of the light. The input light is partially absorbed or transmitted depending on the strength of the electric field. Thus an electrical signal from an antenna can be used to modulate light within the waveguide without the need for intermediary electronics. The residual light, carrying the upstream radio signal, is coupled into the output optical fibre.

The light absorbed in the waveguide induces an RF signal equivalent in frequency to that of the intensity modulated light. This produces an electrical signal at the waveguide contacts which are connected directly to the antenna (an alternative option, considered later in this chapter, is to connect to the antenna via an intermediary power amplifier). The combination of these two effects creates a bidirectional radio transceiver powered only by light. An important feature of these two processes (modulation and detection) is that they take place simultaneously, which means that the link between the central control station and the antenna unit can be full duplex.

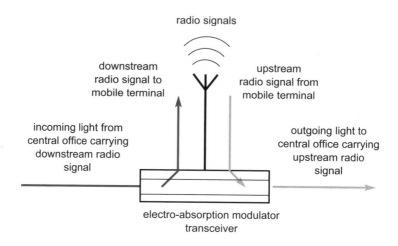

Fig 2.1   An EAM optical radio transceiver.

Figure 2.2 shows the variation in the transmission and responsivity of a typical EAM device with applied DC bias for transverse magnetic (TM) polarised light at a wavelength of 1540 nm. When biased at −5 V, this device simultaneously behaves as a 30 dB extinction ratio modulator and a 0.9 A/W responsivity photodetector. Another important feature of the EAM radio is that it can be operated in a mode that does not use any DC bias. This passive mode of operation has the significant advantage of removing the need for an electrical power supply and hence gives a completely passive optical radio system.

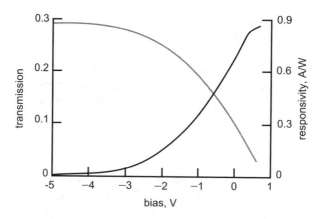

Fig 2.2   EAM transmission and responsivity.

EAMs also have the advantage that they can achieve small-signal modulation performance from DC to 60 GHz, which far exceeds the capability of direct intensity modulation of a semiconductor laser. An additional benefit is that the EAM combines both transmit and receive functions with very little or no external circuitry, and therefore the overall cost of a remote terminal should be less than one that uses a semiconductor laser and a photodiode. In mass production, it is expected that the cost of making an EAM transceiver will be very similar to that of making a semiconductor laser.

Figure 2.3 shows a photograph of an EAM radio transceiver unit built by BT. This unit consists of two fibres, one at either end of the metal case, which are aligned to the optical waveguide of the EAM at the centre of the unit. The optical waveguide transfers light between the input and output fibres. The two microwave connectors provide connection points for the EAM DC bias, when required, and the radio antenna.

Fig 2.3   A prototype radio transceiver unit developed by BT.

The analogue performance of the EAM transceiver was first published by Wake, Moodie and Henkel [4]. They reported the performance of a fully packaged EAM module, with a fibre-to-fibre insertion loss of less than 5 dB, with a multi-quantum well (MQW) absorber layer[2]. This consisted of 17 InGaAsP wells and InGaAsP barriers.

2.2.2   EAM Optical Radio System

Figure 2.4 shows how the EAM is used in an optical radio system. A radio signal to be transmitted is first used to modulate a central semiconductor laser, which is connected via the downlink optical fibre to the EAM transceiver in the remote antenna unit. The uplink fibre is used as a return path to the optical receiver in the central location[3].

---

[2] Mike Burt invented the MQW EAM within BT in 1982.

[3] Using a microwave laser and photodiode to measure the microwave insertion loss of such a bi-directional system, Wake, Moodie and Henkel [4] reported a value lower than 30 dB.

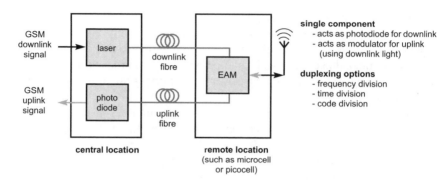

Fig 2.4   EAM-based optical radio system.

Consider, for example, that the laser transmitter in the central location is intensity modulated by a GSM signal. The resulting optical signal is then sent along the downlink fibre and the optical intensity variation at the input to the EAM is an exact replica of the GSM signal.

As previously described, in section 2.2.1, the light that is absorbed in the EAM generates an electric current that varies with the intensity of the received light signal — the EAM acts as a photodetector. The electric current thus produced generates an electrical field that is transferred to the antenna and hence is radiated into the atmosphere. The radiated RF signal is an exact replica of the original GSM signal. The light that is not absorbed in the EAM is returned to the central location via the uplink fibre. Because the remote location antenna is connected directly to the EAM device contacts, any GSM signal detected will directly modulate the intensity of the unabsorbed light as it passes through the EAM — the EAM acts as an optical modulator. Therefore, the intensity of the optical signal received at the central location from the uplink fibre is an exact replica of the GSM signal that was detected by the remote unit's antenna.

If there are several different wireless systems located at the central location (central office) they can be combined together and used simultaneously to drive the optical radio system. This is illustrated in Fig 2.5.

The central office contains the various radio modems that are connected to their respective networks. These modems provide the radio signals that are transmitted to the radio antenna units. Each radio modem is associated with a particular system, for example, GSM, 2G/3G (UMTS), DECT, or WLAN and their signals are combined for onward transmission. A network constructed using the optical radio concept will allow, for the first time a fully integrated wireless network without the need for special terminals or adapter cards. Someone in an optical radio cell will be able to use their existing mobile or cordless phone, their existing wireless computer, or, in the future, their own personal multimedia communications terminal. Optical radio will provide access to all of the users' communications requirements without the need for cabled terminals.

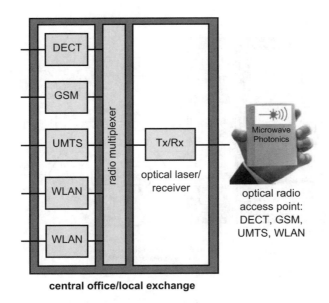

central office/local exchange

Fig 2.5   Optical radio access architecture.

2.3   Benefits of Centralisation for Cellular Infrastructure

A cellular architecture using optical radio is shown in Fig 2.6 and described in more detail in Chapter 1. Here all the signal processing associated with the base-station transceiver system (BTS), usually found in the radio access point, can now be moved to the central office. Consequently, the radio access point becomes a passive device that only consists of an EAM, a bandpass filter (BPF) and an antenna. In practice, some loss of simplicity may need to be traded for increased range (see section 2.6).

The benefits of such a system result directly from the shift of complexity away from the antenna unit to the central office. In other words, centralisation can be used to aid simplification. Some of the benefits of centralisation are listed below.

- Efficient use of resources

  As a result of gains in trunking efficiency, less processing power is required for any given grade of service.

- Relaxed specification

  The electronic processing no longer needs to be ruggedised and protected from harsh environments. Temperature specifications can be greatly relaxed.

- Reduced maintenance cost

  The optical radio antenna unit simply comprises an EAM, a filter and an antenna (where range is paramount, a bidirectional amplifier can be added — see

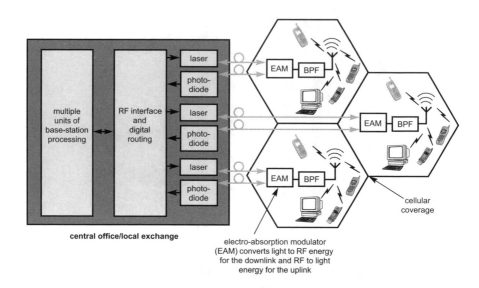

Fig 2.6   Multiplexing several radio protocol systems to a single optical radio access point.

section 2.6). Optoelectronic components, such as the EAM, are extremely reliable — they were originally developed for use in applications where reliability is critical, for example undersea transmission systems. Most maintenance will be performed at the central control point and will therefore cost far less.

- Multi-operator capability

  Electronic processing for different operators may be collocated as part of infrastructure-sharing arrangements. This is likely to be very common in future as operators seek to reduce network deployment costs.

- Air interface transparency

  Network upgrades and enhancements can be performed without any changes to the radiating infrastructure. Simple and easy card replacement at the central control point is all that is required. This makes optical radio a future-proof technology.

- Dynamic capacity allocation

  With centralised processing it is very straightforward to switch capacity between various locations as a function of time. For example, if a WLAN in a conference room suddenly requires more capacity it can be switched from neighbouring cells. When the conference room is empty or the demand drops the resulting spare capacity would become available for re-allocation elsewhere.

2.4 Optical Radio Options

As mentioned earlier in this chapter, there are two main modes of operation for an EAM optical radio system. The simplest is based on using a passive radio antenna unit, and the second, active mode, uses electrical power for EAM biasing and signal amplification, for achieving improved performance.

2.4.1 Passive Mode

The optical signal energy alone is capable of directly generating sufficient radio transmission energy, which has the advantage of removing the need for additional electronic amplification and/or a local power supply. Removing the need for a local power supply enhances system reliability and provides a radio antenna unit that is very easy and cheap to install. The provision of electrical power for outside locations is often a significant cost.

2.4.2 Active Mode

In the active mode, or powered mode as it is also known, a bidirectional amplifier is placed between the electrical contacts on the EAM waveguide and the antenna in order to achieve output powers as great as those that can be attained using conventional electronic systems. However, the optical radio solution still retains the advantages of low cost and transparency to radio system protocols.

The following two sections consider both of these modes of operation in greater detail.

2.5 Passive Optical Radio Access Point

The feasibility of using an unbiased EAM as a remote transceiver to produce a passive, pico-cell radio, antenna unit has been demonstrated in a commercial 2.4 GHz wireless LAN system [5], and is described in more detail below.

2.5.1 Demonstrator System

The demonstrator system was based on the same architectural layout as illustrated in Fig 2.5, except that in this case there was only one radio modem at the central control office.

The radio modem operated at a raw data rate of 1 Mbit/s and had a transmit power of +17 dBm and a sensitivity of − 82 dBm. The optical fibre link had an RF insertion loss of 40 dB for the downlink and 35 dB for the uplink. An antenna with a gain of 8 dBi (with a beam width of about 70°), was used at the radio antenna unit, which gave an effective transmit power at the radio antenna unit of −15 dBm. The antenna used at the mobile terminal had a gain of 2 dBi; consequently the effective receiver sensitivity of the radio modem was −84 dBm.

The pico-cell was demonstrated in a meeting room with dimensions of 6.5 m × 3.5 m. The radio propagation loss within this room was typically 50 dB and this was relatively independent of antenna position. With this value of propagation loss the received power at the radio modem antenna was −65 dBm, so the downlink power margin was 19 dB. The uplink power margin was 24 dB because this path had 5 dB less loss than the downlink. The wireless LAN system operated robustly over a line-of-sight radio range of a few tens of metres for a variety of broadband applications, which included video-streaming.

The optical fibre path length used for this demonstration was less than 100 m. However, lengths of at least 10 km should be possible given that the incremental optical fibre loss is only 0.2 dB/km. This translates into an incremental electrical insertion loss of 0.4 dB/km for the downlink and 0.8 dB/km for the uplink (this difference reflects the way the link is constructed).

2.5.2   Operating Range Considerations

In wireless systems the operational range associated with the downlink is often different from that of the uplink. This is because the radio access point and the mobile terminal can operate with different transmitter powers, antenna gains and also receiver sensitivities. Diversity reception is often included at the access point but not at the mobile terminal, mainly because of size constraints. Clearly, the shorter of the two ranges will determine the range limitation for symmetrical operation. This situation is no different when using an EAM transceiver, especially as the mobile transceiver will, in general, be using conventional electronics and have higher receiver sensitivity than the central location receiver.

In a passive optical radio system (shown earlier in Fig 2.4) there are practical limitations on the amount of RF power that can be produced by the radio access point. Firstly, there are limits to the amount of RF-modulated optical power that can be produced economically when linearly modulating a semiconductor laser (this is the source of power ultimately used by the radio access point). To operate with acceptably low levels of inter-modulation distortion the laser output power has to be held below some critical value (typically, less than 50% modulation depth of the output power). Secondly, there is a threshold power beyond which the EAM's optical waveguide saturates (typically, less than 10 mW). Other limitations arise from the coupling efficiency between the input/output fibres and the EAM waveguide.

Figure 2.7 shows the small signal model of the passive pico-cell. To ensure matched conditions exist between the antenna and the EAM, the EAM load is equal

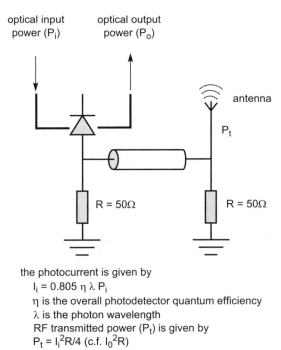

the photocurrent is given by

$I_i = 0.805\, \eta\, \lambda\, P_i$

η is the overall photodetector quantum efficiency

λ is the photon wavelength

RF transmitted power (P_t) is given by

$P_t = I_i^2 R/4$ (c.f. $I_0^2 R$)

Fig 2.7   Passive pico-cell small signal efficiency.

to the characteristic impedance of the antenna (50 W in this case). Because the EAM is a high-impedance source, this arrangement for signal matching further limits the power transfer to the antenna. In practice, the various limitations in overall transfer efficiency to the antenna limit the RF output power that can be achieved to less than 1 mW when the central location semiconductor laser source is operating at an output power level of 20 mW.

The minimum value of received power needed at the radio access point is described by equation (2.2). Although implied, it is not immediately obvious from equation (2.2) that the shot and relative intensity noise sources associated with the optical carrier are not negligible. For this reason the minimum received power for a passive optical radio access point will never be as low as can be achieved with a conventional radio access point.

Nevertheless, despite all of these limitations the passive pico-cell has a role to play in short-range wireless systems such as 3G and WLANs. The receiver sensitivities, Pbr, for W-CDMA modulation at bit rates of 144 kbit/s, 384 kbit/s and 2048 kbit/s when using a passive pico-cell are given with equations (2.1) and (2.2) along with the passive pico-cell characteristics:

- transmit power from BS antenna:

$$p_t = \left(\frac{p_i^2 \cdot g_f^2 \cdot R_m^2 \cdot M_d^2 \cdot Z_m \cdot g_a}{2} \right) \qquad \text{...... (2.1)}$$

- minimum received power of BS antenna:

$$p_r = \left(\frac{p_{br}}{g_a \cdot p_i^2 \cdot g_f^4 \cdot g_m^2 \cdot \eta_m^2 \cdot R_p^2 \cdot Z_m \cdot Z_p} \right) \qquad \text{...... (2.2)}$$

where each parameter is defined with typical values:

optical power, P_i, 1-20 mW,
modulation depth, M_d, 50%,
EAM efficiency, η_m, 0.5 V,
EAM responsivity, R_m, 0.3 A/W,
fibre insertion gap, g_f, −3 dB,
EAM insertion gap, g_m, −7 dBm,
BS antenna gain, g_a, 6 dBi.

The receiver sensitivity, P_{br}, is:

−115 dBm (144),
−110 dBm (384),
−103 dBm (2048).

The maximum value of RF transmission path loss that can be tolerated is given by taking the ratio of radiated RF power to received power (the receiver sensitivity), adjusted as necessary to reflect any antenna gain. The associated propagation distance can then be calculated by using an appropriate propagation model representation of the environment within which the passive pico-cell is deployed.

One such model is the simple dual-slope propagation model illustrated by equations (2.3) and (2.4) which show that the propagation distance is directly proportional to frequency and that it can very quickly fall beyond the Fresnel zone:

$$L = 20\log\left(\frac{4\pi d}{\lambda}\right) \qquad \text{...... (2.3)}$$

for $d < d_F$

$$L = 20\log\left(\frac{4\pi d_F}{\lambda}\right) + \gamma\log\frac{d}{d_F} \qquad \text{...... (2.4)}$$

for $(d > d_f)$

d_F is the Fresnel zone distance: $\quad d_F = \dfrac{4h_1 h_2}{\lambda}$

where:
   path loss slope, l, is 43 dB,
   BS antenna height, h_1, is 5 m,
   MT antenna height, h_2, is 1.5 m.

Although this model is well suited to line-of-sight transmission, its limitations are such that it should not be used when the radio access points are deployed in urban environments. Such environments require other models to be developed. Two very useful empirical models for representing urban micro-cellular propagation have been produced by COST 231 [6] — namely the COST 231 Hata model and the COST 231 Walfisch-Ikegami model.

The Walfisch-Ikegami model has been used with the parameters given in Table 2.1 to derive the maximum operating range that can be achieved when using a passive pico-cell optical radio system to carry a 3G WCDMA, 64 kbit/s data service [7]. Three different deployment scenarios were chosen based on estimations of potential future advances in EAM technology (see Table 2.2). In other words it considers systematic improvements beyond what is possible today.

Table 2.1   Parameters for Walfisch-Ikegami model.

Width of the road	20 m
Height of buildings	15 m
Base-station antenna height	13 m
Mobile height	1.5 m
Road orientation with respect to radio path	90°
Building separation	40 m

The range calculation results are given in Table 2.2 [6]. When deriving these results, in-building penetration loss was modelled as a fixed 15 dB loss, the required uplink Eb/No[4] was assumed to be 2.2 dB and the base-station antenna gain was assumed to be 10 dBi.

Table 2.2   Passive pico-cell range and radio access output power.

	Passive pico-cell			3G microcell
	Current EAM technology	Realistic EAM future technology	Optimistic EAM future technology	
Uplink range	17 m	28 m	34 m	118 m
Uplink power	−23 dBm	−115 dBm	−1.35 dBm	+30 dBm

---

[4] The carrier-to-interference ratio is related to the Eb/No requirement by the processing gain of the CDMA system (the spreading bandwidth divided by the bit rate).

The results produced by the Walfisch-Ikegami model will probably yield a conservative estimate of operating range because there is unlikely to be any significant scattering or diffraction caused by buildings over distances up to about 30 m. It is also expected that the range for urban down-the-street paths would be greater due to not experiencing the 15 dB in-building loss component combined with line-of-sight (LoS) use. Nevertheless, the results illustrate that the passive pico-cell is best suited to short-range communications when compared to a traditional micro-cell. However, note that this simply reflects that the micro-cell is able to use much larger values of uplink power. It is also worth noting that for urban deployments, although the micro-cell can use 1000 times more transmit power than the passive pico-cell, it only increases the operating range by a factor of 3.5 times because of the high path-loss slope beyond the Fresnel zone distance.

2.6   Active Optical Radio Access Point

The operating range limitation of the passive optical radio access point can be overcome by using electronic amplification within the antenna unit. Simply by placing a power amplifier between the EAM and the antenna it is possible to improve on the operating range of a passive pico-cell. Unfortunately, the obvious disadvantage is that this amplifier will need a source of electrical power to be provided. However, one option is to supply this power from the central control point over twisted copper pairs in the same ducts as the backhaul fibres.

With sufficient amplification from the power amplifier it would be possible for an EAM active access point to achieve the same output power as conventional electronic systems. Similarly, by including a low-noise amplifier after the antenna, but before the EAM, it is possible to improve the receiver sensitivity and to also increase the depth of the modulation that the EAM produces on the uplink signal.

Figure 2.8 shows a simple diagrammatic representation of an active optical radio system. Because a bidirectional amplifier is needed to provide amplification in both the transmit and receive directions, microwave circulators, or similar devices, will be needed to provide separate transmit and receive signal paths between the EAM and the antenna. An advantage of using electrical amplification is that lower tolerance EAMs can be used because their lower output signals can be offset by increasing the electrical gain. This is important for WLAN applications where cost is an important consideration. Typically, a power amplifier gain of 40-50 dB is required to achieve a transmit power of 100 mW.

2.7   Cellular Applications of Optical Radio

Figure 2.6 (shown earlier in section 2.3) illustrated a cellular application of optical radio where the BTS is centralised in a local exchange or central office.

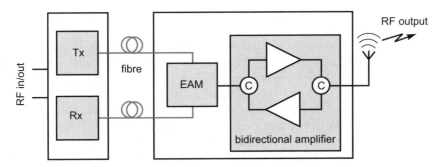

Fig 2.8   Active radio access point for increased range.

EAMs in use today have separate input and output optical ports and therefore need two fibres for connection to the central office — one for the uplink and one for the downlink. This has the disadvantage that it doubles the fibre infrastructure requirements and hence costs.

An elegant alternative system for a cellular application that enables a single fibre to be used between the local exchange and the base-station is shown in Fig 2.9. This becomes possible because new reflection-mode EAMs, achieved by placing a mirror at one end of the optical waveguide, have only one fibre connection. Here one fibre is used for both the upstream and downstream radio signals. Because the optical signal in the EAM now transverses the length of the waveguide twice, allowing double absorption, it is necessary to choose different device parameters.

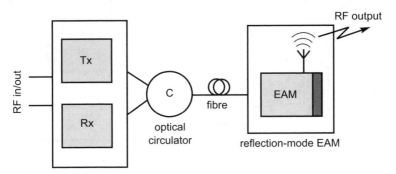

Fig 2.9   Single-fibre option for cellular.

2.7.1   Cellular Architectures Using Wavelength Division Multiplexing (WDM)

The use of reflective EAMs will reduce backhaul cost by halving the number of fibres required. However, as it is necessary at the local exchange to separate the uplink and downlink optical signals, an optical circulator will be needed. Small amounts of residual downlink optical signal will, however, still be fed back into the semiconductor laser.

Such feedback needs to be avoided to prevent dramatic increase in the optical source's relative intensity noise, which would cause a serious degradation in system budget. Therefore an optical isolator may be needed between the semiconductor laser and the circulator to avoid optical feedback.

Blanket coverage of an area of micro-cells will require the use of a large number of backhaul fibres (connecting the antenna units to the central control point) and this could become very expensive. Alternatively, a 'tree', 'bus' or 'ring' architecture could be used in conjunction with WDM technology to reduce the backhaul fibre count.

Various WDM technology options exist (splitters, couplers, add-drop multiplexers, array waveguide gratings, etc), but how they will be used will depend on the type of architecture chosen and the costs involved. One possible type of architecture (see Fig 2.10) is where WDM multiplexers and demultiplexers are used with the reflective EAM antenna units discussed previously.

Another type of architecture (see Fig 2.11), is where WDM techniques are used to enable a number of semiconductor lasers to simultaneously use the same downlink and uplink paths to communicate with their nominated EAM. This is achieved by each EAM having a wavelength selective filter tuned to the optical wavelength being used by its nominated source laser.

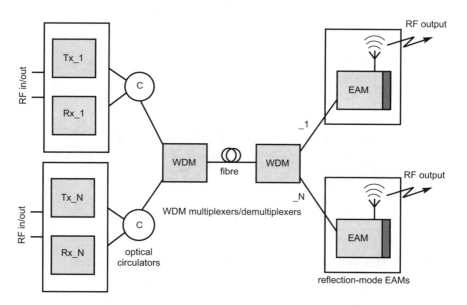

Fig 2.10   Single fibre, multi-wavelength option for cellular.

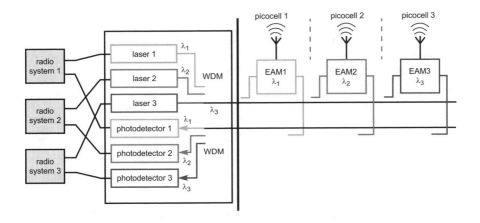

Fig 2.11 Separate uplink and downlink optical sources in the optical hub.

2.8 Summary

Optical radio is a radical new concept for wireless communications networks that allows centralisation of the signal processing normally found in the radio antenna unit. This allows the use of simple low-cost radio antenna units. These can now be designed to be insensitive to the protocols and frequency bands used by current and future radio systems. Centralisation of signal processing allows many other advantages, particularly in management and control.

The passive pico-cell will find uses in short-range broadband radio systems and where line-of-sight is the dominant mode of use. Its wide-frequency range of use (DC to 60 GHz) will potentially allow its use in most of today's radio systems.

An active version of optical radio can achieve the full operational range offered by conventional electronic access points while still retaining the advantage of being future-proof.

These attributes combine to give the very attractive and powerful advantages that have been outlined in this chapter. As we have seen there are many potential application areas for this exciting new technology. If optical radio lives up to only a fraction of its early promise we will witness a major shift in the way future wireless networks are designed and constructed.

References

1   Vanblaricum, M. L.: '*Photonic systems for antenna applications*', IEEE Antennas Propag Mag, **36**(5), pp 30-38 (1994).

2   Microwave Photonics — http://www.microwavephotonics.com/

3   Westbrook, L. D. and Moodie, D. G.: '*Simultaneous bi-directional analogue fibre optic transmission using an electro-absorption modulator*', Electronics Letts, **32**(19), pp 1806-1807 (September 1996).

4   Wake, D., Moodie, D. G. and Henkel, F.: '*The electro-absorption modulator as a combined photodiode/modulator for analogue optical systems*' IEEE (1997).

5   Wake, D., Johansson, D. and Moodie, D. G.: '*Passive picocell: a new concept in wireless infrastructure*', Electronics Letts, **33**(5), pp 404-405 (February 1997).

6   COST 231 TD (86) 109: '*1800 MHz mobile net planning based on 900 MHz measurements*,' (September 1991).

7   Smith, J. and Reed, D.: '*Private communication*', Motorola.

3

WIRELESS LANS — PRESENT AND FUTURE

L Burness, D Higgins, A Sago and P Thorpe

3.1   Introduction

Wireless local area networks (short range broadband radio systems) are rapidly becoming a normal part of the communications access infrastructure. Demand for wireless local area network (LAN) products has grown dramatically over the last two years, and shows no sign of slowing. Indeed it is strengthened by the growth of laptops and personal mobility products. Business users now expect to be able to access the Internet and even their private corporate intranets through other wireless LAN networks. It is hard to remember that less than two years ago this technology was still viewed with some scepticism. Back in August 2001, BT's ability to provide a wireless LAN to support the Internet Engineering Task Force (IETF) meeting in London won the acclaim 'Wow!' from IETF [1]. BT continues to push ahead in this field with the launch of public wireless LAN 'hot-spots' in the Openzone product [2] (this topic is discussed further in section 3.3.1). This technology and its rapid success presents service providers with incredible opportunities and challenges.

The aim of this chapter is to answer the following questions.

- Why is this technology so successful?

- What needs to be developed to maintain the successful growth in use of this technology?

The short answer to the first question is of course that the technology fulfils a real user need, at a cost the user can afford. The answer to the second question can only be derived through considering what other applications might be stimulated through this technology. There is a vision to use wireless LANs as a low-cost access network to provide a nomadic broadband service combined with lower speed ubiquitous cellular coverage. In this situation a user would be able to download large amounts of data (such as music or video) using the high-speed wireless LAN functionality and still be able to roam on to a cellular network for voice and location based services (this is discussed in greater detail in section 3.2). It is no longer sufficient to have Internet access at the office or at home. Increasingly, users expect Internet connectivity and secure access to corporate networks in public places such as airports, train stations, hotels and conference centres.

This chapter firstly provides a brief description of the main technology currently used for wireless LANs — the IEEE802.11 standard which is virtually synonymous with the term wireless LAN today. We then consider current and future applications of wireless LAN technologies.

One key focus is on the issue of interworking between wireless LAN (WLAN) and 2G/3G networks, which is a means of giving BT an opportunity to provide mobile services without owning any regulated spectrum. While this application can be provided using existing technology, a focus on the longer-term applications, such as networks on trains and *ad hoc* networks, helps us identify weaknesses with the current wireless technologies. We then examine technology developments that address these specific problems.

3.2   IEEE802.11b

IEEE802.11b is the prevalent standard across the world. This is down to a number of factors, the most important of which is the compatibility standard (WiFi) introduced by the WiFi Alliance[1] [3] (discussed further in section 3.2.3). The low cost of hardware and its inclusion in many new laptop computers and personal digital assistants (PDAs) has also helped IEEE802.11b become one of the world's most popular wireless networking standards. We will consider the operation of IEEE802.11b in some detail, as this is useful to understand the strengths and weaknesses of current wireless networks.

3.2.1   IEEE802.11 — Physical Layer Characteristics

IEEE802.11b is part of a series of wireless LAN standards, some existing and others under development. These standards differ at the physical layer.

The protocol at the air interface needs to be very resilient to cope with high levels of interference, which will occur from other wireless LAN systems and from the other users of the band. Other wireless LAN systems in the vicinity may well employ a carrier sense mechanism to ensure politeness when attempting to gain access to the medium, but other sources will be less friendly. Many of the different standards are about different physical layer techniques to improve the achievable bit rate, or for different frequency bands.

---

[1] Previously known as Wireless Ethernet Compatibility Alliance (WECA).

3.2.2   Medium Access Control (MAC) Layer Characteristics

In networking terms the MAC layer is responsible for overseeing that stations take it in turn to access a shared transmission medium. The shared medium can take the form of physical cables in the case of Ethernet or the radio channel for wireless networking. MAC layer controls also deal with packet transmission errors and make sure that a lost or broken packet is retransmitted in a timely fashion.

The IEEE802.11 family share the same MAC layer. Access to the radio channel is gained using a collision avoidance (CA) protocol. On a wired Ethernet IEEE802.3 LAN [4], collision detect (CD) technology is employed to recover from two stations transmitting at the same time. Stations sense the signal on the wire as they transmit, and quickly detect when simultaneous transmission is occurring. The protocol requires both to cease output, and to employ a pseudo-random back-off algorithm for retransmission, so that subsequent attempts by both parties to transmit are unlikely to occur at the same point in time. For a radio system, CD is not an option because it would require a station to have both its receiver and transmitter active at the same time, and, if receiver destruction did not result, the transmission would in any case swamp any signal received from the other party. The answer chosen for the IEEE802.11 family of standards is therefore carrier sense multiple access/collision avoidance (CSMA/CA).

Initially the station performs a carrier sense/clear channel assessment (CS/CCA) and, if the radio channel is sensed as idle, it begins transmission immediately. If, on the other hand, the medium is busy, the station waits for the end of the current transmission and then enters a contention period. If the channel is still idle at the end of this period, then the station has chosen the shortest delay and has gained access to the channel in advance of any competing contenders. This procedure occurs for every packet using random contention delays, ensuring equal access to the channel for all stations.

In a radio network, use of CSMA/CA does not guarantee a free medium. Apart from the small possibility of choosing the same contention delay as another station, there may also be another device, out of range of the sending station, which is nevertheless within range of the target and causing interference. This is the classic hidden node problem (Fig 3.1).

This problem is overcome in IEEE802.11 using optional RTS/CTS (request to send/clear to send) handshaking, also known as virtual carrier sense. When a station has a packet to send, it first sends an RTS and waits for a CTS from the target node. This will show that the target node is free to receive the data packet. As a bonus, the hidden node hears the CTS sent from the target and, since both RTS and CTS contain a field indicating the length of the following data packet, the hidden node refrains from any transmissions for the required amount of time. Of course, RTS packets can still collide, but the time lost is much less than if data packets collide and efficiency is improved in a busy network with many access points. Figure 3.2 illustrates the RTS/CTS and CSMA/CA mechanisms.

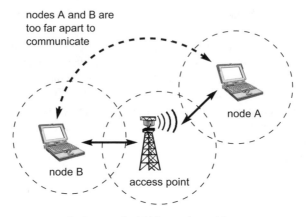

Fig 3.1   The hidden node problem.

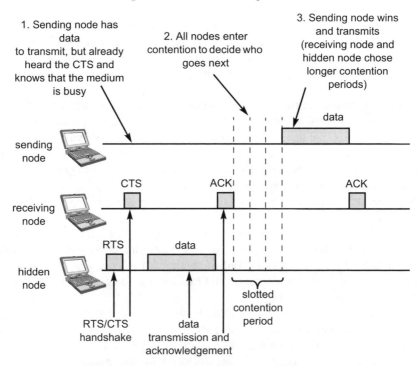

Fig 3.2   IEEE802.11 RTS/CTS and CSMA/CA mechanisms.

As well as the possibility of interference, the nature of the radio medium means that the characteristics of the environment (such as shape and size of room and people moving around in the vicinity) will produce unwanted effects on the signal known as fading. Measures can be taken at the physical layer to mitigate against these effects, but there could still be the occasional loss of data. Higher layer protocols used over a radio medium may not cope well with data loss, and recovery must take place within the wireless LAN medium access layer. This is simply achieved by returning a short positive acknowledgement (ACK) packet for every transmitted MAC data packet. The absence of an ACK indicates that the previous data packet should be resent, either because the packet was lost or because the ACK itself was lost.

3.2.3   IEEE802.11 — Interoperability

Interoperability of wireless LAN equipment was the initial aim of the standardisation activities. Although the multiple physical layers of the original IEEE802.11 prevented interoperability within the standard as a whole, it has been achieved to a great extent in IEEE802.11b.

On top of the mandatory features of the standard, individual vendors will naturally provide optional or proprietary features which will differentiate and enhance their product, while at the same time restricting the choice of vendor when the time comes for additional infrastructure or client cards. Examples of this can be found in proprietary protocol extensions in some access points to increase data throughputs, the various access server security solutions, and voice over wireless LAN solutions, which only operate with wireless LAN access points from selected vendors. Nevertheless, vendors are aware that success in the market place is through volume selling which will be greatly enhanced by good interoperability, achieved by close company co-operation. To address this, the WiFi Alliance was set up in August 1999.

Products from member companies are put through basic interoperability tests to ensure that a client card from one manufacturer will communicate with an access point or client card from another. Successful products are awarded a 'wireless fidelity' or WiFi sticker, and this simple step has contributed greatly to the recent dramatic take-up in the acceptability of wireless LANs. In April 2003, 202 chipset manufacturers, test equipment and wireless LAN vendors and resellers were listed as members of the WiFi Alliance, with 611 WiFi compliant wireless LAN devices [3].

As a final caveat on interoperability, certain areas have not been standardised and are therefore not part of the WiFi Alliance WiFi testing. For instance, handover between access points was not standardised at the same time as IEEE802.11b. When operating in a network which has access points from more than one vendor, it may be that the client device will not be able to roam on to and associate with a second access point that is much closer and offering better performance, without first completely losing contact with the first access point (and consequently losing the connection to the network) (Fig 3.3).

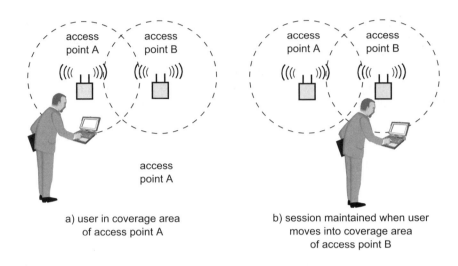

a) user in coverage area
of access point A

b) session maintained when user
moves into coverage area
of access point B

Fig 3.3   Access point roaming in a wireless network.

3.3   Current and Future Applications of Wireless LANs

The success of IEEE802.11b wireless LAN technology is undoubtedly due to the ease of use of the technology, the acceptance of standards, the low cost of the equipment, and the real need for such services. Client-side equipment is widely available in the credit card size PCMCIA (Personal Computer Memory Card International Association) card format suitable for laptops and PDAs, for less than £100. ISA (Industry Standard Architecture) and PCI (Peripheral Component Interconnect) cards are also available from some manufacturers for desktop machines. Access points, costing £100 — £1000 depending on functionality, typically occupy a footprint no bigger than an A5 sheet of paper and can be wall or ceiling mounted, or may be hidden behind ceiling panels or in cable risers within an office.

While initial adoption of wireless LAN systems has traditionally been in vertical markets such as retail, warehousing and manufacturing, current growth is being driven by other market segments. These include enterprise, small office/home office, telecommunications/Internet service provider (ISP) and the public access environment. Wireless peripheral connectivity, wireless Internet and higher throughput compared to cellular mobile networks are the lead drivers for wireless LAN deployment. Voice over IP (VoIP) is also expected to drive this technology in the future.

The introduction of seamless roaming is also believed [5] to be an important next step development for wireless LAN networks — and one that BT is in a strong position to implement.

3.3.1   Wireless Hot-Spots

Cells of wireless LAN coverage at public 'hot-spots' such as airport departure lounges, hotel receptions and shopping malls can provide entertainment, information and Internet access services to members of the public. This type of public access wireless LAN (PAWLAN) offering can also enable a business traveller to obtain a high-bandwidth connection back to their corporate network, enabling access to secure intranet and e-mail services, using a laptop or PDA and a virtual private network (VPN). Such networks are already proliferating in the USA and some parts of Europe, and now also in the UK with BT's own Openzone product [2]. The hot-spots are straightforward to install, with wireless LAN access points typically connecting over an ADSL line back to a main data centre to provide authentication and gateway functions. The supporting network infrastructure is relatively low functionality, which helps minimise the initial cost of such networks. For example, users can only really use such networks for services they initiate — such as Web browsing. There are no mechanisms to help data find a user.

The ultimate success of PAWLAN services is related to the availability of user terminals, the ease of use of the service, and the ubiquity of hot-spots. The introduction by the WiFi Alliance [3] of interoperability testing for terminal equipment has resulted in a change in the user experience, in that Wireless Fidelity (WiFi) certified cards are now guaranteed to work out of the box with access points from other vendors. In turn, there is now a wide choice of manufacturers offering wireless LAN connectivity at reasonable prices, and many people buying laptops or PDAs this year may find they have wireless LAN functionality as a default rather than an option. This may lead them to want to try out this functionality on a public wireless LAN service. For Openzone, a new user can just walk up to a hot-spot, connect to the network (automatically with modern PC operating systems), start their browser and surf the free pages. If they like what they see, they can buy time-limited access on-line with a credit card or purchase a voucher from the partner company (such as the hotel reception).

User credentials and credit card details are protected at the hot-spot through SSL between the browser and the central system, but subsequent traffic is in the clear. Business users (or their company's security policies) are likely to require protection from eavesdropping of their data on both the wireless itself and the wireless parts of the transmission medium. This can be achieved with a VPN, which can be provided as part of the Openzone subscription for business users. Alternatively companies may have an existing VPN solution which can be used over Openzone.

The targets for BT Openzone are for 400 hot-spot sites by summer 2003 and 4000 sites by 2005. It is recognised that providing a large number of sites will encourage use of the service, since there will be a greater possibility that a traveller will be close to a hot-spot at some point in their journey. The free pages accessible at the hot-spots themselves provide details of where coverage is available, and this information is duplicated on BT's Web site. The actual wireless LAN coverage area at a motorway service station or airport is only meant to provide access for users in a specific area such as the restaurant or departure lounge, and it is not intended that there will be continuous coverage across a town or city, for instance. Roaming agreements with other hot-spot operators will eventually open up complementary sites across the UK, Europe and the rest of the world. Covering the areas in between the hot-spots is a role for wide area mobile technology, and the following section on WLAN-3G interworking examines the requirements.

3.3.2   WLAN-3G Interworking

Ovum [5] has identified roaming as a key next-stage development for WLANs. To have long-term success WLANs need to be considered as part of a bigger, more mature data communications system. This is a new area for WLANs that have for a number of years been more academic, experimental, small scale and localised, rather than carrier-grade communications systems.

3.3.2.1   *Why WLANs Must Interwork*

As a part of this push we see a demand for the interworking of WLAN technologies with other systems and technologies. The reason behind the demands for this is twofold, both driven by user needs.

- Micro to macro

  The user has a need for data everywhere, and a lifestyle and work-style that is commonly geographically macro mobile (i.e. from home to office) rather than micro mobile (i.e. along the street). This requires a system that is capable of working in the macro domain as well as offering the advantages of the micro domain. For example, LAN connectivity speed when stationary or moving at a low speed is accompanied by a complementary, probably lower speed, connection when on the move. Although the WLAN is ideal for the micro-area system, as described in this chapter, its suitability to macro domains is not as ideal. Therefore there is a need for the WLAN micro-mobility technology to interwork with a macro-mobility technology.

- Universal systems

  The user does not care about the underlying technology, they want a solution that offers what they want/need. This means that they typically want services that can be offered by systems with multiple technologies and core systems that interwork together to deliver a seamless or near seamless experience, e.g. the interworking of systems such as corporate LANs, Internet, ISP access, and telco access, creating the 'universal system'.

3.3.2.2   *What We Are Interworking With*

So what techniques are we talking about when we say that WLANs must interwork with other systems? Primarily there are three considerations.

- 3G systems

  This is a case of a macro area technology which, due to its high-speed data transfer characteristics, is an ideal companion to the high-speed micro area technology. This is one of the most heavily investigated interworking considerations at the current time.

- General Packet Radio System (GPRS)

  This is also being considered for interworking, as, unlike 3G systems, it is a here-and-now data technology offering macro mobility.

- ISP systems

  Another important consideration is the ISP network. This form of interworking is very different but equally important.

Different solutions exist depending upon the system being interworked, but we also later present a generic approach (see section 3.2.4).

3.3.2.3   *Types of Interworking*

There are three forms of interworking:

- tight coupling;
- loose coupling;
- no coupling.

The third case (no coupling) is that of two completely independent networks operating independently and is what is commonly seen today. This can only be considered as interworking of the lowest level and we shall not consider it further in this chapter.

Tight coupling is a complete integration of the WLAN network into a host network as a new radio access network. The interfaces of the WLAN radio access network would be aligned to those of the existing core network of the host system. Adopting the tight coupling approach has the advantage that the existing mechanisms, for functions such as quality of service (QoS), mobility management, security and authentication, can be used. But the disadvantage is that the WLAN network is now specific to the host network used. This approach would require extensive standardisation activities in the WLAN bodies.

Loose coupling is when the WLAN is used as a complementary access network for interworking with the other network. Within loose coupling, the subscriber databases would be shared, but the user plane interfaces would be independent. For the previous case of Universal Mobile Telecommunications System (UMTS), this would mean that the 3G subscriber database is shared but the user-plane interfaces such as Iu is not, avoiding any impact on the serving GPRS support node (SGSN) and gateway GPRS support node (GGSN). In this case, security, mobility and QoS are addressed independently using IETF schemes.

Loose coupling allows more flexible methods of provision, allowing multi-provider networks with independent operators of the WLAN and interworked network. Additionally it offers increased opportunities for service providers to operate in the same market. This does, however, require co-operation between network operators in order to meet any subscriber end-to-end requirements. In the multi-provider case, a balance will need to be taken between the amount of information shared to ensure efficiency and that to maintain confidentiality of the mechanisms for user identity. There must be trust for charging and revenue accounting across the providers and operators.

3.3.2.4   The Basic Interworking Functions

The work of the European Telecommunications Standards Institute (ETSI) Broadband Radio Access Network (BRAN) group [6] (to date), which has been submitted to the Wireless Interworking Group (WIG), is the most advanced and also offers a genuine solution, from a WLAN perspective, that can work in a generic manner, as we suggested earlier.

The key principle is shown in Fig 3.4 where what we see is the access technology having an interface based on its current interfaces with some interworking functions (I.1). The host network having an interface based on its current interfaces with some interworking functions (I.3). Then a third interface (I.2) which is standard and generic (independent of the access technology or the host network) between these two sets of interworking functions (IWFs).

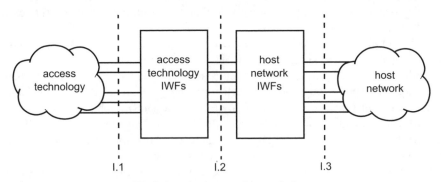

Fig 3.4   An interworking solution.

More detail on this work can be found in Findley et al [7]. Within these functions the most major considerations that must take place are those of the basic AAA functionalities:

- authentication;
- authorisation;
- accounting.

Authentication may be undertaken at any point; however, it will definitely be undertaken initially to extract both authentication and subscription information from the user's home network when an initial association is requested. In the loose coupling model it is important not only to support user authentication and accounting based on existing cellular credentials, but also to be able to have WLAN-only operators and users. It is therefore of importance that the solution provides a single set of AAA protocols which supports both scenarios.

One method of achieving this while still allowing both an IETF and UMTS flavour of authentication is to use a mechanism based on extensible authentication protocol (EAP) as described by Blunk and Vollbrecht [8]. An adaptation of the EAP would be specific to the WLAN solution, e.g. EAP over HIPERLAN (EAPOH) for HIPERLAN or EAP over LAN (EAPOL) for IEEE 802.1x; this does not, however, stop the common interface as on the network side 'Diameter' is used to relay the packets to the home AAA functionality (AAAH). As Findley et al [7] explain: 'This scheme directly supports IETF-flavour authentication, and by use of the currently proposed EAP Authentication and Key Agreement (AKA) mechanism would also directly support the UMTS-flavour authentication.'

Authentication ensures that the user has an acceptable set of credentials to use the access service and that their actions can be accounted for. However, users may have the rights to be able to use different services after association has been established based on their current subscription. For this purpose authorisation is required. Authorisation information would then be transferred (on request) from the policy decision function in the home network to the access point (AP) in the WLAN network. This information is then used to regulate the user's services and application of resources to the required level. The support of authorisation means that users of different classes of subscription, for example the well-touted 'bronze, silver, gold' classes of customer, can be supported.

Accounting is the final AAA function. We have already mentioned that the loose coupling means we can have a multi-provider network, where different sorts of relationships between the providers can be established. In order to fulfil the accounting requirement, information is needed on each user's resource consumption.

These three primary interworking functions allow us to both offer a service, and collect revenue from it. Although we have addressed the most basic functions needed to interwork the seamless vision we mentioned, there are two more important considerations needed, those of QoS and mobility management. These are essential interworking issues to be considered if we aim to get to a mature system capable of a form of seamless operation. Some aspects of how these functions can be handled in WLANs are discussed later in the chapter (see section 3.4).

3.3.3   Home Applications

WLAN is increasingly popular, particularly in America [5] as a home networking solution. America has many households with multiple PCs, and users appreciate the convenience of not having to wire up their house. Home configurations enable computers to talk to, and share, peripherals such as printers and scanners.

Within the UK, the 'BT Voyager' product set [9] includes a wireless ADSL modem, PC cards and USB adapters. This provides a wireless LAN for to up to ten devices, which can also share a common ADSL Internet connection.

3.3.4   Fixed Network By-Pass

Wireless LAN technologies also enable networks on two sites to be interconnected using a point-to-point topology. Users on each site can be on the same subnet or different subnets and the wireless LAN equipment can be set up to provide the appropriate routing.

A point-to-point architecture may not strictly fall under the heading of a 'Wireless LAN', but in practice it is a popular niche application for wireless LAN equipment, and many radio equipment manufacturers provide options to facilitate the point-to-point case. The main advantage to the company operating the network would be avoidance of the need to cable between buildings or to pay a leased line rental. Throughput may actually be slightly improved compared to a 2 Mbit/s leased line solution, but there will be a reduced quality of service in terms of link availability due to the use of a shared medium, namely the licence-exempt radio spectrum. To cover distances over approximately 100 m, high gain directional antennas are typically required and steps must be taken to ensure regulatory maximum power levels are not exceeded. Although usage of current wireless LAN equipment within an office or home environment is generally licence-exempt, anyone intending to deploy point-to-point wireless links should pay regard to any regulatory requirements for the country of use, which may restrict outdoor use, the distance between buildings which can be interconnected, maximum transmitter power levels and the type of traffic to be carried. There may also be a requirement for individual licensing which may attract a fee, especially if the communications service carried over the wireless link is offered to a third party or made available to the public.

Building on this principle, a number of companies are developing products to enable 'wireless mesh' networks to be established. Here wireless routers on rooftops can provide a network connecting not just two, but many buildings. Because of the potential for interference, in the long term this type of network may be more suited to more remote areas, reducing installation costs and times and enabling users to connect to a broadband access point.

3.3.5   Corporate Applications

Many small, medium and large enterprises currently deploy wireless LANs, as part of a wired network infrastructure, to enhance network coverage or to provide mobility to employees within a building or complex of buildings. All users require a client card or wireless-LAN-enabled PC or laptop, and must be in range of an access point (AP) in order to communicate on the network. Cell size will depend on the building construction and the required capacity per cell, which in turn depends on the number of users, but 10-30 m cells are typical indoors. An overlapping cell architecture will provide for continuous coverage and allow users to move freely throughout the building, using the access point roaming feature to have their network connections handed off from one cell to another as they go.

3.3.6   Networks in Motion — on the Train

Already we have seen the launch of public access wireless LANs in hot-spot areas such as railway stations. There is also interest in providing such hot-spot access for users actually on the trains. However, the types of application that may be supported in such a situation are heavily influenced by the actual architecture of the system. More specifically, current hot-spots have a high bandwidth link back into the Internet. This then means that users will have a fast, wireless Internet access service which could be used for Web browsing, mail, or games. Within the train, however, the link between the access point and the rest of the world will be based on a mobile technology such as GPRS or satellite (for cruise ships). Thus users of the train WLAN system will be sharing the limited bandwidth backhaul with other WLAN users. This will introduce a bottle-neck into the system.

In such situations, users might prefer to each have their own GPRS connection to the Internet. However, there is still a large scope for such networks:

- as users of mobile computers tend to be equipped with WLAN cards, using a train WLAN means that they do not need to also have, and change to, a GPRS card — they will still be able to have a good enough service for vital applications such as e-mail;

- the train network may operate a local cache of popular information, which may be slow-changing (pre-recorded radio programmes), and dynamic information, such as newspapers or weather reports, may be downloaded via satellite, or digital audio broadcasting (DAB), or WLAN at stations to an on-train cache — accessing this information via the train WLAN would be much quicker and cheaper for users;

- the train may provide games management servers;

- simultaneous handover of many GPRS sessions may cause performance issues for networks — this would be avoided if a single train mobile router was performing handover (perhaps with dual channels if the extra capacity is required).

One key technical issue to be considered will be address management. Currently, GPRS would assign a single IP address to the train mobile router. The GPRS mobility management then ensures that this remains reachable after handover events. However, we need to determine how the mobile nodes will be reached. One option would be for the mobile router to provide network address translation (NAT) functionality. This, however, might restrict the services that can be offered — it may make it impossible for users to access their corporate network securely. Another approach would be possible if a 3G network could give a subnet prefix to the mobile train router. In the longer term, the advent of broadband UMTS networks should improve the backhaul capacity.

The other technical issue that this system raises is that of quality of service. The train operator could provide an on board games management server. However, many popular games rely on timely co-ordination of events on different terminals, which is not possible in the current IEEE802.11b systems. Consider two user playing some kind of fighting game. The first user moves his avatar behind a wall and then the second user fires a shot. Both terminals attempt to begin transmission; however, because of the MAC implementation, the second node actually transmits first, killing the first user who in fact should have survived. Further, because the nodes are sending many small data packets, the overall throughput of the WLAN is severely reduced. Because of this, a high priority message from a doctor in first class is delayed. These issues relate to quality of service, which will be discussed in more depth later.

3.3.7   *Ad Hoc* Networks

The IETF working group charter defines a mobile *ad hoc* network (MANET) (see Chapters 7, 8 and 9) as '... an autonomous system of mobile routers (and associated hosts) connected by wireless links — the union of which form an arbitrary graph. The routers are free to move randomly and organise themselves arbitrarily; thus, the network's wireless topology may change rapidly and unpredictably. Such a network

may operate in a stand-alone fashion, or may be connected to the larger Internet ...'
[10]. These networks could be used:

- in a meeting room — this network enables users to share documents, with access
  to the Internet being used to obtain additional material;

- to extend the operated network range — a group of users in a rural community
  form an *ad hoc* network, which gives them the ability to reach an ADSL link in a
  nearby village;

- to extend the operated network capacity — an operator rapidly builds a network
  of powered, wireless routers to provide extra capacity for a stadium or city during
  a major sporting event.

There are many issues in the formation and management of such networks. Here
we simply consider those related to the wireless technology.

The use of medium sensing for access control in networks such as IEEE802.11
can cause a problem known as the 'hidden node problem'. This phenomenon
(Fig 3.1) occurs when an access point is able to successfully receive data from two
separate stations, but the stations are unable to receive signals from each other. The
problem with this type of scenario is that one of the stations may sense the medium
as being idle even though the other station is transmitting data on it. When this
happens a collision will occur at the access point as both stations are trying to send
frames on the medium at the same time.

The current fix for the hidden node problem in IEEE802.11b (the 'request to
send/clear to send' initial handshake procedure) relies on the fact that any station
that can possibly interfere with the reception of a packet from node A to B is within
the sensing range of node A. Although true for the basic service set (BSS) type of
networks it does not hold true for multi-hop *ad hoc* networks. A second problem is
the exposed node problem (Fig 3.5).

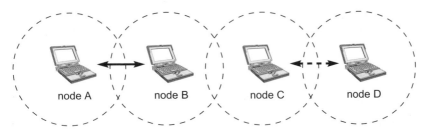

Node C detects the medium as busy
due to the transmission between A
and B. Therefore node C is unable to
transmit to D even though it would not
cause a collision

Fig 3.5   The exposed node problem.

In this situation node A is transmitting data to node B. Node C wishes to send data to node D, so it senses the medium to see if it is able to do so. Because it is in range of node B it detects that the medium is busy and so is unable to transmit its frame until the transfer between A and B is completed. The effect of this is that the legitimate (would not cause a collision) communication between node C and node D does not occur because, from the point of view of node C, the medium is unavailable. There is currently no solution for this problem within the IEEE802.11 specifications.

The hidden node problem and the exposed node problem are the main contributors to the failures encountered when using IEEE802.11b for multi-hop networks. In general, carrier-sensing wireless networks are implemented in such a way that the sensing and interfering ranges are larger than the communication range. In the case of IEEE802.11 it is generally the situation that the sensing range is more than twice the size of the communication range. The large interfering range makes the hidden node problem worse while the extended sensing range intensifies the exposed node problem.

3.3.8   Application Summary

There are many opportunities for network operators to support, and benefit from, wireless LAN 'tails' to their networks. However, there are technical issues with the IEEE802.11 standard, the basis of the vast majority of current WLANs, which will limit its long-term usefulness. As we have identified by considering applications above, these issues specifically are:

- only non-real-time applications can be supported as delay and jitter are uncontrolled and unbounded;

- there is no mechanism to prioritise different types of traffic;

- security remains a concern — in addition to issues caused by user mobility, wired equivalent privacy (WEP) implements relatively weak, static encryption keys without any form of key distribution management, making it quite straightforward for hackers to decipher information being transmitted on an IEEE802.11 wireless network;

- bandwidth is limited, and dependent on the typical packet size;

- the use of licence-exempt spectrum will become more of an issue as the number of deployed access points increases — the different systems will then start to interfere, reducing overall capacity and service reliability;

- the technology is not well suited to *ad hoc* networks.

Within the next section we look in more detail at wireless LAN technologies, and how they are developing.

3.4   Technology Developments

There are developments in all aspects of wireless LANs — for example developments to use different parts of the radio spectrum, improving data access speeds and security, providing quality of service management and mechanisms to facilitate user movement. As indicated within Table 3.1, some of these developments are within the IEEE802.11 standards. Others, also discussed below, use significantly different technologies. Beyond simply improving the user experience, many of these developments will open up new applications and services. For example, some of these technologies operate in the 5 GHz radio spectrum. Since some of this spectrum permits outdoor use we could envision such wireless LANs being used to make a low-cost 'cellular' style network.

3.4.1   IEEE802.11a/h

In addition to the IEEE802.11b standard, there are many other standards in the IEEE802.11 series as illustrated in Table 3.1. The IEEE802.11a standard is currently receiving much attention.

The IEEE802 committee has developed the IEEE802.11a standard [11] for the 5 GHz band. This standard provides data rates of up to 54 Mbit/s using ODFM. Although the transmission range is likely to be less than that of IEEE802.11b. IEEE802.11a equipment is available now, at approximately twice the price of IEEE802.11b equipment.

However, there are a number of problems with the IEEE802.11a standard for operation in Europe. Firstly, Europe has allocated this spectrum to the high performance radio LAN (HIPERLAN/2) system (discussed in section 3.4.2). This is similar to IEEE802.11a at the physical layer, but the standards have significant differences at the MAC layer. Also, the 5 GHz band is shared in Europe with particular services such as satellite uplinks, radar and space science exploration. HIPERLAN/2 took account of this problem during its development. However, the IEEE have now developed IEEE802.11h. This adds two features to IEEE802.11a to make it suitable for use in Europe. The features are:

- dynamic frequency selection — this is a switching of frequency if interference from other systems and services is detected;

- transmit power control — transmission power is reduced to the minimum necessary, as this both reduces interference (into other systems and services) and saves battery power on mobile devices.

International studies have examined the various sharing issues so that official changes to international frequency allocations can be discussed at the World Radiocommunications Conference.

Table 3.1   The different IEEE802.11 standards.

IEEE802.11a	A physical layer standard that provides data rates of 6-54 Mbit/s over the 5 GHz frequency band. Orthogonal frequency division multiplexing (OFDM) is used to provide these high data rates and the 5 GHz band offers less RF interference compared to the 2.4 GHz band. For use in USA (IEEE802.11h is the European solution).
IEEE802.11b	A physical layer standard that enhances the original IEEE802.11 direct sequence spread spectrum (DSSS) to provide 5.5 and 11 Mbit/s data rates on top of the original 1 and 2 Mbit/s, IEEE802.11b uses the complementary code keying (CCK) modulation system to make more efficient use of the radio spectrum.
IEEE802.11c	Bridge operation procedures — provides the required information to ensure proper bridging operations. Manufacturers use this standard when implementing wireless access points to ensure interoperability.
IEEE802.11d	Allow terminals to listen to AP transmissions and automatically set their physical layer attributes (power level, frequency, etc) to satisfy the local regulatory domains. The work of this group is especially important for the 5 GHz spectrum as the permitted use of this band varies considerably from one country to another around the globe.
IEEE802.11e	This standard addresses required modifications to the medium access control (MAC) layer to provide a prioritisation-based quality of service.
IEEE802.11f	Inter-access point protocol (IAPP) — the communications between access points in order to support users roaming from one AP to another.
IEEE802.11g	Physical layer enhancements to the 2.4 GHz band to enable high-speed (up to 54 Mbit/s) data throughput. The important proviso is that the IEEE802.11g equipment can interoperate with standard IEEE802.11b equipment on the same network. IEEE802.11g will use OFDM rather than DSSS to provide the higher data rates.
IEEE802.11h	The issues of using the 5 GHz band within the European regulations. This entails providing dynamic channel selection (DCS) and transmit power control (TPC) to avoid interference with satellite communications.
IEEE802.11i	MAC enhancements for enhanced security — encryption techniques such as advanced encryption standard (AES).
IEEE802.11j	Enhancement of IEEE802.11 standard to add channel selection of 4.9 GHz and 5 GHz in Japan in order to conform to Japanese rules of radio operation.
IEEE802.11k	Enable functionality for higher layers in the stack to get access to radio resource management (RRM) data captured by the PHY layer. Easier access to these measurements will enable simpler management of services (e.g. roaming, coexistence) from external systems.
IEEE802.11m	Maintenance of the IEEE802.11 MAC/PHY specification. Update the standard documentation with technology and editorial corrections.

3.4.2   HIPERLAN/2, HiSWANa

HIPERLAN/2 is a solution for the 5 GHz band [12, 13], offering a number of air interface data rates up to 54 Mbit/s, mobility and QoS for applications such as multimedia, VoIP and real-time video. HiSWANa, a Japanese standard, also operates in the 5 GHz band and has a MAC similar to HIPERLAN/2.

At the physical layer the radio uses OFDM technology, which is well known in DAB and ADSL. OFDM does not use frequency hopping or a spreading code, but it simultaneously modulates multiple narrowband sub-carriers in a channel. In HIPERLAN/2 there are 48 sub-carriers carrying data (plus 4 pilot sub-carriers) in a 20 MHz channel using binary phase shift keying (BPSK), quadrature phase shift keying (QPSK), 16 QAM (quadrature amplitude modulation) or 64 QAM modulation to deliver the required data rate from 6 to 54 Mbit/s. All the sub-carriers are used for one transmission link between a mobile station and an access point. OFDM has been chosen because it offers good protection against multi-path propagation, which affects the performance of any radio link. Although the technique has been known for some time, it is only recent advances in silicon and processing power (to undertake the necessary fast-Fourier transform signal processing) that have enabled OFDM transceivers to be realised practically and at reasonable cost.

In contrast to IEEE802.11, in a HIPERLAN/2 network, control plane signalling is used to set up a path between the client node and the access point before any data is transmitted. Essentially a connection-oriented protocol is used, with a time division multiple access/time division duplex (TDMA/TDD) structure over the air interface. This enables QoS parameters and fair access to the channel to be negotiated, controlled from the access point. *Ad hoc* networking is partially supported through provision of a direct mode for client-to-client communication, but both clients need also to be in communication with an access point in order that the radio resource can be allocated.

The benefits of HIPERLAN/2 are improved quality of service, increased throughput, and less interference. However, the recent economic downturn has seriously affected HIPERLAN/2 implementations. In many ways the standard is in competition with the IEEE802.11a/h standard, and it is not yet clear which will be most successful in the market-place. A globalisation study group and a joint task force have been established to investigate whether and how these standards could coexist, interoperate, or merge to create a single global market for 5 GHz systems.

3.4.3   Ultra-Wideband (UWB)

This is a development that is generating considerable interest within the wireless community — for further details, see Chapter 5. UWB has its origins in military radar applications [14], but the benefits of using UWB to provide multi-user communication have long been recognised since it provides good performance in multi-path environments such as inside buildings [15]. Recent advances in silicon technology are now enabling practical communications systems to be built. The principle is that pulsed transmissions are used to convey information at a high data rate. The transmission is effectively multiple narrowband signals over a very wide bandwidth, perhaps as much as 1.5 GHz. Pulse position modulation is employed with a low power level (typically 50 mW) but a high processing gain, so that the data is recovered even in the presence of high power levels from other services in one part of the spectrum. Of course, although the power level may be low, the pulses affect existing equipment in many frequency bands, with interference into the satellite global positioning system (GPS) being a major concern. The US Federal Communications Commission (FCC) have now agreed to limited use of UWB technology, but the nature of ultra-wideband transmissions still poses a challenge for the regulatory authorities in other countries. When current regulations are written for signals that fit into clearly defined frequency bands and occupy a few MHz of bandwidth at most, how is it possible to regulate transmissions that cover several bands, with an ultra-wide bandwidth by definition?

3.4.4   QoS Solutions

Quality of service is a broad term concerning the perceived value of a service by the user. Aspects of QoS include:

- controlling delay and jitter in the wireless link to support real time traffic, for services such as voice and video, as delay and jitter in a voice stream can cause echo effects as well as choppy sound which breaks when the data is delayed in the network;

- allowing interactive or urgent traffic to receive priority treatment — such services could include 999 emergency signals, or network control traffic.

The IEEE802.11 standard actually has support for transmission of time-bounded data such as voice and video using a different, TDMA-based MAC (the point co-ordination function (PCF) rather than the previously described distributed co-ordination function (DCF)). However, it is very difficult to find it implemented in any of the current IEEE802.11b wireless networking equipment. This scheme is not particularly scalable or efficient, since a single point in the area must poll the stations around it to control access to the medium.

A large number of different QoS schemes have been proposed. A simple example is known as distributed fair scheduling. In this the back-off time is proportional to the size of the data packet and inversely related to the priority of the flow. The inclusion of the packet length in the calculation provides the fairness since flows with smaller packets (real-time voice data tends to be regular small packets) get the chance to transmit more often. This gives flows with the same weighting the same bandwidth regardless of the packet sizes used, and does not starve low priority flows.

Blackburst is a medium-access scheme with the goal of minimising the delay for real-time traffic. It takes a more revolutionary approach to the problem than other MAC schemes and relies on a couple of important requirements. One requirement is that all of the high priority stations try to access the medium with a constant interval (t_{sch}). Secondly, blackburst requires that the stations are able to jam the wireless medium for a period of time to prevent other stations from accessing it. Within this scheme, the low priority stations still use the basic DCF method for medium access.

A high priority station wishing to send real-time data will wait until the medium becomes idle and then enter a blackburst contention period by jamming the channel for a period of time. The amount of time for which the channel is jammed is determined by the amount of time that the station has waited to gain access to the medium. Once the blackburst has been sent the station listens out to see if any other stations are sending longer blackbursts (meaning that they have been waiting longer to access the medium and so should access it first). If the medium is idle, the station will transmit its frame; otherwise it will wait for the medium to become idle again and then go into another blackburst contention period. The use of slotted time and enforcing a minimum frame size on real-time data guarantees that each blackburst contention period yields a unique winner who will have access to the medium.

Once a high-priority station has successfully transmitted a frame it schedules its next access attempt for t_{sch} seconds in the future. Accessing the medium in constant intervals means that the high-priority stations will eventually synchronise and share the medium without collisions. Very little blackbursting is required once the stations have become synchronised. Of the schemes we have seen, this solution:

- provides best performance to high-priority traffic with regard to throughput, especially at lower loads;

- provides high throughput and lowest delays to high-priority stations;

- ensures packet delays have a low upper bound and little jitter occurs;

- provides good medium utilisation;

- ensures no collisions between high-priority stations due to synchronisation — collision rates decrease as number of high-priority stations increases;

- can still, however, starve low-priority nodes.

The scheme under consideration for the IEEE802.11e standard is known as the enhanced distributed co-ordination function (EDCF). In this scheme the minimum contention window (CWmin) can be set differently for the different traffic priorities, providing stations with high-priority traffic a smaller contention window. As well as the contention window, the inter-frame spaces are also set differently for the various classes of service. In this case they are referred to as arbitration inter-frame space

(AIFS), where the AIFS for a given class will be a DIFS plus some (possibly zero) time-slots — the smaller the AIFS, the higher the priority.

EDCF also permits packet bursting to allow for improved medium utilisation. This means that a station can transmit more than one frame without having to contend for the medium again. Once the station has access to the medium, it is allowed to transmit as many frames as possible within a given time limit (TxOpLimit). Other stations are prevented from interrupting the burst because the transmitting station has to wait a shorter inter-frame space (IFS) between packets than normal. A collision causes the packet burst to be terminated. As it is possible for packet bursting to increase jitter within the network, it is important that TxOpLimit is not longer than the time required to transmit a maximum size data frame.

Although an improvement on PCF (better performance and less complexity), this scheme still has some weaknesses:

- low medium utilisation — this is a surprise since packet bursting should improve medium utilisation;

- suffers from high collision rates — this is probably a cause for the low medium utilisation;

- can starve low priority traffic at high loads;

- at high loads packets can suffer from very high delays (>100 ms).

Although, as described, better schemes exist, it is likely that the EDCF solution will become the only standard QoS solution — although proprietary solutions have been implemented and are being used in specialised applications [16].

3.4.5   Interaction between Wireless LAN Technology and IP Technology

In all the preceding discussions, we have been considering how wireless LAN technologies deliver bits from a base station to a user terminal — for example, how to make that bit delivery faster and more reliable. We have also implicitly assumed when discussing applications that these bits are in fact bits of IP data packets, which have been generated somewhere either within an ISP domain, or within the wider Internet, or on an intranet. Indeed, it is practical to assume that the transmission of IP is likely, for the foreseeable future, to be the only networking protocol of interest for wireless LAN systems. While we have so far assumed that the wireless technology can be considered independently from the nature of the bits that it is carrying, this is not actually true. There are a number of ways in which the two technologies can interact — and most of the interactions are not good!

Consider the following scenario. We develop a wonderful wireless link-layer technology that is high bandwidth, supports voice, video and file transfers efficiently, but when it is used as simply a hop in an entirely IP-based communications path, we find that much of the functionality is unused or, even worse, it has a negative impact upon the end-to-end communications. A commonly cited example of this would be the interaction of transmission control protocol (TCP), which manages packet loss on an end-to-end basis with a wireless network that provides local loss management across that segment of the end-to-end path through automatic repeat request (ARQ). Suppose a frame[2] fails to transmit across the wireless link layer, which then tries ARQ. While the link is trying to correctly transmit the frame, the TCP has decided that the packet (assumed to be wholly contained in that frame) must be lost as the delay in receiving the packet is so great, and so it also retransmits the packet. This clearly is very inefficient — we now have two copies of the same packet in transmission. However, it is also clearly inefficient to leave all packet-loss handling to end-to-end protocols when only one particular link is prone to high levels of loss. In this situation, frame drops on that one link could cause large numbers of packets to be retransmitted, end to end, across the network.

This is not the only way in which the wireless network can interact with IP protocols. This section considers some of the interactions, and also proposes a mechanism to control these interactions. Within this section, some standard communications model terminology is used. Layer 2 entities are individual links that make up the communications path. Thus, the wireless LAN is a layer 2 entity. It can be joined to other links, of different physical type, providing layer 3 connectivity. The IP protocols manage this connectivity. Layer 4 refers to the end-to-end communications as managed between the end terminals — the example protocol being TCP.

3.4.5.1   Mobility Management

There are mechanisms to support user mobility at both layer 2 (sometimes called roaming) and layer 3 (for example mobile IP). These mechanisms allow users to move between different wireless access points and still remain active.

Mechanisms at layer 2 are already commonly used. In a typical configuration, multiple base-stations are used to increase the coverage and capacity of the wireless network on a corporate site. Users automatically attach to the base-station with the strongest signal and, should they move around the site, they are automatically transferred to a nearer base-station. These mechanisms are very efficient. They require no complex signalling across the Internet. They are fast. They are transparent to users and they require no additional software or complex configuration.

---

[2] In this section, we will use frame to indicate the data unit that exists across the wireless network, and the term packet will be reserved for the data unit that exists at an IP level. In physical terms, the IP packet will exist inside a wireless network frame when it is being transmitted across the wireless network.

However, a layer 3 solution is also required. Firstly, the layer 2 solutions only provide mobility to computers on the same IP subnet. This typically restricts the scale of such networks to a maximum of 254 portable terminals. Thus, the layer 2 solution is not scalable. Secondly, a layer 3 solution is required for interworking across access technologies[3] and administrative domains.

3.4.5.2   Quality of Service (QoS)

Quality of service has long been recognised as something that requires elements in every layer of the communications model. Lack of QoS in just one segment of the network could totally destroy any QoS that was present within the rest of the communications path. For example, voice packets that are guaranteed a low-delay path through the network could be fatally delayed at the start of their journey as the mobile node attempts to access a wireless LAN that did not support QoS. However, QoS within the wireless segment can also destroy any QoS that was present within the rest of the communications path. For example, the link layer may attempt to provide a high QoS through use of ARQ to reduce the loss rate. This would be fatal for delay-sensitive services, which typically require timely delivery and can tolerate certain amounts of loss. Such applications would probably require no more than forward error correction (FEC) techniques. Other interactions can result from the fact that packets may be scheduled at both the network and link layers. It is then possible that the link-layer scheduler undoes the work of the network layer scheduler. Thus there are two requirements:

- the network layer needs to understand the link-layer QoS;

- the link-layer QoS should support network requirements.

This implies that there is some communication between these two layers.

3.4.5.3   Radio Resource Management (RRM)

Radio resource management is important to allow for the efficient management of what are commonly scarce or valuable resources in the radio layer. Within cellular systems very complex RRM systems are deployed. However, the WLAN is very different, and such complex and therefore expensive options are unfeasible when still trying to maintain the cost advantages of the WLAN. Therefore approaches must be taken in a more localised fashion.

However, we have already mentioned that applications for WLAN networks include interworking with macro area communications technologies.

---

[3] Today, for example, to provide this layer 2 mobility requires that all the wireless LAN base-stations are from the same network manufacturer.

Once we start to consider the interworking of different networks there are clear gains that could be made by the management of resources between them. The way to do this is by using RRM — but this has not, typically, been considered in depth. However, precisely this issue has been considered within the EU Mobile IP-based Network Developments (MIND) project [17] (see also Chapter 11). We will make the assumption here that the WLAN technology and macro area network are interfacing in an IP-based core network (not unrealistic).

Consider RRM which is managed solely at the link layer (L2). With a single technology but multiple domains, it is still possible to have a L2 solution; however, in the 'native-IP' environment this could cause conflicts with the network layer (L3) interactions that will be taking place. Therefore communication with L3 entities becomes important.

When multiple technologies are introduced, different link layers (L2) have to interact with each other. We have an area of commonality, that of the IP Layer (L3), so it is natural to use this layer as the bridge between the technologies through the IP to Wireless (IP$_2$W) interface. At L3, a decision can then be made on the best resource management across the multiple technologies. Similarly in the 'multiple domain to multiple technology' case, L3 decisions are needed not only in order to manage cross-technology RRM, but also to remove inter-domain management conflicts at L3.

3.4.5.4   IP to Wireless (IP$_2$W) Interface

We have seen that many functions can be provided at the wireless layer. This can be essential, or can lead to a more efficient solution. However, we have also seen that there are dangers in this — functions can interact in unpredictable ways. One way to solve these problems is through the use of a well-defined interface between the layer 2 and 3, such a one being the IP$_2$W interface proposed by the EU Broadband Radio Access for IP-based Networks (BRAIN) project (Fig 3.6) [18] (see also Chapter 11). The IP layer uses this to interface to any link layer, and a link-layer-specific convergence layer is used to adapt the native functionality of the specific wireless technology to that offered by the IP$_2$W interface.

In addition to providing support for QoS and mobility management, this interface also handles the normal L2/3 issues such as address resolution, in a standard, technology-independent way.

The concept of a truly dumb link layer will become increasingly outdated as the functionality expected from the Internet grows, so there will be an increasing requirement to maintain the benefits that IP achieved through its clear layering principles. The IP$_2$W interface will enable link-layer technologies to provide advanced support for features such as QoS in a controlled way. It is likely that wireless technologies may develop and implement this type of interface initially, but in truth it is to be hoped that any such implemented interface follows the model above, as this is a truly generic interface that does not care whether the physical layer is wired, wireless or purely telepathic.

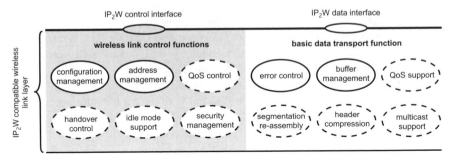

Fig 3.6   BRAIN IP$_2$W interface.

3.5   Summary and Future Outlook

Wireless LAN technology has grown rapidly in importance over the last two years. Indeed, we happily predict that, by the time of publication, much of this material will already be outdated! However, the key points of our discussion will still be valid:

- wireless LAN technology provides both competitive and complementary functionality to 2G and 3G mobile networks;

- there is a wide range of applications in many market segments;

- the markets are still growing rapidly with the increasing penetration of laptop computers and PDAs.

The market segment of wireless networks looks set to continue growing strongly through 2007 and beyond with world-wide end-user expenditure on wireless LAN equipment expected to grow to $3.9bn by 2007 [19]. The actual compound annual growth rate of WLAN peripheral shipments is estimated to be 42% between 2001 and 2007. Figures like these show that wireless is set to make a big impression over the next few years.

The ease of use of the technology has been key in the explosion of use. However, there are concerns [20] that future developments — particularly those aimed at improving security — could be incompatible with the ease-of-use requirement. Indeed, the practical implementation of the security model for many wireless LAN products has been questioned, casting doubt on the use of in-built security. In fact, the default 'out-of-the-box' wireless LAN has security switched off, and many early networks ignored security issues. Reports of 'drive by hacking' have proliferated in the press, where Internet or even intranet connectivity is stolen by uninvited users parked outside enterprise premises. In general, these problems are now easily solved through higher layer solutions, for instance the deployment of Internet protocol security (IPSec) implementations supporting the use of certificates and other forms of strong authentication. For the intelligent user, security should now be a non-issue. BT is strong in the security area and can help build those secure wireless LAN solutions.

The installation base of wireless LAN equipment in new laptop computers and PDAs is likely to continue to grow in the future. In most cases this is aimed at the professional business user. In 2001 over 20% of laptops and PDAs destined for professional use came complete with wireless LAN adapter equipment. This figure is expected to reach 90% by 2007 [19] and is likely to be helped along by the chipset manufacturer Intel who have introduced the 'Centrino' mobile networking concept, building wireless networking technology directly into some of their support chipsets [21]. Intel's move is likely to cause other chipset manufacturers to follow suit and could cause the take-up of WLANs to grow even faster than expected. Furthermore, figures from Analysys [22] suggest that nearly 50% of the estimated 1.5 million users of public wireless LANs by 2006 will be corporate or SME users.

Roaming — the ability to move from one wireless LAN site to another will be one of the key new features of wireless LANs. This can be seen in the rise of BOINGO [23] — an umbrella organisation that shares a common set of security details to a number of different wireless LAN owners, thus providing a kind of roaming functionality. We predict that the ability to roam between different operators and different network types will become increasingly important. A key role for BT in future will be in bringing together WLAN with other access technologies to provide a seamless service to users, with single authentication procedures, simple billing, simple access to network-independent services, VPN over WLAN, and roaming agreements for global services.

Although current wireless LANs offer significant capacity advantages over existing mobile networks, a few Mbit/s shared between a few users will rapidly become a limitation. New technologies and standards that offer increased bandwidths will again become necessary. We can look to technologies operating in the 5 GHz band, IEEE802.11h or HIPERLAN/2, and UWB systems to provide the first stage of advancement.

The ability to support real-time services over wireless LANs will become increasingly important — giving users the chance to support voice-over-IP over WLAN. This will not be possible without improvements in QoS mechanisms for wireless LANs. Further, the inefficiency of current standards when transmitting small data packets will need to be addressed.

Wireless LANs only provide a small part of the future mobile solution. In our discussion on L2/3 interworking, we have considered some of the problems of wireless LANs operating as simply a component within a larger (IP-based) network, and proposed that a well defined L2/3 interface will become more necessary in the longer term, as more (layer 2) networks are combined seamlessly into a larger, access-independent IP network. Another key broader issue, that of personal mobility, is considered elsewhere [24].

This chapter has indicated that there is significant activity in the development and application of wireless LAN technology. While many issues remain, solutions to many others are near to realisation — in many cases the technical solutions exist, the relative strengths and weaknesses of different options are understood (at least from a technical point of view), and it is simply a matter of the standardisation process being completed. This process will now be complicated by the widespread use of the IEEE802.11 solution. In other areas, innovative research may be required for many years. During that time, however, while the technology issues continue to be investigated, we can expect to see many exciting changes in the uses and applications of wireless LAN technology.

References

1   Proceedings of the 51st IETF meeting (August 2001) — http://www.ietf.org/

2   BT Openzone — http://www.bt.com/openzone/

3   WiFi Alliance — www.wifialliance.net/

4   Institution of Electrical and Electronic Engineers: '*IEEE Std 802.3 —Information technology — Telecommunications and information exchange between systems — Local and metropolitan area networks — Specific requirements — Part 3: Carrier Sense Multiple Access with Collision Detection (CSMA/CD) Access Method and Physical Layer Specifications*' (2000) — http://www.standards.ieee.org

5   Ovum Forecasts: '*Global Wireless Markets*', (February 2002).

6   ETSI BRAN — http://www.etsi.org/bran/

7   Findley, D. et al: '*3G Interworking with Wireless LANs*', Proc IEE 3G 2002 Int Conf, London (May 2002).

8   Blunk, L. and Vollbrecht, J.: '*PPP Extensible Authentication Protocol*', IETF RFC 2284 (1998).

9   BT Voyager — http://www.voyager.bt.com/

10 IETF MANET working group charter — http://www.ietf.org/html.charters/manet-charter.html

11 Institution of Electrical and Electronic Engineers: '*IEEE Std 802.11a and errata, Supplement to IEEE Standard for Information technology — Telecommunications and information exchange between systems — Local and metropolitan area networks — Specific requirements: Part 11: Wireless LAN medium access control (MAC) and Physical layer (PHY) specifications: High-speed physical layer in the 5 GHz band*', (1999) — http://www.standards.ieee.org/

12 Johnsson, M.: '*HIPERLAN/2 — The Broadband Radio Transmission Technology Operating in the 5 GHz Frequency Band*', HIPERLAN/2 Global Forum (1999).

13 Khun-Jush, J., Malmgren, G., Schramm, P. and Torsner, J.: '*HiperLAN type 2 for broadband wireless communication*', Ericsson Review No 2 (2000).

14 Engler, H. F. Jr: '*Technical Issues in Ultra-Wideband Radar Systems*', in Taylor, J. D. (Ed): '*Introduction to Ultra-Wideband Radar Systems*', CRC Press (1995) — http://www.uwb.org

15 Win, M. Z. and Scholtz, R. A.: '*Impulse Radio: How It Works*', IEEE Communications Letters, **2**(1) (January 1998) — http://www.uwb.org

16 Spectralink: '*Provision of VoWLAN solutions for special markets including health and education*', — http://www.spectralink.com/products/pdfs/SVP_white_paper.pdf

17 The MIND project — http://www.ist-mind.org/

18 The BRAIN project — http://www.ist-brain.org/

19 Rolfe, A.: '*Wireless LAN Equipment Market: Strong Growth Set to Continue*', Gartner Dataquest (October 2002).

20 Henry, P. and Luo, H.: '*WiFi — what next*', IEEE Communications Magazine (December 2002).

21 Intel Mobile Techologies — http://www.intel.com/products/mobiletechnology/index.htm

22 Analysys — http://www.analysys.com/

23 BOINGO Wireless — http://www.boingo.com/

24 Ralph, D. and Searby, S. (Eds): '*Location and Personalisation: Delivering Online and Mobility Services*', The Institution of Electrical Engineers, London (2004).

4

FUTURE APPLICATIONS OF BLUETOOTH

S Buttery and A Sago

4.1 Introduction

After several years of over-promotion, Bluetooth short-range wireless technology has finally made the transition from 'slideware' to hardware. Bluetooth is described in different ways by different interested parties but the Bluetooth Special Interest Group (SIG) — the organisation that has driven its development since it 'went public' in 1998 — gives Bluetooth the following headline description [1]: 'It works whenever you work, seamlessly connecting all of your mobile devices. Creating unprecedented productivity.'

However, moving away from marketing-speak and into the less dramatic language used by engineers:

- Bluetooth is a standard for a short-range wireless technology;

- the *de facto* standard from the Bluetooth SIG was converted to an IEEE standard in 2002 [2];

- Bluetooth operates in the 2.4 GHz ISM (industrial, scientific and medical) band, which is licence-exempt on a global basis (although a few countries still have some specific local regulations);

- Bluetooth uses a 'fast frequency hopping' radio technique, changing its operating frequency 1600 times a second — this enables it to carry on working even in areas of high interference, an important point, considering that it has to share its radio spectrum with many other devices, including microwave ovens and wireless local area networks (WLANs);

- Bluetooth can support voice and/or data (at rates up to 723 kbit/s);

- the Bluetooth specifications define a number of different 'power classes', with the highest power variant ('Power Class 1', transmitting up to 100 mW) supporting operating ranges up to about 100 metres. However, most of the devices available today use lower power transmitters, restricting the typical range to around 10 m.

The modest range offered by most current Bluetooth products is a deliberate design choice. This arises because Bluetooth is primarily a wireless 'personal area network' (PAN) technology — something that links together all the various devices that you might have about your person. To perform this role, it has to meet three specific design challenges:

- it must be so compact that it can be included in small products, such as mobile phones and wireless headsets;

- it must consume so little power that it will not have a significant impact on the battery life of the devices into which it is built;

- it must eventually be so cheap that vendors will include it within their products as a matter of course.

Bluetooth is well on the way towards satisfying these objectives, with silicon vendors now releasing chip-sets that can deliver Bluetooth functionality on a single chip (plus a few basic external components). So, a complete Bluetooth implementation could now [3]:

- occupy an area of just 90 mm^2;

- consume as little as 25 μA while in 'deep sleep' mode, rising to a peak of ~25 mA when transmitting;

- cost less than $4 a terminal when produced in volume.

It is not surprising, therefore, that large numbers of Bluetooth-enabled products are now being produced. It is estimated that 35 million Bluetooth chip-sets were shipped in 2002, and annual shipments are predicted to rise to 510 million by 2006 [4]. However, an important question for the future of both fixed and mobile telecommunications is: 'How will these products actually be used? Will Bluetooth just continue to be a convenient way of linking together the devices that you carry around (i.e. a 'wireless wire'), or could it have a more fundamental impact on the way we all communicate?' This is the question that this chapter attempts to answer, speculating on alternative future roles for Bluetooth and providing some initial thoughts on how they could be enabled.

4.2   What's Special about Bluetooth?

4.2.1   Bluetooth versus DECT

DECT (Digital Enhanced Cordless Telecommunications) is a highly successful technology which has been deployed in over 100 countries worldwide for delivery of voice communications [5]. In the UK, the majority of cordless telephones on the market now use DECT. With its own dedicated frequency band at 1.88 GHz and an automatic algorithm for avoiding interference from other DECT systems, consumers are able to enjoy high-quality, trouble-free cordless telephony. From the start, the DECT standard has included data networking capabilities and these have been brought together under the guise of the DECT Packet Radio Service (DPRS). This provides a level of device interoperability for data applications, in the same way that the generic access profile (GAP) provides voice interoperability for handsets and base-stations from different vendors. The addition of new modulation schemes has also increased the possible data rate to 2 Mbit/s, and DECT is included in the family of radio standards for third generation radio systems.

Full plug-and-play interoperability is now provided through the introduction of the DECT Multimedia Access Profile (DMAP), which brings support for multimedia services over DECT products through GAP and DPRS. DMAP enabled devices such as PCs, printers and cameras are now possible.

It would seem therefore that DECT has already done much that Bluetooth is trying to do. In practice though, despite the runaway success of DECT voice products, DECT data has never really been a big seller. BT has products in this space, but sales of wireless LANs have already overtaken sales of DECT data products. The complexity and cost of DECT means that it has not been automatically built into business electronic equipment such as laptop computers or into consumer equipment such as cameras and printers. Instead, all these items, and more besides, are now arriving in homes and offices with Bluetooth built in.

4.2.2   Bluetooth versus WiFi Wireless LAN

There has been much talk recently about Bluetooth competing with wireless local area networks (LANs) — from a pure performance perspective, there is not really much of a contest:

- the dominant wireless LAN technology, IEEE802.11b (or 'WiFi'), offers much higher data rates than Bluetooth, delivering usable data rates up to 6 Mbit/s — more recent wireless LAN standards (802.11a and 802.11g) are even more impressive, offering useful throughputs in excess of 25 Mbit/s;

- wireless LANs offer much greater range, operating over distances in the order of 50 metres indoors as opposed to a typical range of around 10 metres for Bluetooth (for the widely deployed Class 2 products).

While these comparisons are technically correct, they completely miss the key point about Bluetooth — it is designed for a myriad of wireless PAN applications, not as a competitor to 802.11b. Moreover, the low cost and size of the Bluetooth radio means it is being included in equipment by default, and over time it is likely to become significantly more widespread than 802.11b wireless LAN. The main reason for this is the fact that Bluetooth will increasingly be included in mobile telephones — currently, the penetration is less than 10%, but this is expected to rise to 75% by 2006 [6]. To put this into perspective, even if every mobile computer sold in 2006 was equipped with a wireless LAN and not Bluetooth (which is a very hard scenario to imagine), forecasts suggest that the sales of Bluetooth-enabled mobile telephones would still be in excess of seven   times the sales of wireless-LAN-enabled mobile computers [6].

Whereas, in the short term, the use of wireless LANs will be dominated by business users, the Bluetooth user base will consist of people from all market segments — from true mass market right through to high-end corporate. This is what is special about Bluetooth — the sheer quantity of enabled devices. With Bluetooth heading towards 70% penetration of the mobile telephone market, this suggests that (in time) over half the population of the UK will be carrying around a device capable of broadband wireless communications — surely this must create many new service opportunities! The next section of this paper attempts to address this — what roles could Bluetooth play in the future?

4.3   Possible Future Roles for Bluetooth

Over the next few years, the prevalence of Bluetooth devices could enable the technology to broaden into a number of applications beyond its original 'PAN' role of 'cable replacement'. A number of these are examined below.

4.3.1   *Ad Hoc* Networks

Bluetooth's ability to form *ad hoc* connections to other users means that it can be used for a wide range of 'social networking' applications. Wherever people meet, they can use Bluetooth for sharing files (e.g. business cards, photos) and for linking together devices for activities such as multi-player games. One of the companies leading in this area is Nokia, who have recently announced that their N-Gage product will support multi-player gaming via Bluetooth connections [7].

4.3.2   Home Base-Station

A number of vendors are now developing Bluetooth 'home base-stations'. There are a number of ways in which these could appeal to customers — and just three of these are suggested below.

- What 3G could have been

  The media frenzy surrounding third generation mobile (3G) a few years ago was sparked by the vision of a mass market of users being able to access high bandwidth, video-rich services while on the move. As this vision has been passed down the line from the marketing teams to the designers and finally to the unfortunate souls who actually have to worry about the cost of deploying networks, this vision has faded somewhat. While there can be no doubt that the large, full-colour screens of 3G telephones will bring about some exciting new services, economic reality dictates that — in the near future — we will not be able to stream hours of decent quality video to our handsets without running up a fairly hefty bill. However, Bluetooth could change all this — at least in the home.

Plugging a Bluetooth base-station into a broadband network connection will enable flat-rate, high-bandwidth services to be delivered to handsets, virtually anywhere in the house. Interestingly, this may not need to be constrained by the ~10 m range limitation that was mentioned earlier. As the base-station will be mains-powered, it can:

— operate in high power mode ('Class 1'), increasing the range on the 'downlink' (base-station to handset);

— run its radio receiver in a way that uses more mains power but improves its performance (i.e. it improves the 'receiver sensitivity').

- Cutting mobile telephone bills

  Various surveys (for example, Strategy Analytics [8]) have examined the way in which mobile telephones are used and, rather surprisingly, have concluded that around 30% of mobile telephone calls are made from the home. There are many different reasons for this [9], but, for many people, it is purely down to convenience — they have the mobile telephone with them (with all their 'favourite' numbers programmed in), and that old wired telephone is just that bit too far away to go and pick up, even when a call is more expensive on the mobile.

  A Bluetooth home base-station could come to the rescue in this scenario, allowing the customer to retain both their inertia and their money. The customer can make the call from their mobile but the handset (if so designed and configured) can send the call out via Bluetooth, rather than via the cellular radio link. The Bluetooth base-station can then connect the call over the fixed network.

- Mobile coverage enhancement

  Although the UK as a whole has very good mobile coverage, there are still many places where in-building coverage is poor. Solutions are now being developed that enable the mobile telephone network to extend seamlessly over broadband networks and Bluetooth access points — thereby enabling customers to bring mobile coverage into their own homes.

4.3.3   Enterprise Mobility Solutions

This is essentially an extension to the last proposition — developing the simple 'home base-station' concept into a multi-cell (and possible multi-site) solution that can offer an integrated mobile voice and data solution for business customers. However, implementing such a solution requires a number of challenges to be met — including support for handover from one Bluetooth access point to another.

4.3.4   mCommerce

As Bluetooth is built into more and more mobile telephones, some very interesting mobile eCommerce (mCommerce) opportunities are created. The thinking behind this statement is as follows:

- in the absence of the long-awaited 'eWallet', mobile telephones are the only device that people currently carry around that actually combines the guarantee of money (via the presence of a SIM card associated with an account) and the ability to communicate;

- before Bluetooth, it was quite hard to make use of this — what could you do with a mobile telephone that was better than ringing someone up and giving them your credit card number?

- with Bluetooth, things look very different; without realising it, we finally have an eWallet — via pre-pay or post-pay, users load money on to their mobiles, and then can wander around dispensing this wirelessly via Bluetooth, and, with the right support systems in place, such a solution could enable a whole range of 'micropayment' applications, such as:

  — buying from vending machines;

  — buying low-value tickets (e.g. for parking or underground);

  — micro-betting;

  — small value 'cashless' purchases beneath the threshold of credit and debit cards.

Obviously, the security implications of any such service need careful thought, but this approach has a good starting position from which to address these issues.

4.3.5   Public Access 'Hot-Spot' Technology

Several companies around the world have launched public access 'hot-spot' wireless LAN services. The benefits offered by this vary according to your perspective:

- the attraction of this to an existing wireless LAN user is clear — even when they are out and about, if they are at a hot-spot they can achieve data rates approaching those they get in their home or office, while still using their normal data terminal (typically a notebook computer);

- to a fixed line broadband operator, public wireless LANs offer a way to extend their broadband service out into the mobile space, thereby differentiating their product and creating an attractive 'value added service';

- mobile operators are currently endeavouring to roll out 3G networks, and many initially regarded hot-spots as just something else that could erode the business case that underpins this expensive deployment — however, more recently, some mobile operators have started to shift their positions, and are now realising that wireless LANs can actually be sold as part of the 3G vision, providing lower cost data 'picocells' to complement their new cellular technology (W-CDMA, wideband code division multiple access).

However, while everyone can see the attractions of a wireless LAN hot-spot service, getting one to work commercially is not without its difficulties. Figure 4.1 attempts to represent the main challenge — the fact that you only get someone to actually use your service when:

they need the service enough to pay for it, and,

they are carrying a suitably equipped terminal, and,

they have the opportunity to use the service, and,

they are passing through an area where coverage has been provided.

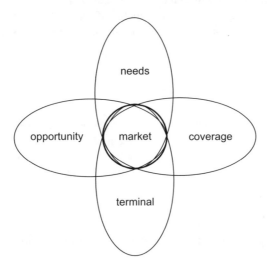

Fig 4.1   The market for public access wireless services.

Aficionados of logical statements will instantly spot the problem here — 'ANDing' a number of 'tests' together like this means that the chances of success become small, especially in the short-term when:

- many people have not yet realised they need this service;

- the majority of notebook computers in use today do not currently have wireless LAN functionality (although an increasing number are now being shipped with the necessary chip-sets built in);

- a user would currently have to have the time, space, security and seating (or remarkable PC balancing ability) to fire up their notebook and log on before they could actually use the service;

- in the early stages of roll-out, no operator is going to have extensive coverage.

Fortunately, Bluetooth has the potential to address a number of these problems. Most importantly:

- it extends the 'public wireless access' concept right into the consumer market space, enabling mass-market applications (e.g. downloading MP3 files on to portable players);

- it could greatly increase the potential user base for these public wireless services;

- it would enable entirely different usage patterns — users could access the service from small PDAs, smartphones, MP3 players, etc, without the need for the space or time to sit down and boot up their notebook computer.

Obviously, Bluetooth's low range means that it is never going to be the right technology to cover, say, every square metre of an airport, but it could be a very attractive way for operators to open up their public wireless services to a much larger potential user base.

4.3.6   Information Services

Tourist boards, transport providers and other organisations are increasingly providing information services over high technology systems such as 'multimedia kiosks'. While these clearly have a higher 'wow-factor' than the traditional, dog-eared leaflets, they do face a number of challenges, including:

- comparatively high cost;

- comparatively high 'footprint';

- the need for input and output devices that are both easy to use and difficult to vandalise;

- the need for sufficient security to stop unauthorised removal;

- the difficulty of finding a reliable and cost-effective way in which customers can take information away with them (e.g. local history, video clips, audio tours, maps, address lists, etc).

If, in a few years time, the penetration of Bluetooth is such that the majority of people will be carrying around an enabled terminal, this should enable Bluetooth to play a major role in this area.

While most information points will have to remain accessible to those without Bluetooth, a large number can be specifically enabled to support control via Bluetooth:

- menus can be passed to the handset, and the information-selection process can be controlled from there, thereby enabling many users to interrogate one information point at the same time (without having to queue up to use a keyboard or touch-screen),

- information can be downloaded into the user's terminal for them to take away.

4.4   Barriers

Some of the barriers that will have to be overcome before Bluetooth can take on these wider roles are discussed below.

4.4.1   Bluetooth Interoperability

There is a distinct feeling of 'Groundhog Day' about the launch of new short-range wireless technologies — every time the same things happen in just the same way:

- a truly impressive burst of industry effort starts to create a new and exciting standard;

- each equipment manufacturer (with the whole-hearted encouragement of solution and service providers) rushes to develop equipment that meets the standard, but is just that bit better than all the other products that are being built;

- consequently, the standards are finalised with just enough flexibility built in to support this differentiation;

- as a result, all of the 'standards-conformant' products made in the first year or two stubbornly refuse to interwork with other vendors' equipment, and the trade press have a field day pointing this out to their readers;

- eventually, the main interested parties get together and find some way to sort this out and, typically about three years after launch, everything is working fine, and the market really starts to take-off — the customers are happy, albeit slightly bemused that this was not sorted out in the first place.

So, following in the well-worn tracks of DECT and IEEE802.11b, Bluetooth is now entering the final stage of this process. The trade press are still carrying stories about Bluetooth interworking problems [10], but these issues are being overcome through the Bluetooth Qualification Programme [11]. Bluetooth qualification is the certification process required for any product using Bluetooth wireless technology before it can be put on the market carrying a Bluetooth logo. Over 2000 member companies in the Bluetooth Special Interest Group (SIG) have a royalty-free licence to use Bluetooth technology in products listed on the Bluetooth Qualified Products List (QPL).

There are over 900 licensed Bluetooth end products, subsystems, components and development tools. Products are added to the QPL by 34 specially authorised Bluetooth Qualification Bodies (BQBs), including four UK companies. The actual testing can be carried out by any competent body, except for Category A tests on radio conformance testing (covering the radio frequency, baseband, and physical test specifications) and protocol and profile conformance testing (covering the baseband, link manager, L2CAP and profile conformance test specifications). For these Category A tests, there are just 16 authorised Bluetooth Qualification Test Facilities.

Companies developing new products can participate in a Bluetooth UnPlugFest. Ten of these had taken place up to 2003. Basically it is a chance for developers to lay their products bare behind closed doors in the presence of developers and equipment from other companies, to tweak parameters and iron out any bugs. To save embarrassment there are never any published results from UnPlugFests and participants are sworn to secrecy. This approach, requiring voluntary co-operation between rival companies, is becoming increasingly prevalent in the field of product development.

As a further caveat on interoperability, although UnPlugFests happen and the Bluetooth Qualification Programme exists, there are no guarantees regarding particular configurations that may be encountered in real-life environments. How does the equipment cope with multiple units configured in a single network? How would that network co-exist with an adjacent network? These performance aspects are what give equipment from one vendor an edge over ostensibly similar products from a rival vendor.

4.4.2   Bluetooth Profiles

Bluetooth interoperability builds on the concept of profiles. These create agreed sets of functions that can be used to support specific applications, and are essential for devices to interwork in the way that their owners want. Currently, as a large number of profiles are recognised by the Bluetooth SIG, it would be impractical for every Bluetooth device to support every profile. Clearly, however, two devices which do want to communicate must have the necessary profiles in common for the tasks they wish to undertake.

Some of the better known profiles are listed below:

- Generic Access Profile (basic device discovery and connection features);
- Service Discovery Profile (allows devices to find other compatible devices);
- Cordless Telephony Profile;
- Intercom Profile;
- Serial Port Profile;
- Headset Profile;
- Dial-up Networking Profile;
- Fax Profile;
- LAN Access Profile;
- File Transfer Profile;
- Object Push Profile;
- Synchronisation Profile.

Profiles that have been recently developed, or are still under development, include:

- Advanced Audio Distribution Profile;
- Audio Video Remote Control Profile;
- Basic Imaging Profile;
- Basic Printing Profile;
- Extended Service Discovery Profile;
- Generic Audio Video Distribution Profile;
- Hardcopy Cable Replacement Profile;
- Hands-free Profile;
- Human Interface Device Profile;
- Personal Area Networking Profile;
- SIM Access Profile;
- Common ISDN Access Profile.

While these profiles are clearly a good thing, the idea of profiles should be borne in mind when considering Bluetooth's use for more radical applications. Unless such new applications are supported by the profiles that are included in a large base of deployed terminals, there is extra work to be done through the SIG and with hardware manufacturers before suitable terminals can be made available. So, there is a real requirement on any company intending to develop a radical new application for widespread use to make sure that there are profiles defined that can support it. Progressing any new ideas for profiles through the Bluetooth structure of Expert Groups, Policy Committees, Study Groups and Working Groups can be both time consuming and expensive. Even then, actually getting the profiles into terminal devices may require tie-ups with hardware manufacturers.

4.4.3   Applications

Even if all Bluetooth terminals were perfectly compatible at both a hardware and a profile level, they still would not be able to support all the applications mentioned in the last section. This is because many of these will require specific application-layer software to carry out the necessary functions, and so either terminals will have to be shipped with this software built in, or there will need to be some user-friendly way of downloading new applications as and when they are required. When the terminal is a portable PC, this downloading is not likely to be a problem. However, when the terminal is a mobile handset or an MP3 player, it could be considerably more difficult.

4.4.4   Device Limitations

Bluetooth will be built into a wide range of devices, and many of these will not be able to take part in all of the advanced mobility-oriented applications mentioned here. For example, it is difficult (although strangely pleasing) to envisage the oft-discussed Bluetooth-enabled fridge being used for e-mail retrieval at an airport. So, it is important not to take the various Bluetooth forecasts at their face values when trying to estimate potential future customer volumes — different applications could have very different numbers of devices that could exploit them.

4.4.5   Usability and Service Integration

The other barriers described in this section are largely technical in nature. However, it is important not to neglect the human aspects of this — as Bluetooth is aimed at the mass market, it must be extremely simple to buy, configure and use. To illustrate the importance of this, consider the 'public access hot-spot' application mentioned in the last section. For this to be a success, it must:

- work with Bluetooth-enabled terminals without any complex configuration;
- be as easy to use as any alternative access method (e.g. a data-enabled mobile telephone);
- offer significantly improved service and/or significantly reduced prices;

- be usable under an existing billing and service relationship — indeed, ideally, the access via Bluetooth should appear to be just a faster, cheaper 'mode' of a user's cellular service.

If all of the above tests are passed, then the question ceases to be: 'Why would you use Bluetooth for public access?' and becomes: 'Why wouldn't you use it?' Usability will inevitably be one of the issues that dictates the success (or otherwise) of various Bluetooth products. In some respects, it would make the user's life easier if there were a high level of consistency across the user interfaces of different Bluetooth products from different vendors. However, the commercial imperative for differentiation and the diverse nature of the various Bluetooth terminals means that, in practice, there will always be wide variations.

Ease of use is now one of the top issues on which the Bluetooth SIG is focusing. It has recently announced its '5-Minute Ready' programme [12], which encourages manufacturers to develop Bluetooth products that consumers can use within five minutes of taking them out of the box. Simple tools to aid developers in this quest include an official lexicon of Bluetooth phrases translated into 34 languages, to ensure users always receive written instructions in a consistent fashion, regardless of the country of origin.

4.5   Summary

After several years of hype, Bluetooth is now finally starting to have a commercial impact. Today, Bluetooth is really just used as a 'wireless wire' — providing a convenient way for people to connect the various devices that they carry around. However, with Bluetooth predicted to be included in over 70% of new mobile telephones (by 2006), Bluetooth's role could grow considerably, even to the extent where it changes how people think of wireless communications. In the future, Bluetooth could play a significant role in applications such as:

- *ad hoc* networking;

- mCommerce;

- information services;

- public wireless hot-spots;

- home and office mobile base-stations.

The last two applications are potentially the most important. Bluetooth has the potential to become a genuine second wireless link into a mobile terminal, enabling users to get the best of both worlds (i.e. wide-area connectivity from cellular technology, enhanced by the localised availability of higher data rate and lower cost services via Bluetooth). However, to realise this vision, the industry will have to ensure that:

- terminals capable of supporting this vision are produced;

- interoperability reaches the high standards that technologies such as GSM have shown to be possible;

- services are presented to customers in such a way that the benefits are obvious but the underlying technologies are not.

References

1   Bluetooth — http://www.bluetooth.com/

2   IEEE standards — http://standards.ieee.org/catalog/

3   Texas Instruments — www.ti.com/

4   In-Stat/MDR — http://www.instat.com/

5   DECT Forum — http://www.dectweb.com/

6   Gartner Dataquest — http://gartner.com/

7   Nokia — http://www.n-gage.com/n-gage/gd_features.html

8   Strategy Analytics — http://www.strategyanalytics.com/

9   Oftel: '*Consumers' use of mobile telephony*', Q10 (August 2002) — http://www.oftel.gov.uk/

10 Computer Reseller News (UK): '*UK: Comment — getting to the root of Bluetooth's problems*', (June 2002) — http://www.crnmedia.co.uk/

11 Bluetooth Qualification — http://qualweb.opengroup.org/

12 Planet Wireless — http://www.blue.telecoms.com/

5

ULTRA-WIDEBAND AND ITS CAPABILITIES

X Gu and L Taylor

5.1   Introduction

Ultra-wideband is a relatively new technology. The term UWB was introduced by the US Defense Advanced Research Project Agency (DARPA) only in the late 1980s [1]. Until recently, its development has been targeted at radar- and location-based applications. This is because the short pulse nature of the signal transmission results in very high-resolution timing information. However, ultra-wideband can also transmit large amounts of data (10s-100s Mbit/s) over a very wide frequency spectrum of a few GHz with a restricted power level <10 nW/MHz over short distance of a few metres. On 14 February 2002, the Federal Communications Commission (FCC) [2] lifted the restriction on the use of UWB technology for non-military applications. New devices that incorporate the technology are now beginning to emerge as wireless LANs. With refinement of the technology, UWB systems will eventually be able to transmit data over a very high speed ranging from 400 Mbit/s to 500 Mbit/s [3]. UWB circuits need very little power to achieve these data rates (within the region of tens of mW), which is between one tenth and one hundredth of the power required by devices such as mobile telephones and existing wireless LANs for the equivalent data rate. This is ideal for battery-powered devices.

Unlike traditional narrowband systems, UWB generates short pulses and uses these pulses for data modulation. Therefore, UWB is alternatively referred to as impulse, carrierless or baseband transmission. However, it has now been realised that UWB does not have to be impulse or carrierless [4]. This is because the FCC report and order [2] only defines UWB as a signal that occupies more than 500 MHz in the 3.1-10.6 GHz spectral mask (see section 5.3 for frequency mask for UWB signals). The new regulation is based on frequency band with power limitations, not on the types of data modulation and multiple access schemes. Therefore, the new regulation has motivated UWB developers to adopt more flexible technologies (not just the impulse-based approach) to fully realise the potential of UWB. This chapter will explain how UWB works, using both the impulse-based approach and a newly proposed multi-band approach [3, 4]. The chapter will also describe potential markets and applications. The final part of the chapter addresses important issues on standards and regulations.

5.2   Single-Band UWB

UWB can be described as a radio technology that transmits and receives extremely short pulses with a controlled duration typically from a few tens of picoseconds up to a few nanoseconds. The energy spectrum of the resultant waveform is extremely wide from near DC to a few GHz. The extent of this wideband feature enables the system to have a very low power spectral density (PSD) across the whole spectrum. The low operation frequencies enable UWB systems to penetrate materials such as walls, something that becomes harder for high frequencies. There is no carrier involved for data transmission and receiving, which means that the system is less complex to manufacture than traditional radio. Information transmission can be achieved using any standard digital modulation scheme, such as amplitude, time and phase.

Because of the wide bandwidth of the pulses, the UWB signal can overlap with the spectrum used by existing narrowband systems, such as a global positioning satellite system (GPS) and satellite radios. Regulation requires that UWB should not interfere with other short-range narrowband radio systems operating in a dedicated spectrum. The bandwidth and the centre frequency of the signal are determined by the pulse width and shape. It is therefore essential, when designing a system, that the interference issue is considered. On the other hand, UWB radio operating at several GHz can also receive interference from the existing in-band systems or proposed future systems.

Despite the difficult problems of coexistence with other radio systems, UWB has many interesting features and unique characteristics that give this technology great promise for many new applications. Specifically, the wide bandwidth (or short pulses) of UWB signals provides excellent multipath propagation properties and has the potential for transmitting very high data rates.

To understand how UWB works, it is essential to understand some basic concepts and system parameters. These concepts and parameters will help in understanding not only UWB itself but also regulatory issues. We start with a generalised UWB signal without defining any modulation and channelisation schemes. This signal can be extended for inclusion of data modulation and channelisation. A pulse train of a generalised UWB signal can be represented as a sum of pulses which takes the form:

$$s(t) = A \sum_{i=-\infty}^{+\infty} \sum_{j=0}^{N_s - 1} v(t - jT_f), \ (i-1)T_b < t \leq iT_b \qquad \text{...... (5.1)}$$

Here A is the amplitude of the pulse. N_s is the number of pulses required to transmit a single information bit. T_b is the bit duration, where $T_b = N_s T_f$. T_f is the frame time, also known as average pulse repetition period. The reciprocal of T_f, called the pulse repetition frequency, has a closed effect on system design. $v(t)$ is the basic UWB pulse of duration T_v. The following sections illustrate each of the components in equation (5.1) and their relationships in more detail.

5.2.1   Ultra-Short Pulses

The starting point of UWB is to generate the short pulses with which the system communicates. Let us start with an initial waveform of a Gaussian pulse which has the familiar form given below [5, 6]:

$$p(t) = e^{-t^2/\tau^2} \qquad\qquad ...... (5.2)$$

Taking the first derivative of this equation yields a Gaussian monocycle, which has the form:

$$v(t) = \alpha t e^{-t^2/\tau^2} \qquad\qquad ...... (5.3)$$

where α is the parameter related to the amplitude of the pulse. A typical waveform for a $\tau = 0.5$ ns width pulse is shown in Fig 5.1.

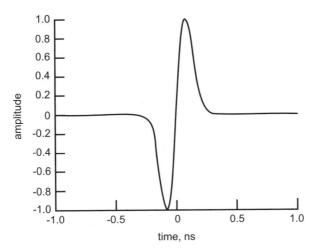

Fig 5.1   Gaussian monocycle pulse.

It can be seen from the waveform in Fig 5.1 that there is a single zero crossing point. Taking additional derivatives yields waveforms with additional zero crossing points, one additional zero crossing for each additional derivative. According to Fourier transform theory, as we take additional derivatives, the relative bandwidth decreases, while the centre frequency increases (for a fixed value of τ). The equivalent of taking derivatives is filtering. This gives us a choice of waveforms to be used as short pulses for UWB, which will depend on the system performance and application requirements. The standards bodies such as FCC [2] and NTIA [7] may also influence the choice of the waveform. For example, if the requirement were to eliminate the signal energy for low frequencies to protect GPS and GSM systems, which operate in relatively low frequency bands, at least one derivative of the Gaussian waveform should be considered, as by taking derivatives, the spectrum tends to move to the higher frequency bands.

A Gaussian monocycle sequence, or 'pulse train' can then be generated for a data modulation purpose. The pulse train is acting somewhat like a 'carrier', which can be used for the purpose of modulation and transmission. In the frequency domain a pulse train with regular time intervals will produce energy spikes, which might interfere with conventional radio systems. These energy spikes should be minimised to reduce the interference level. Section 5.2.3 introduces a method of reducing energy spikes.

5.2.2   Data Modulation

The regular monocycle pulse train contains no information and produces energy spikes. In order to transmit information, the monocycle pulse train needs to be modulated by data. Information transmission can be achieved using a number of ways, including amplitude, time and phase modulation of the UWB pulses. The modulation needs to reduce energy spikes, thereby minimising the PSD as required by the regulation. The choice of the modulation schemes also affects the bit error performance.

Three of the popular modulation schemes proposed for UWB transmission are pulse position modulation (PPM), pulse amplitude modulation (PAM) and phase shift keying (PSK). For binary data modulation, PSK is also known as bi-phase modulation. These modulation schemes, each discussed below, can be compared with each other in terms of spectrum characteristic and bit error performance.

5.2.2.1   *Pulse Position Modulation*

Pulse position modulation is based on the encoding information by modifying the time shift between the pulses.

Figure 5.2 illustrates a pulse position modulation scheme, where the pulse frame length T_f (pulse repetition period), the symbol length T_b, and the chip length T_c are also indicated. The introduction of the chip length T_c is for the purpose of multi-user communication, which will be illustrated in section 5.2.3. As illustrated in Fig 5.2, pulse position modulation changes the time of transmission of every monocycle in a data symbol by a time shift δ. For a binary data sequence, each bit in the data stream is sampled by N_s monocycles. In the system shown in Fig 5.2, transmitting three monocycles represents each data symbol. When the data symbol is 0, the transmission of data symbol starts a nominal position of T_c. Because of the time shift δ, pulse position modulation distributes the signal energy more uniformly across the spectrum. For the PPM modulation example shown in Fig 5.2, there are three frames per data symbol ($N_s = 3$, the number chosen for the purpose of illustration only). The other two modulation schemes PAM and PSK will use the same assumption. In real applications, there are a few hundred frames per data symbol. This increases the robustness of signal reception, because detection of only a few frames per symbol for a receiver requires very high receiver sensitivity.

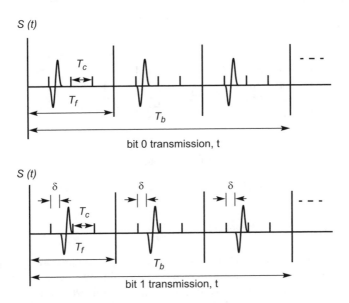

Fig 5.2   Pulse position modulation.

The frame length is only slightly longer than the pulse width, which is limited by the graphic illustration. In real systems, the ratio of the frame length to the duration of a monocycle is much larger, resulting in a low duty cycle pulse. Usually the chip duration is larger than the pulse width because the pulse width is very short to ensure wide frequency occupancy.

5.2.2.2   *Pulse Amplitude Modulation*

Pulse amplitude modulation is based on the principle that the amplitude of the impulses is encoded by data. Digital PAM is also called amplitude-shift keying (ASK), alternatively referred to as on-off keying (OOK) for two-level PAM. Because there is no carrier involved for UWB transmission, the modulated signal waveform is simply the baseband signal. Figure 5.3 shows a two-level amplitude modulation.

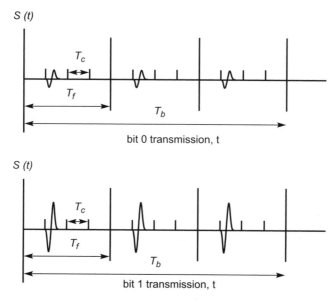

Fig 5.3   Pulse amplitude modulation.

5.2.2.3   *Phase Shift Keying*

Phase shift keying is also referred to as bi-phase modulation, if only two phases are used for modulation. In the case of bi-phase modulation, the impulse is sent, for example, at zero degrees for transmitting '1' and 180 degrees for transmitting '0'. For UWB signal transmission, where the signal level is relatively weak (low signal to noise), analysis of M-ary PSK shows that $M = 2$ and $M = 4$, i.e. BPSK and QPSK offer best bit error performance [8]. This is to say that bi-phase modulation requires the least energy per bit for any given noise level. Figure 5.4 shows the bi-phase modulation scheme.

5.2.3   **Multi-User Communications**

A number of multiple access schemes have been proposed to enable a channel-sharing purpose for multi-user networking. Most recently, a new method called multi-band [3, 4] for UWB transmission has been proposed by several developers, which implicitly includes multi-user communications. The major access schemes are:

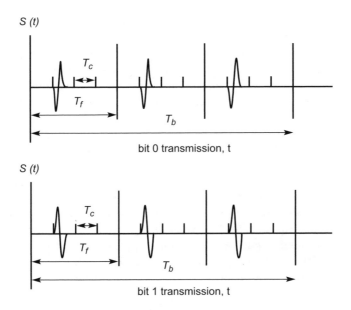

Fig 5.4   PSK bi-phase modulation.

- code division multiplexing;
- frequency hopping;
- frequency division multiplexing;
- time division multiplexing;
- hybrid.

Among these schemes, code and frequency division multiplexing have been most discussed. Code division multiplexing used for UWB is equivalent to time hopping. We discuss the time hopping scheme first. The multi-band UWB will be discussed in section 5.3.

This scheme is designed mainly for relatively low data rate transmissions, typically up to 40 Mbit/s with low duty cycle pulses. Time hopping UWB has been well presented by Scholtz [9]. This scheme can use any of the data modulation methods illustrated in Figs 5.2, 5.3 and 5.4; the monocycle pulse train has the form:

$$s(t) = \sum_j v(t - jT_f) \qquad\qquad ......(5.4)$$

This pulse train may look the same as any other pulse train, which means there is no multiplexing involved for the purpose of sharing a channel. The frame time T_f can be a few hundreds to a thousand times the monocycle width, which results in a signal with very low duty cycle. The low duty cycle feature of the pulse provides a possibility for channel coding, i.e. for multiplexing or channelisation. A frame can be divided into hundreds of time-slots and the pulse generated from the *k*th user occupies only one of these time-slots. The rest of the slots can be allocated to other users. Take the first user as an example. If the pulse generated by the first user occupies the same slot for all the frames, the pulses are equally spaced. Such a system becomes vulnerable because the first user could easily collide with another user — in which case a large number of pulses from two users could be received at the same time.

This problem can be resolved by signing pseudo-random noise (PN) codes to users that share the channel, a concept similar to a conventional CDMA system, except that, in a UWB system, a PN code is signed to every user so that the actual transmission time of each user of each pulse can be shifted over a large time frame. The control of the time-slot can be realised by an appropriate design of the PN code sequence. Therefore, in a UWB multiple access system, each user would have a distinct PN code sequence $\{cj^{(k)}\}$, which is also called a time hopping code. To receive the transmitted signal, a receiver must operate the same time hopping code. The codes $\{cj^{(k)}\}$ are periodic codes with period N_p, that is, $c_{j+iN_p}^{(k)} = c_j^{(k)}$ for all integers *j* and *i*. Each code element is an integer in the range:

$$0 \le c_j^{(k)} \le N_h \qquad \qquad ...... (5.5)$$

Multiplying the PN code provides an added time shift to each monocycle in the pulse train, with monocycle *j* undergoing an added shift of $cj^{(k)}T_c$ sec.

Figure 5.5 shows an example of channel coding using time hopping with pulse position modulation. User 1 transmits '1' and is signed by the PN code sequence $\{2,1,3, ... \}$, so the monocycle from user 1 takes the 2nd slot in the 1st frame, 1st slot in the 2nd frame, 3rd slot in the 3rd frame and so on. The position of the monocycle will shift from slot to slot, hence the name 'time hopping'. User 2 and 3 follow the same steps. So there are three time hopping codes used in Fig 5.3. They are $\{2,1,3, ...\}$, $\{3,2,2, ... \}$ and $\{1,4,1, ... \}$ for users 1, 2 and 3, respectively. The parameter, N_hT_c/T_f, can significantly influence the performance of the system. It indicates the fraction of the frame time T_f over which time hopping is allowed. If N_hT_c/T_f is too small, the system would still be vulnerable to collisions. The capacity of the system can be improved with a large ratio of N_hT_c/T_f and well-designed PN codes.

Using time hopping can have two effects on the system. The first effect is to distribute the monocycles non-uniformly over time to avoid data collisions. The second effect is to spread the signal spectrum further. The time hopping code is periodic with period N_p. After signing a PN code to the *k*th user, its waveform becomes:

$$s^{(k)}(t) = \sum_j v\left(t - jT_f - c_j^k Tc\right) \qquad \qquad ...... (5.6)$$

with period $T_p = N_pT_f$. Therefore, the signal spectrum contains a train of delta impulses with a frequency period that is inversely proportional to the time period T_p

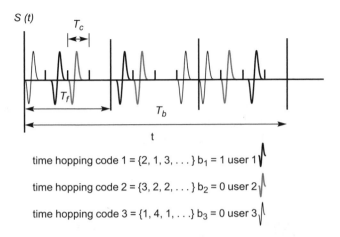

Fig 5.5 A frame sequence showing three users sharing a channel assuming PPM modulation. The PN codes are {2,1,3, ...}, {3,2,2, ...} and {1,4,1, ...} for users 1, 2 and 3, respectively. A time shift of δ is added for transmission 1.

(using the Fourier transform theory of a periodic signal). The spectral resolution increases with the increase of the time period. Because the time period of a coded signal T_p is larger than the time frame T_f, time hopping can effectively reduce the power spectrum density. The spacing between the frequency spikes is now $1/T_p$, which is much less than the original spacing of $1/T_f$. This means the energy is spread into more frequency lines, resulting in a lower spectral density.

Because of the reduced spectral density, by using time hopping in the frequency domain a signal can be made virtually indistinguishable from white noise. In the time domain, each user could be signed by a distinct PN code and there are thousands of channel codes available for the system to choose. In the receiver, without knowing of the time hopping code, the signal is virtually undetectable. This makes the signal inherently difficult to detect or intercept other than by a matched correlation receiver.

5.2.4   Reception and Data Demodulation

Reception depends on a number of conditions, including the modulation scheme and multi-user scheme. It is not realistic to describe all of them. For example, a PAM receiver with time hopping, or a PSK receiver with frequency hopping and so on. Here we only consider a theoretical approach on how to deal with data reception and bit error performance, using PPM and time hopping as an example. To receive time hopping signals with minimal spectral density, it is essential to have an optimal receiver structure. The optimal receiver is a correlation receiver that multiplies the received signal with an embedded template signal and then integrates the output to produce a decision statistic.

When there are K users in a multiple access system, the receiver signal $r(t)$ can be modelled as:

$$r(t) = \sum_{k=1}^{K} A_k s_k(t - \tau_k) + n(t) \qquad \text{...... (5.7)}$$

where A_k is the attenuation of the kth signal, τ_k is the propagation delay, and $n(t)$ represents white noise. We consider $s_1(t)$ to be the desired signal and all other $K-1$ signals are considered as interference signals. Under hypothesis of perfect synchronisation with $s_1(t)$, the correlation receiver computes the following decision statistic:

$$r = \sum_{j=1}^{N_s} \int_{\tau_1 + jT_f}^{\tau_1 + (j+1)T_f} r(t)w\left(t - \tau_1 - jT_f - c_j^{(1)}T_c\right)dt \qquad \text{...... (5.8)}$$

where:

$$w(t) = p(t) - p(t - \delta) \qquad \text{...... (5.9)}$$

is the template waveform and δ is the delay associated with PPM (modulation index). Substituting equation (5.7) and equation (5.9) into equation (5.8) we have:

$$r = \pm N_s + / + \eta \qquad \text{...... (5.10)}$$

where η is a Gaussian random variable with zero mean and variance $(N_0 N_s^2)/E_b$, N_0 is the noise power spectral density, E_b is the energy per bit and I is the accumulated multiple access interference from all $(K-1)$ interfering signals given by:

$$I = \sum_{k=2}^{K} \sum_{j=1}^{N_s} \int_{\tau_1 + jT_f}^{\tau_1 + (j+1)T_f} s_k(t)w\left(t - \tau_1 - jT_f - c_j^{(1)}T_c\right)dt \qquad \text{...... (5.11)}$$

Once the probability density function for the interference is computed one can derive the expression of probability of bit error. The system performance and multi-user capability can then be assessed based on the probability of bit error. We shall not discuss this any further. More details about performance issues can be found from an IEEE special issue on UWB [10], which has several papers discussing methods of performing the bit error calculation.

5.3   Multi-Band UWB

The recent FCC regulation [2] has motivated a number of leading UWB developers to promote a new approach — multi-band UWB [4]. The 'new' way of looking at UWB is based on the available spectrum, not as a particular technology. Based on the FCC regulation [2], permission is given for unlicensed use of 7.5 GHz of available spectrum. The operating spectrum in the USA is 3.1-10.6 GHz, while in Europe there is expected to be a different operating frequency of 3.0-6.0 GHz. The limitation of power is -41.3 dBm per MHz which is the same level as unintentional radiation from common electronics devices such as laptop computers. The frequency and power limit are for indoor and handheld systems. The FCC further defines a UWB device as having fractional bandwidth greater than 20% or occupied bandwidth of at least 500 MHz at all times during the signal. Fractional bandwidth is the ratio of the signal bandwidth to the notional centre frequency. Because of this rule, the 'new' way of defining UWB is that it does not have to be impulse, or carrierless, but can be any technique that generates signals occupying at least 500 MHz of bandwidth within the spectrum mask placed by FCC. Hence the 7500 MHz of unlicensed spectrum can be considered to provide a number of UWB 'bands' that can be exploited in innovative ways.

5.3.1   Spectral Mask

The mask for permitted PSD for outdoor operation in the USA shows a steep reduction of out of band emissions below the lower 3.1 GHz limit in order to protect the sensitive GPS operating band at 1.9 GHz (see Fig 5.6).

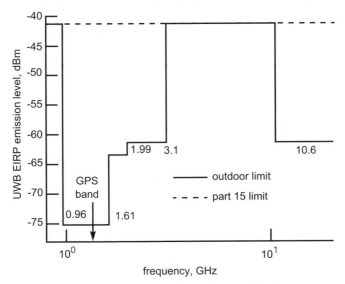

Fig 5.6   Spectral mask for outdoor applications.

The spectral mask for indoor emissions is more lenient (see Fig 5.7). It should be noted that this is one area where European and US regulations are expected to differ. European regulations are expected to define a more practical slope for out-of-band limits rather than the ideal rectangular form adopted by the FCC.

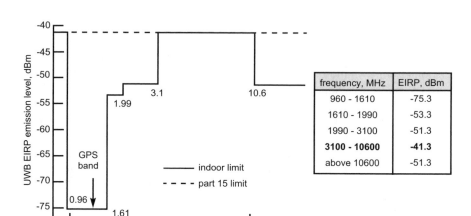

Fig 5.7   Spectral mask for indoor applications.

In both cases, the main point of note is that the permitted emission level within the allocated frequency band for UWB devices is the same as the unintentional radiator limit, i.e. -41.3 dBm/MHz. There are, however, no limits set in the USA on maximum signal levels, only on average PSD and no specific ruling has been made concerning signal modulation techniques. To put this in context, a low cost, integrated UWB radio will be limited in maximum signal swing that can be generated on the semiconductor technology used to implement it, i.e. typically considerably less than 2 V.

5.3.2   New Ways of Looking at UWB

The FCC Report and Order (R&O) [2] not only provides a legal basis for UWB devices and systems to be brought to market for the first time, but significantly changes the way UWB signals should be considered.

Prior to the FCC ruling, UWB signals were frequently defined as signals having a fractional bandwidth of greater than 25%. Fractional bandwidth η is defined as:

$$\eta = 2\frac{f_H - f_L}{f_H + f_L} > 25\% \qquad\qquad \text{...... (5.12)}$$

where f_L and f_H are the frequencies measured at the -10 dB emission points. Following the FCC R&O, any signal occupying at least 500 MHz of bandwidth is considered to be a UWB signal, as well as the more conventional fractional bandwidth definition (although the R&O uses 20% instead of the previously accepted 25% value for η). Figure 5.8 compares the previous and new definitions in terms of occupied bandwidth and fractional bandwidth over the frequency range selected by the FCC for UWB operations.

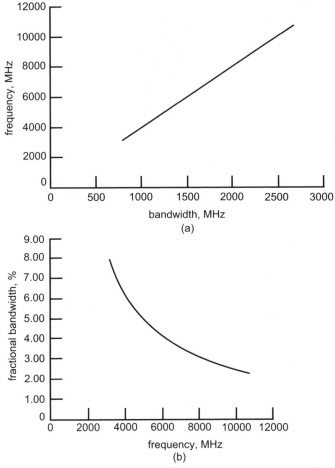

Fig 5.8   Occupied bandwidth and fractional bandwidth over 3.1 to 10.6 GHz —
(a) previous definition, (b) new definition.

It can be clearly seen that the FCC ruling permits significantly narrower emissions than previously considered to be UWB. We can view this ruling in another way. Consider the charts shown in Fig 5.9.

Figure 5.9(a) shows a burst of RF energy constrained in time to be approximately 4 ns wide. Any signal which is shorter than this will occupy a bandwidth greater than that shown in Fig 5.9(b).

Notice that the spectrum permitted to be exploited by UWB signals in the FCC R&O allows many such UWB signals to exist simultaneously, since any one signal only occupies a small fraction of the available spectrum. Using these ideas leads to

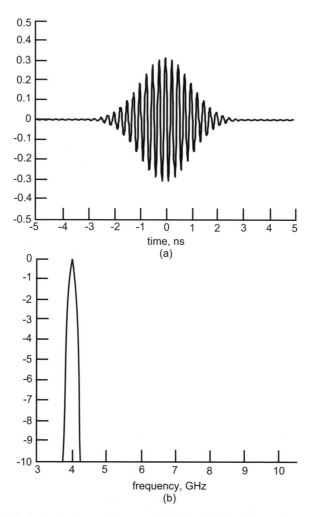

Fig 5.9   Time and frequency display of UWB short pulses.

an interpretation of UWB as a means of accessing available spectrum and not as a specific technique for generating RF energy such as impulse radios, or carrierless radios, or indeed any other limited definition. A UWB signal is simply a burst of RF energy constrained in time to be less than 4 ns wide, with a power spectral density within the permitted limits.

This new way of thinking about UWB has led several leading companies [11-16] to develop multi-bands as a better way to utilise the UWB spectrum. The multi-band idea illustrated by Staccato Communications [3, 4] is similar to the concept of orthogonal frequency division multiplexing (OFDM) and divides available spectrum into several bands for UWB signals to share.

The current FCC regulation allows UWB devices to use a transmission bandwidth of 7.5 GHz and the newly proposed multi-band technology divides that spectrum into multiple 500 MHz bands that can be added, or dropped if interference is detected from other systems. As shown in Fig 5.10, there are multiple bands allocated to UWB signals operating at different frequencies. Each of these signals can be transmitted simultaneously without interference as they occupy different frequency bands. Data modulation for each UWB signal can be achieved using standard digital modulation techniques enabling very high data rates to be attained for the complete system.

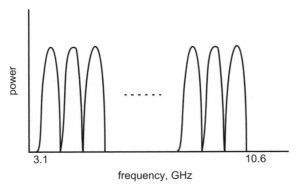

Fig 5.10   Multi-band operation of UWB.

There are several advantages to the multi-band approach. The major advantage would be its increased level of coexistence with other systems operating in its range. The multi-band systems have ability to adjust to in-band interference by selectively using available bands. This is important since UWB is likely to co-exist with WLANs such as IEEE802.11a, which works at 5 GHz. The multi-band approach is also scalable on data rates — more bands can be used to accommodate higher data rate systems, while fewer bands can be used for low rate systems. Multi-band systems can also therefore scale their power consumption. Multi-band can also provide flexibility to adjust the operating band in different spectrums in order to accommodate regulatory changes. It is suggested that the multi-band approach has a drawback of being more complicated and more power consumptive than a single band system but this may be mitigated by using bands in sequence rather than in parallel and coding additional data in sequence bits. Overall, multi-band is a more effective approach to provide scalability, flexibility and coexistence.

The major implication of introducing the multi-band approach is that many companies have already invested heavily in developing single band systems (mainly impulse-based approach), and there could be issues in the standardisation process on how to co-ordinate these two different approaches, i.e. the single band and multi-band systems.

Recent developments include significant industry support for a multiband approach based on an OFDM symbol structure [17]. The Multiband OFDM

Alliance proposes a signal structure over 3 or 7 × 528 MHz bands. Although the work is still in progress, this proposal may well be accepted as the first UWB standard for the indoor applications envisaged by the FCC ruling.

5.4 Potential Markets and Applications

This section briefly describes some applications and new capabilities of UWB [11-22]. More detailed descriptions on applications of UWB can be found elsewhere [18, 19, 23].

The potential markets for UWB can be broadly classified into imaging systems, vehicular radar systems, and communications and measurement systems.

- Imaging systems

  — ground penetrating radar, e.g. finding buried objects for fire and rescue, scientific research, commercial mining and construction;

  — wall imaging, e.g. locating objects within a 'wall' for fire and rescue, commercial mining and construction;

  — medical imaging, e.g. medical diagnostics for licensed health care practitioners.

- Vehicular radar systems

  — vehicular radar, e.g. collision avoidance, improved suspension systems.

- Communications and measurement systems

  — high-speed business networking, for short-range indoor data and video communications (20-100 Mbit/s+);

  — wireless device interconnection, e.g. wireless USB (up to 480 Mbit/s at close proximity);

  — home networking, for wireless connections of all fixed and mobile home appliances such as PCs, PDAs, game consoles, televisions, videos and stereos;

  — storage tank measurement devices, subject to certain frequency and power limitations.

The following sections describe, in a little more detail, applications in high-speed WLANs, precision geo-location, and radar systems.

5.4.1 Exciting New WLAN Offering up to 500 Mbit/s

WLANs (see Chapter 3) are becoming increasingly popular in both the commercial and domestic environments. IEEE802.11b has been dominating the commercial wireless networking area. Digital cordless telephone and home networks are starting to offer solutions for wireless interconnection for the domestic environment. HIPERLAN/2 and IEEE802.11a appear to offer higher data rates for future applications. Table 5.1 summarises the standards and their operation range.

Table 5.1 WLAN standards and operation range.

Standard	Frequency (GHz)	Speed
Digital cordless telephone	1.8	552 kbit/s, approx 200 m
Home RF, Release 2	2.4	10 Mbit/s, approx 50 m
IEEE802.11b	2.4	11 Mbit/s, approx 50 m
IEEE802.11g	2.4	54 Mbit/s, approx 10 m, greater range at lower rates
HIPERLAN/2	5.15	54 Mbit/s, approx 15 m, greater range at lower rates
IEEE802.11a	5.15	54 Mbit/s, approx 15 m, greater range at lower rates
Home RF, Release 3	2.4	100 Mbit/s, short range
IEEE802.15.3a (in progress)	3.1-10.6	110 Mbit/s at 10 m, 200 Mbit/s at 4 m, higher speeds optional at shorter distances

Thus we can see that UWB will give greater throughputs particularly in the home environment.

5.4.2   Applications for Precision Geo-Location Positioning

UWB radios can be used to locate objects in a similar fashion to the method used by the global positioning satellite systems (GPSs) [6]. GPS uses a constellation of satellites to transmit a radio signal that carries timing information. A GPS receiver can calculate its location by comparing timing information from each satellite. UWB devices can be used to measure both distance and position with or without a reference infrastructure. Details of example UWB location systems can be found elsewhere [18, 24].

UWB positioning systems could provide real-time indoor and precision tracking for many applications. GPS could be an excellent system for outdoor applications, while UWB could be an outstanding candidate for indoor systems. Because of the pseudo-random characteristic of the pulse, the UWB system is robust in multi-path environments, such as inside buildings. Multi-path is a major problem for GPS systems attempting to operate inside buildings.

It is worth noting that UWB can also provide some outdoor applications including personnel and target tracking for increased safety and security, and precision navigation capabilities for vehicles.

5.4.3   Low-Power and Radar Applications

UWB technology has been used in the past in ground penetrating radar applications. New applications are being developed for through-wall sensing and imaging systems [2, 6]. These applications can be used for personnel and machinery in various situations, for example, a rescue situation. It requires a low operating frequency to enable the signals to propagate effectively through walls. UWB radar would require a precision time gate [6] in order to detect signals of interest. This timing gate also has the added benefit of a dynamic range, allowing targets at longer ranges and lower signal levels to be detected. UWB imaging devices could also be used to improve safety for personnel. For example, the device could be used for an electrician to detect electrical wiring and pipes hidden inside walls. Vehicle safety could be improved using UWB radar with collision avoidance.

5.5   UWB Regulatory Situation

International regulations for UWB devices have been driven primarily by the activities of the US Federal Communications Commission and its public comment process, enabling interested parties to lobby and respond to intended rulings. The FCC issued its First Report and Order after several years of study by the National Telecommunications and Information Administration (NTIA), with a follow on R&O (FCC 02-48, UWB Report and Order) [2]. The NTIA was tasked with assessing the interference potential of UWB emissions on existing wireless systems and services and establishing suitable measurement techniques for UWB emissions. In Europe, the results of independent spectrum studies carried out by CEPT SE24 are expected to be published before the end of 2003 leading to a CEPT Decision and a corresponding ETSI standard in 2004. The ETSI work is being carried out under TG 31a. The regulatory situation in the Far East is less clear, although considerable interest in UWB technology exists. An experimental 'UWB friendly zone' has been established by the Singapore authorities to encourage research into UWB techniques, systems and applications.

Until the European results are published, the main regulatory data remains that of the FCC. The NTIA interference studies, taking into account expected interference levels, bandwidth of the spectrum occupied by a UWB signal and the sensitivity of existing services (particularly critical safety ones), has led to an initial allocation of the frequency band 3.1-10.6 GHz. This allocation places intentional UWB emissions above key sensitive services, notably the global positioning service in the USA. The FCC identifies three classes of application — imaging systems (including ground penetrating radars), vehicular systems, and communications and measurement systems.

European regulations are expected to permit intentional UWB emissions in the frequency band of 3-6 GHz. Hence it is possible to see a common frequency band for the operation of UWB devices, although emission levels in Europe are expected to be lower than those permitted in the USA.

The Radiocommunications Agency [25] has been co-ordinating UWB regulation and standard processes for the UK. The agency sponsored a one-day colloquium on UWB technology in July 2002. This event presented the state of the art in UWB and its potential applications and addressed implementation and regulation issues. The Radiocommunications Agency presented a UK approach to emerging technologies [26]. The Agency's approach has been mainly to work closely with other European administrations and the CEPT to share information and determine and resolve issues. The Agency is developing an informed UK national position for constructive participation at the international level. The standards bodies including FCC in USA and CEPT and ETSI in Europe presented their approaches on regulatory and standard issues on UWB. UWB development and consumer electronics companies, including Time Domain Corporation, Thales, Siemens, Sony and Philips, gave their views and presentations on UWB technology and applications. Details of their approaches to UWB technology are available on-line at the Radiocommunications Agency web site [25].

In summary, the FCC and expected European regulatory situations are:

- UWB signal defined as:
  - — fractional bandwidth greater than 20%, or,
  - — occupied bandwidth greater than 500 MHz;
- operating frequency:
  - — USA: 3100-10 600 MHz,
  - — European: 3000-6000 MHz;
- emission limits:
  - — USA: -41.3 dBm/MHz EIRP,
  - — European emission limit expected to be lower;
- application classes:
  - — imaging systems,
  - — vehicular systems,
  - — communications and measurement systems (indoor and handheld);
- other restrictions and measurement procedures in the FCC R&O;
- signal and modulation:
  - — no specific modulation or pulsed modulation defined,
  - — no specific signal characteristics defined.

5.6   Summary

UWB holds out great promise for potential applications in communications, radar and imaging systems. The reason UWB can offer these benefits is that UWB waveforms are of extremely short time duration. In communication applications, short pulses can be used to provide extremely high data rate transmissions for multi-user applications. The same short pulses enable high resolution for both radar and positioning applications.

Analysing short pulses is the key to understanding the unique properties of UWB systems being brought to the market. The type of the short pulses can also have an impact on system design. All standard digital modulation schemes can be considered for data modulation. Several multiple access schemes can be used for channelisation for multi-user networking purposes.

The FCC's ruling provides a legal basis for developers to bring their UWB devices and systems to markets. The three main areas of market for UWB applications are:

- imaging systems, including ground penetrating radars, through-wall surveillance and medical devices;

- vehicular radar systems;

- communications and measurement systems (indoor and handheld).

The authorisation granted by FCC has also changed the way UWB signals should be considered. This has led to a new way of looking into UWB — the multi-band approach. Regulatory and standardisation activities must determine if there are technical issues in the use of both approaches.

Studies to establish whether UWB will cause harmful interference to other radio systems are still under way, and the means of calibrating the interference are yet to be standardised. Therefore, the process architecture of co-existence of UWB with other systems is still to be defined, although some companies have tested and proved individual products. It is still too early for telecommunications operators to express their opinions on how to incorporate UWB into their existing network infrastructure, but it does appear that this technology holds a great potential for a vast array of new applications.

References

1   Taylor, J. D.: '*Introduction to Ultra-wideband Radar Systems*', CRC Press, Boca Raton, Florida (1995).

2   Federal Communications Commission (FCC): '*New public safety applications and broadband Internet access among uses envisioned by FCC authorisation of ultra-wideband technology*', — http://www.fcc.gov/Bureaus/Engineering_Technology/News_Releases/2002/nret0203.pdf — and FCC 02-48, UWB Report and Order (released 22 April 2002).

3   Staccato Communications, White Paper: '*IEEE 802.15.3a 489 Mbit/s Wireless Personal Area Networks, Achieving a Low Complexity Multiband Implementation*', (January 2003) — http://www.staccatocommunications.com/

4   Staccato Communications, White Paper: '*New ultra-wideband technology*', (October 2002) — http://www.staccatocommunications.com/

5   Multispectral Solutions, Inc: '*Response to FCC Notice of Proposed Rule Making ET Docket No. 98-153: Revision of Part 15 of the Commission's rules regarding ultra-wideband transmission systems*', (September 2000).

6   Time Domain: '*PulsON® Technology Overview*', White Paper (July 2001) — http://www.timedomain.com/

7   National Telecommunications and Information Administration (NTIA) — http://www.ntia.doc.gov/

8   Proakis, J. G.: '*Digital Communications*', 3rd edition, McGraw-Hill, p 272 (1995).

9   Scholtz, R. A.: '*Multiple access with time-hopping impulse modulation*', in Proc MILCOM, pp 447—450 (October 1993).

10 IEEE special issue: '*Ultra-wideband radio technology in multi-access wireless communications*', IEEE J on Selected Areas in Communications (December 2002).

11 Intel — http://www.intel.com/

12 Staccato Communications — http://www.staccatocommunications.com/

13 Time Domain Corporation — http://www.timedomain.com/

14 General Atomics — http://web.gat.com/

15 Wisair — http://www.wisair.com/

16 Multiband OFDM Alliance — http://www.multibandofdm.org/ieee_proposal_spec.html

17 Alereon Inc — http://www.alereon.com/

18 Time Domain Corporation: '*Applications of Ultra Wideband Technology*', (2001) — http://www.timedomain.com/

19 Fontana, R. J.: '*Multispectral Solutions — Recent Applications of Ultra Wideband Radar and Communications Systems*', — http://www.multispectral.com/pdf/UWBApplications .pdf

20 Multispectral Solutions Inc — http://www.multispectral.com/

21 Pulse-Link Inc — http://www.pulselink.net/

22 XtremeSpectrum — http://www.xtremespectrum.com/

23 Beaumont, D.: '*FCC UWB decision opens new wireless chapter*', Planet Wireless — http://www.blue.telecoms.com/

24 Parker, T.: '*Two-Way Time Transfer*', — http://www.boulder.nist.gov/timefreq/time/twoway.htm

25 Radiocommunications Agency, United Kingdom — http://www.radio.gov.uk/

26 Barron, B.: '*Overview of UK approach to UWB technologies*', — http://www.radio.gov.uk/topics/uwb/uwb-overview.pdf

6

AD HOC WIRELESS NETWORKS

R Gedge

6.1   Introduction

Ad hoc networks[1] offer a radical alternative to existing cellular and fixed networks
for providing communications. These networks form on the fly from the
communications devices themselves without needing any infrastructure or
centralised control. Devices communicate directly with each other and by forming
chains of transceivers they relay information through other devices in order to reach
the final destination (see Fig 6.1). The devices also learn about their peers and then
use this intelligence to route information via the optimum path taking into account
such things as processing power, battery capacity and alternative network connections such as broadband or GPRS. The network therefore works as a symbiosis.

Fig 6.1   Example *ad hoc* wireless network.

---

[1] Also known as symbiotic networks.

An *ad hoc* approach to creating such networks is not new. Packet radio and multi-hop networks have been the subject of research for nearly 30 years and have been deployed by the military for some time. However, technology has now advanced to a point where these networks can be deployed in large numbers at low cost.

Wireless connectivity has become an integral feature of many portable devices such as laptops and PDAs, and in the coming years this trend is certain to extend across a vast array of everyday devices ranging from pens to cars and into fixed structures such as buildings and streetlights. In a future with 'devices everywhere', wireless capability will be present in sensors, clothes, food packaging and a host of other everyday objects.

Ad hoc networks are potentially highly disruptive to network operators, possessing the capability to bypass existing networks, enabling a new generation of privately owned and community networks. These networks also present significant opportunities for expanding the fixed network, since certain strategic nodes could function as access points (or gateways), utilising the fixed network to provide high-speed connectivity for *ad hoc* networks that serve geographically distributed communities.

This chapter discusses the history of *ad hoc* networks, possible applications and the part WLANs can play. It goes on to describe some of the routing challenges and the *ad hoc* test network that has been built by BT.

6.1.1   Definitions

Many terms are used to describe these *ad hoc* networks — the definitions below are those employed in this chapter.

- *Ad hoc*

  Devices form a network as needed on an *ad hoc* basis. The nodes may or may not be able to relay information on behalf of others.

- Multi-hop

  This is a network where information is relayed from one device to another. Each packet may undergo many hops before reaching its destination.

- Peer-to-peer

  This is a network where devices can communicate with each other without requiring a single controlling entity. It includes session establishment and quality of service.

- Intelligent routing

  This is a network where information is relayed to the destination by the best possible path using accrued knowledge such as node location and gateway capability. This may involve multiple hops or going via a fixed infrastructure if available.

- Parasitic

  This is a network where devices make use of each other's resources in order to extend their own capability. This could include data rate, power or by relaying their information via other nodes.

- Symbiotic

  This is a BT term for a parasitic network where the perceived benefits of sharing your connectivity outweigh the negative aspects associated with a parasite.

6.1.2   History

Ad hoc networking began in the early seventies when the US Department of Defense (DoD) sponsored the Packet Radio Network (PRNET) research programme. This evolved into the Survivable Adaptive Radio Networks (SURAN) programme in the early 1980s [1]. These programmes aimed to provide packet-switched networking to the mobile battlefield in a hostile environment with no infrastructure and with soldiers, tanks, aircraft, etc, forming the nodes in the network.

In the 1990s came developments such as laptop computers with infra-red ports and cheap wireless LAN PC cards. The infra-red ports allowed the laptops to communicate with each other directly and the IEEE adopted the term '*ad hoc* networks' for the IEEE802.11 Wireless LAN standard.

The DoD sponsored Global Mobile information systems (GloMo) and Near-term Digital Radio (NTDR). GloMo aimed to provide office-environment Ethernet-type multimedia connectivity any time, anywhere, in handheld devices. NTDR is now in use with the US Army.

A number of standards activities have followed this interest in *ad hoc* networking. Chief of these is the IETF's Mobile *Ad Hoc* Networking (MANET) working group [2], which has tried to standardise routing protocols. The IEEE has continued to work on its IEEE802.11 standards and IEEE802.11b, a, and h, and Bluetooth offers further technologies that can be used in *ad hoc* communications.

6.2   Applications of *Ad Hoc* Networks

The first and most widely deployed application of *ad hoc* networks has been by the military for battlefield systems. The battlefield is characterised by being insecure with no fixed infrastructure available. Devices are moving constantly around the battlefield (carried by soldiers or in tanks for example). This and a high attrition rate mean that the network is undergoing constant change. The inherent ability of the networks to self-organise and self-heal is of obvious interest to the military. A major, and as yet largely unexplored, issue with these systems is security [3].

Other applications for *ad hoc* networks include disaster scenes where there is a sudden and unpredictable need for an increase in connectivity greater than that provided by the existing infrastructure (which might be affected by the disaster anyway). Conferences, campuses and sports stadia are also settings where visitors may wish to communicate with each other for a short time where it may not be cost effective to provide permanent infrastructure. Several business colleagues may wish to collaborate with each other, for example, at a hotel or at an airport.

All the above applications are characterised by the lack of infrastructure with communications taking place in a stand-alone *ad hoc* network. However, these networks may also be used to allow connection to an existing infrastructure. Those devices out of range of a wireless access point can route their packets through neighbouring devices in order to access a fixed broadband connection. Indeed the original UMTS specification proposed a symbiotic-type capability (ODMA [4]) for reaching devices out of range of a base-station, for diverting traffic to alternative base-stations and offering peer-to-peer services.

BT's initial focus is to investigate if *ad hoc* networks can extend the reach of a broadband-connected WLAN access point. Is this a realisable proposition from both a technical solution and perhaps more importantly from a commercial or business model point of view? This scenario could be used to extend the reach of BT broadband to rural areas where it is not economic to provide individual connections to each customer. For full information on the role of community networks using similar technology, see Chapters 7 and 8.

Finally, specific *ad hoc* network infrastructure or relay nodes could be put in place (or seeded) in order to support these networks. These devices need not be connected to any existing network infrastructure and could be situated in places such as on rooftops or streetlights. Therefore, with a known density of seeded relays it could become possible to better predict the probability of connection. Thus rather than rely on the presence of user devices to allow relay chains to be formed, devices can be explicitly placed into the environment with the sole intention of acting as relays.

6.3   IEEE802.11 WLANs and *Ad Hoc* Networks

The widespread use of IEEE802.11b WLAN technology now offers a low-cost, mass-deployment opportunity for *ad hoc* networks. Out-of-the-box WLANs offer simple peer-to-peer *ad hoc* networking although most are used in what is called managed or infrastructure mode in conjunction with an access point (see Fig 6.2).

Extending the *ad hoc* mode to enable relaying provides the basis for a low-cost *ad hoc* network. Several companies are now starting to develop such solutions [5-8] based around the IEEE802.11b standard.

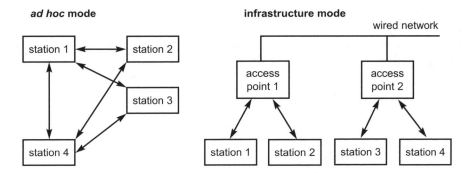

Fig 6.2 *Ad hoc* and infrastructure modes.

BT Exact has constructed an IEEE802.11b symbiotic test bed, based on the best research from around the world, to investigate how well this technology works in practice and to look at the opportunities for BT — this is described in section 6.5. This and other work in the area has shown that IEEE802.11b is not optimal for multi-hop networks [9]. Some of the issues highlighted are:

- interference and capacity;

- MAC layer conflicts;

- absence of quality of service (QoS).

6.3.1   Interference and Capacity

A major issue with IEEE802.11b networks is the 2.4 GHz unlicensed frequency band in which they operate. This unlicensed band, which can be used by anyone providing equipment conforms to laid down specifications [10], has spawned the widespread use of WLAN technology and the current innovation surrounding it. In cellular networks frequency spectrum is allocated to network operators who carefully share this functionality, ensuring appropriate coverage and capacity. With IEEE802.11b networks, anyone can operate in this 2.4 Ghz band without regard for other users and can therefore cause congestion and interference. It is interesting to note that Bluetooth devices operating in the same band, being of a later design, are more resilient to IEEE802.11b signals than vice versa.

The interference between nodes operating on the same network also causes a significant loss of capacity for *ad hoc* networks. Simple analysis shows that, for a single end-to-end session, when data is forwarded between nodes operating on the same radio channel, the throughput can be estimated from:

$$throughput = \frac{channel\_capacity}{number\_of\_hops} \qquad ......(6.1)$$

Actual measurements on a 6-hop IEEE802.11b network in clear line-of-sight conditions confirm this (see Fig 6.3). Note the theoretical maximum throughput is 6 Mbit/s due to the high MAC layer overheads [11], but even this is higher than the measured 5.2 Mbit/s.

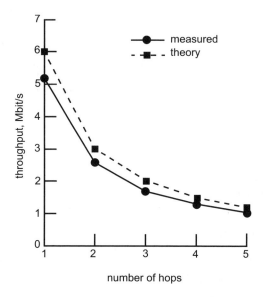

Fig 6.3   Throughput of a multi-hop IEEE802.11b network.

It was anticipated that at the maximum multi-hop communications range shown in Fig 6.4 it would be possible for simultaneous transmissions to occur between those nodes that were not in direct range of each other. For example, node 3 could transmit to node 4 at the same time as node 1 is transmitting to node 2. In these conditions the throughput was expected to rise due to spatial diversity. However, due to the carrier sense collision avoidance reservation system used by IEEE802.11b, the carrier sense range can be two or more times the communication range (see Fig 6.4).

The throughput curves shown in Fig 6.3 are therefore truly indicative of a real system operating in a clear environment.

6.3.2   MAC Layer Conflicts

The IEEE802.11b MAC layer was designed to provide robust communication under a range of operating conditions. There are, however, several features which do not work well with *ad hoc* routing algorithms. Two such features are packet acknowledgement and broadcast message.

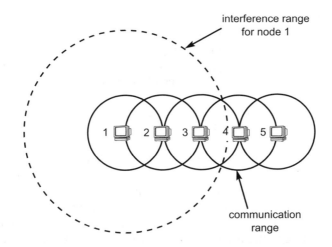

Fig 6.4   IEEE802.11b communication versus interference range.

6.3.2.1   Packet Acknowledgement

In IEEE802.11b every data packet sent from one node to another is acknowledged by the return of a small data 'ACK' packet to the sender informing them of correct reception. In an *ad hoc* network a node may need to forward the received packet on to the next node in the chain; the originating node, on overhearing the forwarding, can assume that the packet was received correctly and not require a separate 'ACK' packet, thus improving efficiency. There are also circumstances where the radio link, due to propagation conditions, may become unidirectional between 2 nodes. In an IEEE802.11b network the packet will be lost, as no acknowledgement can be issued; however, in an *ad hoc* network the acknowledgement could be routed back via another node.

6.3.2.2   Broadcast Message

In an IP network there is a broadcast capability to enable sending messages to all nodes on the subnet simultaneously. Typically this is used for ARP look-up [12]. In IEEE802.11b, broadcast messages are sent using the lowest data rate (and hence lowest signal-to-noise ratio required to successfully receive), in order to reach as many nodes as possible. In many *ad hoc* routing algorithms, this broadcast is used to determine which nodes can be reached directly and those which required a hop through another node. Once a route has been established the data is sent, although in IEEE802.11b the MAC layer may try to send this at the highest data rate. Due to the higher signal-to-noise ratio required for successful reception at the faster data rate, this message may not be received, and therefore the MAC layer reduces the data rate until reception is achieved — at the limit this will be at the same rate as the broadcast message. However, while this is happening the routing algorithm will believe that the route has failed and will re-initiate a route discovery process and the whole process starts again. A simple way round this problem is to force the data rate to stay at the lowest rate, but with a resultant loss of throughput.

6.3.3   Quality of Service

IEEE802.11b has no quality of service capability, although work is in progress to address this issue [13]. This means that voice over IP [14] or other services which require a guaranteed bit rate, cannot be well supported. The issues that were mentioned in section 6.3.1 make this job even more difficult to achieve.

In an *ad hoc* network, a route from A to B may go via several other nodes. If one of the nodes routing traffic decides to shut down, then a new route must be found. This may not be possible or take a long time. Speech circuits operating under these conditions are, therefore, untenable. However, once the density of nodes increases beyond a certain level, then an alternative route is always available and the problem becomes solvable once again.

6.4   Routing Challenges

6.4.1   Overview

One of the challenges of an *ad hoc* network is maintaining routes to other nodes in the network. Nodes may be on the move and the radio propagation environment is constantly changing requiring continuous updates in order to maintain end-to-end data paths. How to do this without flooding the network with signalling traffic is the topic of many published research papers. It is impossible to provide full justice to this topic here, but a good starting point is Perkins [15] who describes the well-known algorithms and issues.

In summary, there are two approaches to routing — proactive and reactive. In proactive systems the nodes continually search for routes and attempt to maintain a complete routing map of all nodes currently in the network. This approach tends to generate high signalling traffic but offers very low latency when data is sent from one node to another as the route is already known. In reactive systems no routes are discovered until a node requires to send data. At this point a route discovery process is started to find the end node. This approach tends to generate lower signalling traffic but higher latency, as a route must be found before data can be sent. Some examples of routing schemes are described below.

6.4.1.1   Destination-Sequenced Distance-Vector (DSDV) Protocol

This is a modification of the distance-vector routing algorithm to address poor looping properties and faster convergence for mobile networks through the use of sequence numbers. Each router maintains a vector containing the 'distances' to all the other nodes that it knows of.

Initially, this list should contain only its direct neighbours. It then distributes this vector list to all neighbouring routers. Upon receiving such a list, the router finds the shortest distance to all known destinations. If an update to its vector list occurs, the router redistributes its updated vector list to its own neighbouring routers. This continues until there are no longer any changes required in any routers' vector list and the algorithm has thus converged.

Further updates are either periodic (to inform its neighbouring routers that it is still alive), or triggered (when a change has occurred to the vector list, such as a link has failed).

6.4.1.2   Dynamic Source Routing (DSR) Protocol

Source routing is where the source node determines the complete sequence of nodes through which to send the packet to reach the destination. This list of nodes is included in the packet header. DSR offers the additional benefit in that the route is dynamically determined by the nodes in the network without any prior configuration necessary.

6.4.1.3   Ad Hoc On-Demand Vector (AODV) Protocol

This is similar to DSR, except that routing is dependent on the route table entries in the intermediate nodes, rather than carried within the packet header.

6.4.1.4   Zone Routing Protocol (ZRP)

Zone-based routing attempts to divide the network into 'zones'. This attempts to create a hierarchical structure into the network, such as is evident in the wired world. This contrasts with proactive and reactive routing protocols, which view the network as a flat structure.

The benefit of doing this is to reduce the overhead when determining routing information, since it is no longer necessary to flood the entire network with routing packets.

6.4.1.5   Temporally Ordered Routing Algorithm (TORA)

TORA is designed to minimise reaction to topological changes. A key concept in its design is that it decouples the generation of potentially far-reaching control message propagation from the rate of topological changes. Such messaging is typically localised to a very small set of nodes near the change without having to resort to a dynamic, hierarchical routing solution with its attendant complexity.

6.4.2   Routing Summary

In reality there is no one routing protocol that suits all scenarios. In a network with a static topology a reactive protocol may be best, whereas a network where nodes are continuously moving may best suit a proactive protocol. In practice, networks change from one state to the other and adaptive systems are being considered [15].

In is probable that until a protocol is standardised or adopted as a '*de facto* standard', networks will exist as islands and communicate with each other through standardised network gateways such as the Internet or GPRS. Nodes may start running multiple routing protocols, choosing one as required for best performance or best compatibility with nodes around it, although this is unlikely to be possible for small low-capacity devices.

6.5   BT Exact's Symbiotic Test Bed

6.5.1   Overview

To be able to exploit this technology, BT Exact is building its 2nd generation symbiotic network. The latest test bed operates under Linux and has been constructed using components from research projects around the globe. The test bed will support application trials and potentially provide the core for future product development. Work is now in progress to develop a proof-of-concept demonstrator that provides low-cost broadband WLAN access. A schematic of a node in the test bed is shown in Fig 6.5.

The nodes can be constructed from a variety of hardware components. Currently nodes have been constructed using:

- PC laptop;
- Compaq IPAQ with PCMCIA sleeve;
- Sharp Zaurus;
- PC desktop;
- BT Exact WRX1.

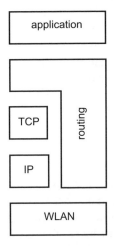

Fig 6.5   Symbiotic node schematic.

The hardware is standard except for the BT Exact WRX1. This unit was constructed at BT's Asian Research Centre in Malaysia.

6.5.2   WRX1

The WRX1 is a general-purpose processing unit (Fig 6.6) capable of performing a number of roles within the test bed network. It consists of a PC104-sized PC board with a power supply, compact flash solid-state storage, PCMCIA card and a custom-built enclosure suitable for short-term external mounting. The unit has been designed with 24 × 7 operation in mind from the outset, has two fixed network connections allowing the unit to be deployed as a broadband gateway and, with suitable software, as a firewall, and can also function as a standard IEEE802.11b access point using the HostAP software [16].

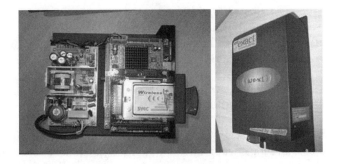

Fig 6.6   The WRX1 processing unit.

6.5.3   Routing

Several routing protocols are being evaluated for the network. Currently AODV from the National Institute of Standards and Technology [17] and DSR from the University of Queensland [18] are being used. The work from Uppsala University in producing their *Ad Hoc* Protocol Evaluation (APE) test bed [19] has also been a useful source of information and ideas.

6.5.4   Demonstrations

The simple network shown in Fig 6.7 was recently demonstrated at Adastral Park. Interesting observations included how the routing changed as groups of people moved around the building altering the propagation conditions. Doors opening and closing also caused new routes to appear for very short periods of time.

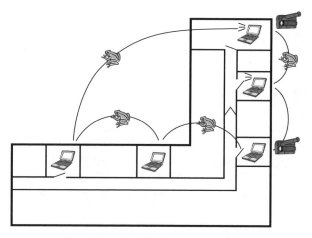

Fig 6.7   The BT Exact test network.

   BT is not alone in this technology area; much research is being carried out around the world and also being spun out into new companies. The team is compiling a database of known activity, with the most interesting projects being evaluated for potential exploitation routes.

6.5.5   Network simulation

It is difficult to build and test large-scale networks that contain 100s of nodes, mainly due to cost and logistical issues, and in these situations the best technique is to use software simulation. To ensure the results from the simulation are as close to reality as possible a number of steps have been taken:

- to emulate rather than simulate wherever possible;

- to calibrate the simulation results against practically measured results using actual networks;

- to use industry-standard simulation tools rather than in-house products that may have little credibility in the outside world.

The simulation tool chosen was OPNET Modeller [20]. This product comes with a full emulation of the IEEE802.11b MAC layer together with TCP/IP and a full range of monitoring and display tools. In addition, a radio layer can be added to better simulate the effects of a radio channel. The radio simulation has been further enhanced by BT to better characterise the radio effects at the operating frequencies.

This work is currently in its early stages but initial results from the simulation have shown great disparity from the practically measured results. Further work has reduced this disparity to an acceptable degree but shows that the calibration phase is absolutely essential if the simulation results are to be trusted.

6.6   Summary

BT Exact's symbiotic test bed has shown that research into *ad hoc* networks has reached a point where networks can be built and deployed in the laboratory using freely available components. The performance using IEEE802.11b is not optimal and improved solutions are required for other aspects such as security, address allocation and network management. However, commercial products are now becoming available, as are free public-domain solutions, and it is time to start looking for the 'killer application'. Work being undertaken by other teams within BT on motivational aspects (why would you allow others to route through your terminal) (Chapter 9) and business modelling (Chapter 8) will further refine the likely uses.

References

1   Freebersyser, J. and Leiner, B.: '*A DoD perspective on mobile ad hoc networks*', in Perkins, C. E. (Ed): '*Ad Hoc Networking*', Addison-Wesley, pp 29-51 (2001).

2   IETF MANET working group — http://www.ietf.org/html.charters/manet-charter.html

3   Zhou, L. and Haas, Z. J.: '*Securing ad hoc networks*', IEEE Network, **13**(6), pp 29-30 (November-December 1999).

4   Opportunity Driven Multiple Access (ODMA) — http://www.3gpp.org/ftp/tsg_ran/TSG_ RAN/TSGR_03/Docs/Pdfs/RP-99269.pdf

5   MeshLAN by Mesh Networks — http://www.meshnetworks.com/

6   SONBuddy by Greenpacket — http://www.greenpacket.com/

7   LocustWorld MeshBox — http://www.locustworld.com/

8   SkyPilot Networks — http://www.skypilot.com/

9   Xua, S. and Saadawi, T.: '*Revealing the problems with IEEE802.11 medium access control protocol in multi-hop wireless ad hoc networks*', Computer Networks, **38**, pp 531-548 (2002) — http://www.elsevier.com/

10 Radio Communications Agency: '*UK Interface Requirement 2005, Wideband Transmission Systems Operating in the 2.4 GHz ISM Band and Using Spread Spectrum Modulation Techniques*', — http://www.radio.gov.uk/publication/interface/index.htm

11 Faimberg, M. and Goodman, J.: '*Maximising performance of the wireless LAN in the presence on Bluetooth*', Third IEEE workshop on Wireless Local Area Networks (2003).

12 Address Resolution Protocol — http://www.redfoxcenter.org/

13 IEEE802.11 and QoS — http://grouper.ieee.org/groups/802/11/

14 Swale, R. P. (Ed): '*Voice over IP: Systems and Solutions*', The Institution of Electrical Engineers, London (2001).

15 Perkins, C. E. (Ed): '*Ad hoc Networking*', Addison-Wesley (2001).

16 HostAP software — http://hostap.epitest.fi/

17 NIST AODV — http://w3.antd.nist.gov/wctg/

18 University of Queensland DSR — http://piconet.sourceforge.net/thesis/

19 APE test bed — http://apetestbed.sourceforge.net/

20 OPNET Modeller — http://www.opnet.com/

7

SCALABILITY, CAPACITY AND LOCAL CONNECTIVITY IN *AD HOC* NETWORKS

S Olafsson

7.1   Introduction

A general definition of *ad hoc* networks is a collection of mobile nodes, which communicate with each other over a wireless channel [1]. By definition, *ad hoc* networks have no fixed infrastructure. All network functions have to be co-ordinated in a distributed manner between the network nodes. Central to the idea of *ad hoc* networks is the concept of multi hop, where each node can act as a router and forward packets on behalf of other nodes towards their destination. This is in radical contrast to conventional cellular systems where each mobile device communicates directly to a base station, which controls all transmission and routing functions [2].

One essential feature of *ad hoc* networks is that generally nodes can be mobile. They can and do move in an unpredictable manner resulting in dynamic network topology. This can lead to the emergence of temporary clusters of communicating nodes, which later may disappear as the nodes disperse. Due to the movement of nodes the association of any two nodes can disappear, to be replaced by new temporary connections. Communication between nodes depends critically on their distance as well as their power control and communication protocols [3]. This dynamic nature together with the distributed control requirements leads to new challenges not present in cellular or fixed network design.

Another aspect of *ad hoc* networks is how their willingness to be active impacts on the resulting network topology. As power is a scarce resource for small mobile devices some nodes may prefer not to participate in the *ad hoc* network beyond their own requirements. They may, for example, prefer not to forward packets on behalf of other nodes or to undertake any processing as a part of some local distributed control procedures. This complicates matters even further, as now the network topology does not only depend on applied power by individual nodes or their mobility but also on their decisions on the level of network participation.

The decision of a node not to participate or link to a local cluster impacts on the network topology and therefore the ability of active nodes to communicate across geographic distances. In reality, the network topology is therefore at any given time different to what it would be if all nodes were active and available to assist their neighbouring nodes. The unpredictable behaviour of nodes adds to the dynamics of the resulting network topology in addition to that caused by their movement alone. At any given time the network is simply defined by all the nodes, which can communicate within a certain geographic area. Due to the node dynamic and the random participation the network size as well as the network topology are subject to continuous random changes. Paths connecting source and destination nodes may be disrupted at any time requiring rerouting across different sets of active and available intermediary nodes.

Clearly it is in the interest of all network nodes to be able to communicate with any other node of their choice. Faced with node dynamics and unreliable node participation, the simplest way for active nodes to provide the required network connectivity is by adjusting their power levels. By increasing the power nodes can send data longer distances and therefore reduce their dependence on intermediate nodes relaying their data. However, this comes at a cost as higher power levels increase the interference with neighbouring nodes [4]. The benefits of higher power levels and therefore enhanced connectivity need to be contrasted with the negative effects they have on neighbouring nodes. Some of these are listed in Table 7.1.

Table 7.1   Some advantages/disadvantages of higher power levels.

High power levels	
Advantages	*Disadvantages*
Better network connectivity	Increased interference
Fewer hops	Reduced battery life
Lower bit rate	Higher costs
Less time delay	
Better usage of network resources	

All nodes in an *ad hoc* network need to balance the advantages of applying high power levels with the consequent disadvantages. In the presence of a central controller this may be a straightforward task. However, due to the distributed nature of the network, and therefore lack of global co-ordination, implementing efficient power control may be hard to achieve. It has been suggested that the task of optimally balancing the power levels of nodes in *ad hoc* networks can be seen as a two-step process [5]. Firstly, the subset of all the nodes that request to send data and can do so simultaneously needs to be identified. Secondly, those nodes that can send data need to adjust their power to suitable levels. The first step of this process does not necessarily imply optimal power levels. However, the first step is undertaken by applying a centralised algorithm, which is clearly not an option in *ad hoc* networks. The development of a decentralised algorithm is still an unsolved problem.

Another feature of *ad hoc* networks, which has no parallel in wired networks, is the relationship between power control, network topology and routing algorithms. This observation leads to an interesting, and not fully exploited, interaction between layers in the control stack. As power control in a mobile network would generally be a function of the physical layer, routing functions are a matter for the network layer. As power impacts on topology it will also impact on routing strategy as a matter of necessity. In addition, nodes may want to select routes that contain nodes with sufficient power as opposed to those running at low power levels. This will require a close co-ordination between the physical layer and the network layer.

How the increase in power levels increases the interference caused at other nodes depends critically on the type of antennas with which the *ad hoc* nodes are equipped. Omnidirectional antennas tend to generate circular interference zones, or footprints. They can therefore cause substantial interference to receiving nodes that may be sideways, or in the opposite direction to the node to which they are sending data. This effect can be very drastic, particularly in cases where the radius of the footprint may be two or even three times the distance the nodes can reach for a given power level. Under these circumstances the benefits of antennas that can focus their beams towards the node with which they are communicating can be very substantial [6]. In this case the circular footprint of the omnidirectional antenna is reshaped to a narrow beam, which causes little side or backward interference. We discuss this issue at some length in this chapter.

Applications of *ad hoc* networks are likely to be very diverse. They include communication between mobile devices within conferences, companies, campuses and homes. Also, fixed and mobile wireless sensor networks, multi-hop extensions to WLAN systems or spontaneous building of communicating nodes (hot spots) provide examples of *ad hoc* networks. Furthermore, *ad hoc* networks will provide the connectivity between different future networks. These will include, body area networks, personal area networks, Bluetooth and other larger area networks. There are numerous other applications some of which are discussed in [7]. However, before these scenarios can be realised a number of technical issues need to be addressed and solved. We summarise some of them.

- Reachability

    To be able to communicate globally nodes need to be able to reach other nodes, either directly or via multi-hops across other nodes in the network. Reachability, and therefore global network connectivity depends on node density and the power nodes apply when communicating [8]. The *ad hoc* network is globally connected if any node can communicate either directly or via multi-hops with any other node in the network. From the above discussion it is clear that global network connectivity should preferably be achieved at minimum possible power level. It has been suggested that this minimum possible power level should be the same for all nodes in the network [4]. However, this is only the case when the nodes are reasonably evenly distributed over some defined area. Usually that is not the case, and in these situations global connectivity requires node-dependent power control. This approach has been discussed in [9].

- Interference

  As in all wireless systems, nodes in *ad hoc* networks are subject to interference from other nodes. The degree of disturbance depends on the applied power and the number of neighbouring nodes. This fact is independent of the access protocols used. Only the degree of interference is protocol-dependent. Also, the type of antennas used do severely influence the extent of the nodal interference. For example, focused beams for data transmission cause far less interference than the more circular footprints of omnidirectional antennas. A way forward for more efficient *ad hoc* networks lies in the usage of smart antennas, which locate and track the receiving node in a dynamic environment. There are still several technical problems to be overcome before smart antennas can be made sufficiently 'smart' to work efficiently and reliably in dynamic *ad hoc* environments. However, as we shall show in this chapter, the benefits of using focused antennas are substantial and quantifiable.

- Power management

  Long life, light weight batteries are still not a reality. Presently this fact provides one of the main limiting factors for a large-scale implementation of *ad hoc* networks. It also has a profound effect on the way in which different protocol layers need to interact to be able to perform their tasks. Power control is traditionally a job for the physical layer, and it will, to a very large extent, continue to be that way. However, the networking layer, whose responsibility it is to find and record network routes, needs full awareness of the applied power. It may also need to be able to modify some of the functions controlled by the physical layer. This layer interaction needs to be fully exploited in efforts to make the whole protocol stack more efficient.

In this chapter we discuss mainly the problems of network capacity and local network connectivity. The probability of nodes being connected to their neighbours affects the local connectivity. This probability will depend mainly on the power applied by individual nodes as well as their density and distribution. In general, statements about local connectivity do not imply global connectivity, that is, the condition that any two nodes in the network can communicate with each other. In fact, there are, to date, no rigorous models available which enable one to make statements about global connectivity. Only some approximations based on percolation arguments have been considered [10]. All statements we make about connectivity do therefore refer to local connectivity, i.e. the probability that any randomly selected node can, at a given applied power level, communicate with its nearest neighbour.

We study the complementarity between local connectivity and network capacity, meaning that as the conditions for local network connectivity improve, for example, by nodes increasing their sending power, the totally available capacity in the network is reduced. By totally available network capacity we mean simply the fraction of network nodes which can simultaneously be active, that is, send or receive data from neighbouring nodes. Due to mutual interference, the activity of some nodes will reduce the capability of other nodes being active at the same time.

Assume that the maximum transmission capacity of all nodes in the network is the same and C bit/s. If all N nodes can simultaneously submit data at this rate the total network capacity is NC. However, if only a fraction f of all nodes can simultaneously be active, the total capacity is reduced to fNC. We show how f can be estimated as a function of node distribution and applied transmission power.

Finally, we analyse how the average rate achievable by nodes in the network scales with the mean number of hops required to connect source node and destination node. In the case where the two communication nodes are selected randomly from within a network of identically and independently distributed nodes, the ratio of average rate over maximum rate decreases proportionally to $N^{-3/2}$. However, whenever the geographic length between two communicating nodes grows at a slower rate than the network itself, the utilisable rate scales better. We consider the case when the number of hops between communicating nodes grows proportionally to the logarithm of the total number of *ad hoc* nodes that constitute the network.

7.2   Mean Geographical Distances

To analyse the conditions for local connectivity one needs to make assumptions about the distribution of nodes in a given geographical area. We make the assumption that nodes are randomly distributed over a two dimensional circular area. This is certainly a simplifying assumption, which will be modified in a separate publication.

Consider N *ad hoc* nodes distributed uniformly and independently across the two-dimensional circular area with radius R. The constant node density is ρ, which gives the average number of nodes per square unit, $\rho = N/\pi R^2$. We are interested in finding the distance distribution between two points placed randomly and independently in the area. The probability density function for the distance function on a unit disk given by Kendall and Moran [11]:

$$d(x) = x/\pi(4\arctan(\sqrt{4-x^2}/x)-x\sqrt{4-x^2}); \quad 0<x<2$$

The expression:

$$\langle L \rangle = R \int_0^2 x d(x) dx = (128/45\pi)R \approx 0.9R$$

represents the mean physical (geographical) length between two nodes selected randomly and independently from within a circle of radius R. Note that this quantity only depends on the geometry of the circle and does not take the density of nodes into account. For performance evaluation we need to express the mean geographical length of connection between two nodes in terms of the mean number of hops required to get from source to destination. For this we need to consider the node density and the distribution of the shortest distance between nodes.

7.3   Connectivity

We address the question of connectivity for a large group of nodes randomly distributed over the two-dimensional plane. Fundamental for the question of network connectivity is the statistics, that is, the distribution of the shortest distance between any neighbouring nodes. In the limit of an infinite circle the nearest neighbour distance r for the nodes in the plane is given by Cressie's expression [12]:

$$f(r) = 2\pi\rho r \exp(-\rho\pi r^2)$$

where ρ is the density of nodes. From this we find the cumulative distribution function as:

$$F(r) = \int_0^r f(r')dr' = 1 - \exp(-\rho\pi r^2)$$

This is the probability that the distance between neighbouring nodes is less than or equal to r. The probability that a node is **not** connected to any of its neighbours is $1 - F(r) = \exp(-\rho\pi r^2)$.

To demonstrate that the neighbour density distribution also gives a reasonable approximation for the distribution of nodes in a finite-sized square we performed a number of simulations to compare theoretical and empirical results. The results are presented in Fig 7.1. Because of this reasonable fit we can use the following expression for the mean shortest distance:

$$\langle r \rangle = \lim_{R \to \infty} \int_0^R r f(r) dr = \frac{1}{2\sqrt{\rho}}$$

and the variance is:

$$\sigma_r^2 = \lim_{R \to \infty} \int_0^\infty (r - \langle r \rangle)^2 f(r) dr = \frac{4 - \pi}{4\pi\rho}$$

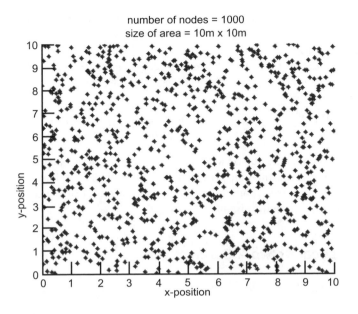

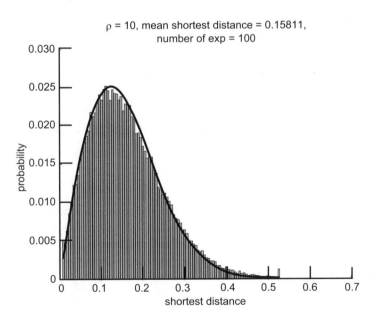

Fig 7.1 Distribution of 1000 nodes in a 2-dimensional square of size 10×10 (top). The empirical distribution of the nearest neighbour distance is presented by the bar chart and the theoretical distribution by the solid line (bottom).

Generally we will be interested in using the required number of hops between two nodes as a distance measure, rather than the geographical distance. The number of hops will of course depend on the density of nodes. Using the results from the previous section we find that the mean length between two selected nodes, measured in the number of mean shortest distance units, is:

$$d = \frac{\langle L \rangle}{\langle r \rangle} = 2\alpha R \sqrt{\rho} = 2\alpha R \sqrt{\frac{N}{\pi R^2}} = 2\alpha \sqrt{\frac{N}{\pi}}; \quad \alpha = 0.9$$

From this we conclude that on average there are $d + 1$ nodes involved in relaying data from source to destination provided that the radius R is kept constant. The number of hops increases proportionally to the square root of the total number of nodes, distributed within the circular of radius R.

7.4   Local Network Connectivity

For any two nodes to be able to communicate directly they have to be sufficiently close. For the moment we assume that each node is able to reach any other node directly provided that it is within a reach distance r_e. The reach r_e depends on the power applied by the nodes.

We assume that the path loss in the access medium is described by the following expression [13]:

where γ is a constant, P_0 the transmitted power and $\beta \geq 2$ is a constant. If the minimum power required to receive a packet is given by P_c then the reach r_e has to satisfy:

$$r_e = \left(\gamma \frac{P_0}{P_c} \right)^{1/\beta}$$

Assuming that the distribution of the shortest distance between nodes is given by the function $f(r)$ in section 7.3, one finds that the probability that a randomly selected node has a distance to its closest neighbours of less than r_e is $F(r_e) = 1 - \exp(-\rho \pi r_e^2)$. By making the assumption of statistical independence [8] the probability that none of the N nodes are isolated is:

$$Pr(local\ connectivity) = \left(1 - \exp\left(-\rho \pi r_e^2 \right) \right)^N$$

We refer to this as the network being locally connected (see Fig 7.2). However, it is important to realise that this condition is not sufficient for global connectivity, that is, the condition that any node in the network can communicate with any other node in the network. The condition of global connectivity is stronger than the condition of local connectivity.

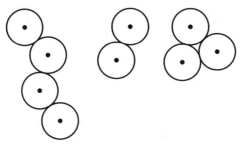

Fig 7.2   The network of nine nodes is locally connected as every node is connected to at least one neighbour. However, the network is not globally connected as nodes belonging to one cluster cannot communicate with nodes belonging to another cluster.

To deal with the question of global connectivity one introduces the percolation probability p_c, as the probability that an arbitrary node in a network belongs to an infinite cluster of connected nodes. Unfortunately, no exact formula for the percolation probability in two dimensions is presently known. Only bounds have been calculated [10]. We will therefore work with the condition for local connectivity and view it as a necessary condition for global connectivity.

- **Example** — consider a system of radius $R = 10$. The number of nodes randomly distributed over the available area is $N = 100$. Figure 7.3 presents the probability that the network is locally connected as a function of r_e.

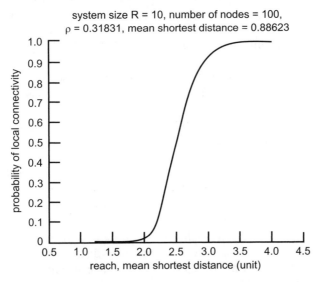

Fig 7.3   Probability of local network connectivity as a function of node reach.

From the graph we realise that the reach r_e has to be approximately three times the mean shortest distance to provide a high probability of local connectivity. How-ever, with a reach of that length, active nodes tend to interfere with other nodes in their vicinity with the effect that these cannot simultaneously send or receive data.

7.5   Interference

Assume the maximum (stand-alone) capacity per node is C. However, due to interference the effective capacity per node can be substantially reduced. In this section we demonstrate how efforts to maintain effective capacity per node tend to reduce the probability of local connectivity and vice versa.

We assume omnidirectional antennas. Therefore, when a node is active no other node can simultaneously be active within a circle of radius r_i, referred to as the interference range. The activity of each node therefore makes it impossible for all nodes in an area $A_i = \pi r_i^2$ to be active at the same time. We assume that N_a is the maximum number of nodes that can simultaneously be active. Then $\pi R^2 = N_a \pi r_i^2$ and therefore $N_a = (R/r_i)^2$. With N the total number of nodes, the fraction of simultaneously active nodes is given by the expression

$$f_a = \frac{N_a}{N} = \frac{1}{N} = \left(\frac{R}{r_i}\right)^2$$

If $r_i = R$ then $f_a = 1/N$, that is, only one node can be active at any given time. From the condition $f_a = 1$ we have that $r_i = R/\sqrt{N}$. We therefore assume that the interference r_i takes values in the range from $R/\sqrt{N}$ to R. In terms of the node density, the inference range varies between $r_{i,\min} = 1/\sqrt{\rho\pi}$ and $r_{i,\max} = \sqrt{N/\rho\pi}$.

The probability of local connectivity increases as the reach increases. However, increasing the reach reduces the fraction of those nodes, which can simultaneously be active. The minimum interference range, $= 1/\sqrt{\rho\pi} = 0.56/\sqrt{\rho}$ is just above the mean shortest distance $\langle d \rangle = 0.5/\sqrt{\rho}$. Therefore, as is clear from Fig 7.1, when $r_e \leq 0.56/\sqrt{\rho}$ the probability of local connectivity is negligible. This is demonstrated in Fig 7.4.

Clearly the assumption of a 'circular' disturbance zone around each active node is fairly simplistic and only approximately true in the case where there are no obstacles in the way; but it provides a starting point for quantitative analysis.

Differently shaped disturbance zones can be created by antennas, which focus their beams towards the nearest neighbour they want to communicate with. In some cases this focused beaming will minimise the disturbance caused to other neighbouring nodes. What impact these focused antennas will have on the overall network performance will be discussed in next section.

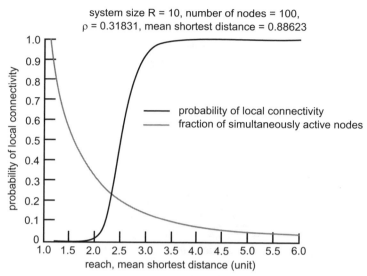

Fig 7.4 The decreasing graph presents the decline in the number of simultaneously active nodes as a function of reach. The increasing sigmoid shaped curve presents the probability of local network connectivity as a function of reach.

7.6   Focused Antennas

In this section we will discuss how the application of focused antennas (smart antennas) can increase the available network capacity substantially without reducing the local network connectivity. Instead of modelling the precise shape of the antenna beams we make a simplifying assumption, which only demonstrates the beneficial effects of applying beam-focusing antennas. This will be studied in more detail in subsequent publication.

We assume that the interference domain around each active node has the shape of an ellipse. The area is therefore given by $A_i = \pi a_i b_i$ where $2a_i$ and $2b_i$ are the lengths of the major and minor axes respectively. With N_a the maximum number of nodes that can simultaneously be active, we set $\pi R^2 = N_a \pi a_i b_i$. If we assume that $b_i = a_i/n, n \geq 1$, then the fraction of simultaneously active nodes is given by

$$f_a = \frac{n}{N}\left(\frac{R}{a_i}\right)^2$$

where a_i varies between the maximum and minimum values:

$$a_{i,\,min} = \sqrt{\frac{n}{N}}\,R = \sqrt{\frac{n}{\rho\pi}} \quad \text{and} \quad a_{i,\,max} = R\sqrt{n} = \sqrt{\frac{nN}{\rho\pi}}.$$

From the expression for f_a we notice that the fraction of nodes that can simultaneously be active grows linearly in $n = (a/b)$. This is demonstrated in Fig 7.5,

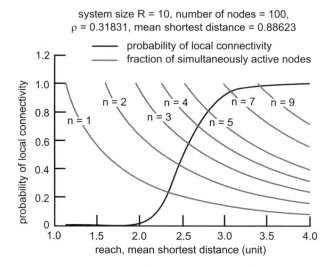

system size R = 10, number of nodes = 100,
ρ = 0.31831, mean shortest distance = 0.88623

Fig 7.5   The declining lines present how the fraction of simultaneously active nodes is reduced as a function of reach, for different parameterisations of the elliptical disturbance zone.

which plots f_a as a function of r_e for different n values, together with the probability
of local connectivity as a function of r_e.

The capacity improvement as a function of the ratio $(a/b) = n$ is presented in
Fig 7.6.

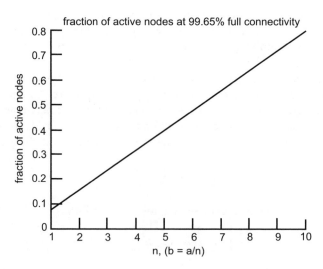

Fig 7.6   The fraction of simultaneously active nodes as a function of the ratio $(a/b) = n$ at
the 99.65% full connectivity level.

For example for $n = 10$, 79.72% of all nodes can simultaneously be active. This is a considerable improvement from only 7.97% in the case of $n = 1$, that is, when the disturbance has the shape of a circle.

Previously we have assumed that the interference and the reach are of the same magnitude. This assumption is generally not correct. In fact it is possible that the interference exceeds the reach by a factor of two or more. Figure 7.7 demonstrates what effect an interference twice the reach, $r_i = 2r_e$ has on the complementarity between local connectivity and network capacity.

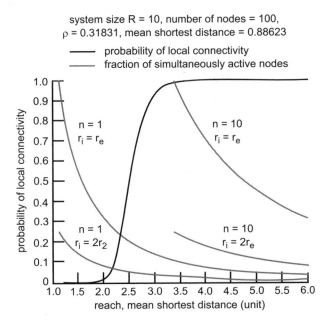

Fig 7.7   The declining graphs show the available capacity in the case when the interference is equal to the reach for the two cases $(a/b) = 1$ and 10 and for the case when the interference is twice the reach again for $(a/b) = 1$ and 10.

From this it is clear that the ratio $\alpha = (r_i/r_e)$ is important for the capacity available in the network. The details of this and other relationships are the subject of another publication.

7.7   Multi-connectivity

Previously we have only discussed the probability of local network connectivity as a function of density and reach. In other words we have considered the condition for each node in the network having at least one neighbour it can communicate directly with. However, that approach assumes that all the nodes are available for forwarding packets for other nodes at all times. In reality this will not be the case. As discussed, light-weight long life batteries are still not a reality. Battery power is therefore valuable and nodes in *ad hoc* networks may not have the right incentives to stay switched on for long periods of time. This fact reduces the probability that the network stays locally connected, even though that may be the case when all nodes are on and available.

In view of the above comments we need to consider the possibility of introducing some redundancy into the network so that even when some fraction of nodes may be switched off the network still stays at least locally connected. It can be shown, see for example [8], that the probability that a node with reach r_e has m neighbours is given by the expression:

$$P(d = m) = \frac{\left(\rho \pi r_e^2\right)^m}{m!} \exp\left(-\rho \pi r_e^2\right); \quad m = 0,1,2...$$

Therefore, the probability that a randomly selected node with reach r_e has up to $m - 1$ neighbours is:

$$P(d \leq m - 1) = \sum_{k = 0}^{m - 1} P(d = k)$$

From this one finds the probability that all N nodes in the *ad hoc* network have at least m neighbours:

$$P(d \ for \ all \ nodes \leq m) = \left(1 - \sum_{k = 0}^{m - 1} \frac{\left(\rho \pi r_e^2\right)^k}{k!} e^{-\rho \pi r_e^2}\right)^N$$

Notice that for $m = 1$ this expression reproduces the probability of locally connected network given in section 7.4.

In the graph in Fig 7.8 we have plotted the probability of each node having m neighbours with $m = 1,2,3,4,5$ together with the network capacity for different ratios between the major and minor axis both as functions of the reach.

It is clear that for a given node distribution the power (reach) has to be increased substantially to move from single to multi connectivity. Therefore, to secure multi-connectivity, the network capacity will reduce substantially compared to the case of just single local connectivity.

From Fig 7.8 we see that to provide close to 100% probability of local connectivity of degree 5 the capacity in the case of circular antennas has been reduced to less than 10%. However, in the case of elliptic antennas with a ratio (a/b) = 10 the capacity has been reduced to only 50%. These figures demonstrate clearly the substantial benefits provided by focused antennas.

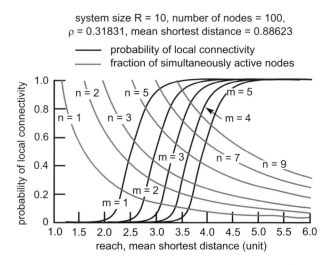

Fig 7.8 The tinted declining curves represent the capacity available for different ratios $n = (a/b)$. The solid sigmoid curves represent different degrees of local connectivity, ranging from $m = 1$ to $m = 5$.

7.8 Valid Node Arrangement

In the previous section we have demonstrated by quantitative analysis how the increased power level applied by communicating nodes in an *ad hoc* network improves the network connectivity and therefore the probability that randomly selected nodes can communicate. However, as we have shown, increasing power levels also has a detrimental effect on the overall performance of the network.

There is also an issue whether nodes should seek to reduce the number of hops involved in the relaying of packets from source to destination by increasing transmission levels. This will not require the same number of intermediate nodes for relaying the data but other neighbouring nodes would generally suffer from the interference caused by the higher power levels required. That, on the other hand, would render them inactive for the time slot under consideration. However fewer nodes need to use their own limited power for the relay of packets.

A node receiving data from a neighbouring node can only do so provided that there is no other node active within a certain minimum distance D_c. This distance does depend on the applied power levels and the type of antennas used as has been discussed in previous sections. We have also assumed that nodes that are receiving data cannot simultaneously send data and vice versa. Given these constraints only a subset of all nodes can send and receive data at any given time. These two sets of sending and receiving nodes build a set of active links, where in each instance the data flows from the sending node to the receiving node (see Fig 7.9).

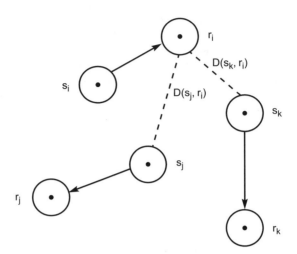

Fig 7.9   The graph represents a possible arrangement of sending and receiving nodes. No node can transmit and receive at the same time, i.e. $r_i \neq s_i$ for all i. Furthermore the distance between all sending nodes and receiving nodes has to satisfy, $D(s_j, r_i) \geq D_c$ for all i and j.

We assume that node s_i wants to send data to node r_i. Then the signal to interference plus noise ratio (SINR) at the receiving node is given by the expression [2]:

$$S_{s_i, r_i} = \frac{P_{s_i, r_i} G_{s_i, r_i}}{\sum\limits_{k \neq i} P_{s_k, r_k} G_{s_k, r_i} + \sigma^2_{r_i}}$$

where is the gain for the connection from the sending node s_i, to the receiving node r_i. σ_{r_i} represents the noise level at the receiver's location. The denominator $\sum_{k \neq i} P_{s_k, r_k} G_{s_k, r_i}$ represents the unintended signal at node r_i when node s_k is sending a signal to node r_k. For the receiving node to be able to receive the signal correctly its signal to noise ratio has to exceed some critical level [2], $S_{s_i, r_i} \geq \beta_{r_i}$. Rearranging this equation results in the following power being applied by the sending node s_i:

$$P_{s_i, r_i} \geq \frac{\beta_{r_i}}{G_{s_i, r_i}} \left\{ \sum\limits_{k \neq i} P_{s_k, r_k} G_{s_k, r_i} + \sigma^2_{s_i, r_i} \right\}$$

From this expression it is obvious that the ratio $G_{s_k, r_i} / G_{s_i, r_i}$ influences strongly the required power levels for node s_i to communicate with r_i. This observation makes a very strong case for focused (smart) antennas, which, by tracking the nodes they beam to, reduces substantially the interference to other nodes as captured by the term. As this term gets smaller, node s_i can reach the required threshold value at a substantially reduced power level, reducing at the same time the interference to neighbouring nodes.

7.9   Scalability for Node Rates

The question of how the performance of *ad hoc* networks scales is an important question, which needs to be addressed. The parameters in the network topology or traffic characteristics that have an impact on the network performance need to be linked within a generic model whenever possible. In this section we consider briefly the effect that connection length, measured in terms of number of hops, has on the effective send rate available to nodes in the network.

We consider the case of an *ad hoc* network consisting of N nodes. Each node acts as a source node which wants to communicate with a destination node. The mean geographical distance between source and destination is L and the mean shortest distance between neighbouring nodes d. We assume that each node can effectively (on average) send at the rate λ_e. The total average packet rate for the whole network is then [4]:

$$\lambda_{\text{tot}} = \lambda_e \left(\frac{L}{d}\right) N = \lambda_e N n_h$$

Furthermore we assume that the maximum rate at which each node can submit at is $\lambda_{\max}$. Then the total number of packets that can be transmitted at any given time is:

$$\Lambda = N_a \lambda_{\max} = \left(\frac{A}{\pi r_i^2}\right) \lambda_{\max}$$

The following inequality must hold $\lambda_{\text{tot}} \leq \Lambda$, from which we derive:

$$\left(\frac{\lambda_e}{\lambda_{\max}}\right) \leq \frac{1}{d} \left(\frac{n}{N}\right) \left(\frac{R}{r_i}\right)^2$$

If we make the assumption that the mean geographical distance between randomly selected nodes is related to the system radius as $\langle L \rangle = \alpha R$, then we find:

$$\left(\frac{\lambda_e}{\lambda_{\max}}\right)_{d \sim \sqrt{N}} \leq \frac{\sqrt{\pi n}}{2\alpha N^{2/3}} \left(\frac{R}{r_i}\right)^2$$

This is equivalent to the mean shortest distance being $d = (2\alpha/\sqrt{\pi})\sqrt{N}$ (therefore the sub-index $d \sim \sqrt{N}$). If, however, the mean geographical distance of connections grows at a slower rate than the system size itself, then the mean available rate scales better than is indicated by the above expression for $\lambda_e/\lambda_{\max}$.

We make the assumption that the mean shortest distance does not grow proportionally to $\sqrt{N}$ but to the logarithm of N, $d = \beta \log(N)$. Then the ratio $\lambda_e / \lambda_{max}$ behaves like:

$$\left(\frac{\lambda_e}{\lambda_{max}}\right)_{d \sim \log(N)} \leq \frac{\pi}{\beta N \log(N)} \left(\frac{R}{r_i}\right)^2$$

This is considerably better scaling, as is demonstrated in Fig 7.10, which plots the following expression as a function of N:

$$\frac{(\lambda_e / \lambda_{max})_{d \sim \sqrt{N}}}{(\lambda_e / \lambda_{max})_{d \sim \log(N)}} \sim \frac{N^{3/2}}{N \log(N)} = \frac{\sqrt{N}}{\log(N)}$$

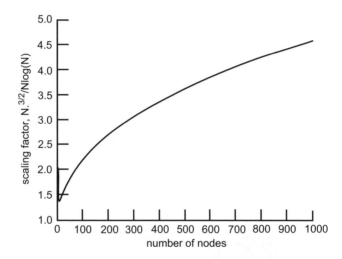

Fig 7.10   When number of hops increases proportionally to $\log(N)$, the mean effective bandwidth per node scales better compared to the case when it increases proportionally to $\sqrt{N}$. In the case of 1000 nodes, the effective rate is, by approximately a factor of four, larger in the $\log(N)$ case.

It is reasonable to assume that the geographic distance between communicating nodes does not grow at the same rate as the network itself. This follows from the other assumption that nodes are more likely to communicate with neighbouring nodes, rather than nodes far away. This improves the scaling behaviour of the network as is demonstrated in Fig 7.10. One way to improve the scaling of the *ad hoc* network is to insert fixed nodes in the network. This technique has been discussed in Sharma et al [14].

7.10   Summary

Presently there is a great interest in the future potential of *ad hoc* networks. However, before they take off on a large scale a solid understanding of their performance and coverage is essential. In this chapter we have discussed in detail the overall available capacity in a network of randomly distributed nodes as well as the conditions for the network being locally connected. We have demonstrated that capacity and local connectivity are complementary quantities in the sense that by improving the probability of local connectivity one decreases the overall available capacity in the network and vice versa.

We have shown that by applying antennas which focus their beams towards the nodes with which they directly communicate, the overall capacity of the network can be increased substantially without reducing the level of local connectivity. This issue is likely to become very important, particularly in the case of dynamic nodes.

Nodes in *ad hoc* networks rely on the local presence of other nodes to route and relay their data across the network towards their destination. Due to the limited duration of battery power in mobile nodes many may prefer to be switched off when not engaged in their own tasks. Clearly when a large number of nodes are switched off the local connectivity of the network may be substantially reduced. To counteract this we have studied the necessary conditions for introducing redundancy into the network, that is, the probability that each node can directly connect to $m > 1$ neighbouring nodes. We have shown how this more stringent requirement reduces the available capacity even further compared to the case where $m = 1$ only.

Finally we consider how the average available rate per node scales with the increasing number of nodes within an area with fixed radius R. We show that if the mean geographic length of connections grows linearly with the system size then the available rate scales poorly with the increasing number of nodes. In this case the mean shortest distance between nodes grows proportionally to the square root of the number of nodes in the network. This poor scaling is improved considerably if the mean shortest distance grows at a slower rate, for example proportionally to the logarithm of the number of nodes.

References

1   Pursley, M. (Ed): Special Issue on Packet radio Networks, IEEE Proceedings, **75**(1) (1987).

2   Lee, W. C. Y.: '*Mobile Communications Engineering*', 2nd edition, McGraw-Hill (1997).

3   Gupta, P. and Kumar, P. R.: '*The Capacity of Wireless Networks*', IEEE Transactions on Information Theory, **46**(2) pp 388-404 (March 2000).

4   Narayanaswamy, S., Kawadia, V., Sreenivas, R. S. and Kumar, P. R.: '*Power Control in Ad Hoc Networks: Theory, Architecture, Algorithm and Implementation of COMPOW Protocol*', — http://black.csl.uiuc.edu/~prkumar/ps_files/compow_ewc_2002.pdf

5   ElBatt, T. and Ephremides, A.: '*Joint Scheduling and Power Control for Wireless Ad Hoc Networks*', IEEE INFOCOM (June 2002).

6   Olafsson, S.: '*Modelling Capacity and Local Connectivity in Ad hoc Networks*', International Conference on Wireless Networks, Las Vegas (June 2003).

7   EURESCOM Study: '*Homogeneous Infrastructure for Hot Spot Scenarios*', pp 1244 — http://www.eurescom.de/public/projects/P1200-series/p1244

8   Bettstetter, C.: '*On the Minimum Node Degree and Connectivity of a Wireless Multihop Network*', MOBIHOC'02, Lausanne (June 2002).

9   Kawadia, V. and Kumar, P. R.: '*Power Control and Clustering in Ad Hoc Networks*', IEEE, INFOCOM 2003-08-05

10 Meester, R. and Roy, R.: '*Continuum Percolation*', Cambridge University Press, Cambridge (1996).

11 Kendall, M. J. and Moran, P. A. P.: '*Geometrical Probability*', Charles Griffin & Company Limited, London (1962).

12 Cressie, N. A. C.: '*Statistics for Spatial Data*', John Wiley (1991).

13 Prasad, R.: '*Universal Wireless Personal Communications*', Artech House Publishers, (1998).

14 Sharma, S., Nix, A. and Olafsson, S.: '*Situation-Aware Networks*', Communication Magazine (July 2003).

8

THE ROLE OF *AD HOC* NETWORKS IN MOBILITY

A Readhead and S Trill

8.1   Introduction

In this context, mobility is the ability to access communication services any time, any place, anywhere — the Martini moment. WLAN *ad hoc* networks would offer users mobility within range of a base-station. If they work together, lots of *ad hoc* networks could offer lots of users quite extensive mobility.

This possibility has already been foreseen by Nicholas Negroponte[1], who has a vision of a mesh of interconnected WLANs 'like lilypads where users hop from one to another'. This could come about by the spread of *ad hoc* networks.

Ad hoc networks are created when two or more devices are connected together to form a network. The ready availability of wireless devices has made this sort of network comparatively easy to set up. People can buy the equipment they need to build wireless networks off the shelf, plug it in and go[2]. If they are unsure, there are books written specifically for non-technical readers telling them how to do it [2].

8.2   *Ad Hoc* Networks

A true *ad hoc* network may have little or no external connectivity and be limited to hopping from node to node within a relatively small area. As a result we believe that true *ad hoc* networks will have limited appeal. Because of this we expect the *ad hoc* networks that develop in the UK to have some form of external connectivity.

---

[1] Professor Negroponte gives credit for the water lily analogy to Alessandro Ovi, Technology Adviser to European Commission President, Romano Prodi.

[2] LocustWorld are selling a hardware MeshAP for £250 or $440 or 355 euros. Prices drop to £220 or $395 or 320 euros for orders of 10 or more. The system utilises over-the-air self-updating software and cryptographic mesh authentication. The system simply plugs into the user's PC and connects them into a nearby IEEE802.11b 2.4 GHz wireless network (if there is one). If they have no Internet connection, they can use one via the MeshAP, or if they do have one, they can share this via the same mesh system [1].

This means that the distinction between an *ad hoc* and a community network may be less clear in practice than it is in theory.

For *ad hoc* networks to become a significant part of the network landscape, we believe the following three factors must be present.

- Motivation

  The users must want to communicate with each other and/or have a reason to co-operate with each other. Probably the most obvious reason to co-operate is to gain access to connectivity because without this the network will have little to offer. Some users may want to band together to negotiate for better connectivity. Some will want to do it just 'because it's there[3]'.

- Technically possible

  The instinct to tinker is present in most small boys and the 'freedom to tinker' [3], e.g. with a Pringle tin [4], is an important element in innovation. In this case we already know the technology works. It may not offer the quality of service network operators normally provide but it is good enough for many purposes.

- Cost effective

  When people are doing something as a hobby, the cost in time and money is less of an issue. However, for it to become a mass-market service, and move beyond the early adopters, the technology needs to be cheap and simple to deploy. The good news is that it is moving in that direction already.

From an economic point of view, the technical details do not matter. If the right conditions exist, people will do this, and by trial and error find the most cost-effective means of doing it. We believe these conditions do exist. We believe *ad hoc* networks will develop in the UK.

For many people in the telecommunications industry the broadband ideal is fibre to the home (FTTH) and an ultra-wideband (UWB) network within the home. However, if *ad hoc* networks grow faster than the speed of fibre roll-out, the border between the network operator owned infrastructure and end user owned infrastructure could move much closer to the exchange. If so, this could have interesting implications for traditional network operators as ownership of the last mile has been perceived as an important part of the value chain to own and control. For example, an *ad hoc* network might allow its users to hop to the best network or connection point or share network connections. If so, part of the management of how the network is used will move to the edge of the network.

For the telecommunications companies, *ad hoc* networks are private networks outside their control. The telcos will only control the external connectivity. This is an important lever because without external connectivity the usefulness of an *ad hoc*

---

[3] The response of George Leigh Mallory (1886-1924) on being asked, 'Why do you want to climb Everest?'

network would be very limited. On the other hand if users band together, they will be in a much better position to get what they want from network operators.

Some simple scenarios follow.

- Example 1

  An Internet user sends an e-mail to a neighbour. The message travels from their PC down their telephone line to their ISP, then to their neighbour's ISP, then down their neighbour's telephone line to their PC. The marginal cost to the Internet users will be zero, if both have flat-fee Internet access. The extra revenue to the network operators will also be zero, but they have a little extra network traffic.

  In an alternative scenario, both invest in some wireless kit so they can communicate directly. Our conclusion is that there is little point in doing this because (a) they probably do not e-mail each other that often, and (b) as the marginal cost of using a fixed line to do this is zero, there can be no cheaper alternative.

- Example 2

  An Internet user wants to send an e-mail to somebody who lives some distance away — same zero marginal costs and revenue as above. An *ad hoc* network cannot do this at all because it does not have the reach. To do this you must have some kind of connectivity to the Internet.

- Example 3

  An Internet user wants to download a big file from a Web site on the other side of the world — same zero marginal costs and revenue as above. Again, an *ad hoc* network cannot do this because it requires connectivity to the Internet.

- Example 4

  An Internet user wants to share quite a large file with a neighbour. Sending the file over their fixed line Internet connection would not cost anything *per se*, but it would tie up the connection for quite a long time. It may be much quicker to send over a wireless link.

  Therefore, in our view, an unconnected *ad hoc* network is unlikely to be much use unless people need to share files with other people who live relatively nearby.

  However, if one or more individuals on the *ad hoc* network have external connectivity, then a whole new range of possibilities opens up, because Metcalfe's Law states that '... the value of a network goes up as the square of the number of users' [5]. They now have a good reason to want to belong and therefore a good reason to co-operate, if that is the price of belonging.

- Example 5

  An Internet user decides to allow other Internet users to share their always-on but not always-in-use Internet connection, by means of a wireless *ad hoc* network. For example, a business with Internet connectivity in use from 9.00 am to 5.00 pm Monday to Friday might decide to make its Internet connection available to nearby and passing wireless users during the evenings and at weekends. The rise of Warchalking[4] suggests this is already happening — with and without the owner's consent. Additional cost to the Internet connection owner will be zero, if they have flat-fee Internet access and adequate security to prevent unauthorised access to their own computers. Extra revenue to the network operators will also be zero, but they could have significant network traffic.

- Example 6

  A group of Internet users pool their always-on but not always-in-use Internet connections. For example, each user has fixed line Internet access and they arrange to manage their (joint) traffic to make the best use of their (joint) access. Additional cost to the Internet users will be zero, if they have flat-fee Internet access. Extra revenue to the network operators will also be zero, but they may have additional network traffic.

- Example 7

  A group of Internet users get together to buy and share a high bandwidth Internet connection. This is probably most likely, but not necessarily only, where there is no fixed line broadband service available. This scenario is very similar to a community network (see section 3). The cost to the individual Internet users may be no more than the cost of a separate Internet connection, but the bandwidth may be higher. The revenue to the network operator may be less than the sum of the revenues it would get from individual Internet connections. This may be partly offset by lower local network build costs but it also means that the fixed line network operator is losing control over the last mile. In addition, traffic volumes may be higher as the users take advantage of the higher bandwidth.

- Example 8

  A broadband wireless Internet service provider (WISP) sets up business in an area offering higher bandwidth Internet access than is offered by the fixed line network operator[5]. The WISP may need to buy a big pipe from the fixed line network operator but the fixed line network operator will lose the revenue it

---

[4] This activity consists of labelling buildings and pavements with a series of cryptic symbols to signpost the presence of a WLAN and to provide information about how to gain access. The practice has become so popular that a set of standardised symbols can now be found 'chalked' upon buildings around the world [6].

[5] E-xpedient [7] offers Internet access speeds of up to 100 Mbit/s primarily to business users over a combination of fibre and wireless in (parts of) 14 cities in the USA.

might have received from the end users[6]. Instead of a number of small customers, the national network operator will now have just one large customer. Where there is a choice of network, the negotiating power will cross the table. This will drive prices down. This will benefit consumers.

These scenarios do not pretend to describe all the ways in which people will use *ad hoc* networks — because we cannot know how people will use the technology. That will be revealed over a period of time. We have assumed that people will want to communicate with other people on other networks, so they will want connectivity to other networks.

In these scenarios, the last mile or so[7] will be made up of equipment owned by third parties. This equipment can take a number of forms. The most likely is some form of local wireless network, as this avoids the need to install equipment or cables on other people's property and helps distribute the cost among the users. The form this network takes does not matter to the fixed line network operator. It is not the fixed line network operator's infrastructure. It is not the fixed line network operator's expense. It is also beyond the fixed line network operator's control.

8.3   Community-Based Networks

Because of the advantages of working together, we believe that the most common form of *ad hoc* network will be a community-based network, i.e. groups of people working together.

There are already a large number of community-based networks around the world. One directory on the Internet [8] lists 14 in Australia, 18 in Europe and 30 in North America — and is probably already out-of-date. Some of these are networks where the costs of running the network is shared among the users. Others go further and aim to provide free access to the Internet for all.

8.3.1   Free networks

Free network access for all is a noble aim and if the network is purely local, it may be possible as each user will bear their own costs and only communicate with other users similarly equipped. As already explained, in our view such networks will be of limited appeal because of their limited reach. Without access to a large network, such networks will have little to offer.

---

[6] A typical rental for a high capacity leased line for the backhaul is circa £10k pa. If this supports 50 users (the normal contention ratio for residential ADSL customers is 50:1) the lost revenue could be circa £15k pa. If this supports 100 users the lost revenue could be circa £30k. If this supports 150 users the lost revenue could be circa £45k.

[7] Broadband suppliers in the USA are currently testing wireless access points with a range of up to 25 miles, albeit with higher power levels than would be legal in the UK.

If some of these users have Internet connections, it may be possible to share this access with those who do not. However, there will be access costs and the network will rely on some paying the costs so all can enjoy the benefits. In certain circumstances (e.g. in development areas) it may be possible to apply for grants (e.g. from the government or other interested parties), but these grants are normally intended to cover capital costs and another way will need to be found to cover running costs.

Unrestricted access to the Internet is a commendable aim, but the problem with providing unrestricted access to anything is the danger that some people will abuse the unrestricted use of the resource and end up spoiling it for all[8]. If access is unrestricted, there may be no way to stop 'bandwidth hogs' behaving badly.

In addition, the free networks will not have any funds to pay for network support and maintenance, but something of this nature will be required. If the free network relies on a few key individuals for these skills, the network may have to cease operations if they leave the group.

For these reasons, we don't think the free and unrestricted access network business model is sustainable in the long term.

However, there are a number of free networks in the UK that seek to prove us wrong. For example, Arwain [9] in Cardiff seeks to provide free (or not-for-profit) high-speed broadband access to all sections of the community including the young, sick, elderly, unemployed and disadvantaged. If Arwain can provide free access for these groups by utilising bandwidth that others have paid for, but are not using, then they may have found a way to provide free access for worthy causes. This may be possible on a small scale but it is unlikely to be possible on a large scale. It is much easier to organise such altruistic co-operation when a network is small (where everyone is known to everyone else), than when the network is large and impersonal.

8.3.2   Shared Cost Networks

Shared cost networks start with a much better business model. As the costs are shared, there are funds to pay for Internet access and other expenses. Because these are proper commercial transactions they are more stable. In addition, there are a number of ways the members can keep costs down. They will not have large overheads. They may provide their own first-line technical support. They will not need large advertising and marketing budgets. Instead, all the income can be spent on providing the service. Therefore, the cost of belonging to such a network is likely to be less than that offered by a commercial operation[9] — but may also offer more.

An example of a sort of community network is the network around Cambridge. In this case there is a monthly subscription and commercial support. A company

---

[8] Forster-Lloyd, W. (1794-1852): 'The Tragedy of Freedom of a Commons'.

[9] According to www.wlan.org.uk their studies suggest that the not-for-profit ethic can provide end-users connections for as little as 7% of current commercial rates.

called Invisible Networks manages and supports the network[10]. The community provides some self-help and assistance to other community users.

Although the network has commercial support, it is not like other commercial networks. Just because the community has outsourced the management of the network to a commercial company, it is still a community network, because of the strong community element. By outsourcing the management of the network, the community has made sure the network is an entity in its own right. Members can come and go but so long as the network remains viable it should carry on regardless.

In this case, the main driver for the community network was to provide broadband access in an area not served by fixed line broadband operators. But the network also offers a much faster speed within the network. For this reason, this type of wireless service could appeal to subscribers who are within reach of a slower fixed line service.

The prices of one of the companies involved in the network around Cambridge, Carnet [10], are comparable to commercial prices but offer more bandwidth (see Table 8.1). In addition, £2.00 per month of the subscription is paid to a not-for-dividend community company. By purchasing the residential service, customers become members of the community company and have a say in how the money that goes to the company is spent. If they wish to do so, the community company could use these funds to provide or subsidise Internet access to local worthy causes.

Table 8.1   Comparable service charges.

Product Name	Type	Speed	Install	Monthly
BT Openworld Home 500 Plug & Go	RADSL	512/256	Free	£29.99
Carnet Standard Service	WiFi	512/512	£169.36	£25.52

Notes:
1. All prices exclude VAT.
2. We have tried to compare like with like but no two services are exactly the same.
3. BT Openworld also offers engineer-assisted installation for £210.00 and line only installation for £63.82 and other services. For details of the current service offerings see www.btopenworld.com/broadband.
4. Carnet offers other services too. For details of the current service offerings see www.bbb.uk.net.
5. The Carnet installation charge includes service set-up, wireless card or Internet antenna and cable.
6. Price per month for the Carnet service falls to £20.41 if paid 15 months in advance.
7. Carnet claims to offer a minimum of 512 kbit/s Internet connectivity in both directions.

By retaining control of the network, the community can make sure this resource is not abused. Carnet have an 'acceptable use policy' that includes the statement that

---

[10] In 2003 Invisible Networks appeared to run into some problems and on 9 October 2003 Microcomm Systems trading as 'Mesh Broadband' (http://www.mshbroadband.co.uk) announced that it had purchased the business and assets of Invisible Networks (http://www.invisible.uk.net) allowing the community broadband networks developed by the Cambridgeshire-based company to continue.

'... users of the service must not make excessive use of the service to the detriment of other users and a limit of 30 Gbit/s per month usage'.

If a large number of these community-based networks develop they could find it cheaper and faster to communicate directly with each other over wireless links, rather than route local traffic via a slower fixed line link[11].

For these reasons, we think shared cost networks offer a more sustainable business model than free networks. They will have the means to meet any expenses they incur and the ability to manage the network resources for the common good.

Such community networks may be able to offer national, or even global, reach by co-operating with each other, the way other clubs do. For example, a member of a yacht club is normally welcome to use the facilities of another yacht club, as a visitor, once they have identified themselves as a member of a bona fide yacht club. The Internet offers the ability to verify membership details promptly, so there should be little or no delay in gaining access to the facilities, in another location.

8.4   Commercial Alternatives

Two commercial alternatives to *ad hoc* networks are WISPs and 3G mobile operators.

8.4.1   WISPs

There may already be approximately 70 broadband WISPs operating in the USA. Some WISPs have already ceased operations for economic and technical reasons but other companies have taken advantage of the shake-out in the telecommunications industry to acquire useful assets very cheaply, giving them a much lower cost base[12], but a WISP will probably still cost more than a community based not-for-profit wireless network.

BOINGO [11] is a roaming aggregator, which aims to simplify the complexity of dealing with multiple wireless network providers by offering a single account that works across multiple networks. It provides the marketing, technical support, end-user software and billing, while hot-spot operators build and operate the WiFi networks. Therefore, it passes on the risk of running a public hot-spot to businesses,

---

[11] For example, Cambridge appears to be in the process of being ringed by wireless broadband networks. Cambridge Ring West covers the West of Cambridge. Fens Broadband covers the North of Cambridge. Cambridge Ring South East covers the South East of Cambridge. This may only leave the North East [10] and the South West to complete the ring. Carnet claims the wireless part of their network operates at an effective 5 Mbit/s.

[12] Metricom Inc had invested over $1.3 billion to develop the Ricochet technology and construct the Ricochet network in 21 major US cities before it had to suspend operations in August 2001 for financial reasons. In November 2001, Aerie Networks Inc (through a wholly owned subsidiary called Ricochet Networks Inc) purchased Metricom's Ricochet technology, patents and assets for (only) $8.5 million.

e.g. hotels, cafes, which subscribe to their programme. It operates mainly in North America and in Europe. BOINGO claims to have over 1000 hot-spot locations covering 300 cities and 43 states in the USA. The BOINGO network includes full or partial coverage in most major US airports, service in the lobbies of many hotels and coverage at lots of cafés, coffee shops and free networks (see Fig 8.1). There are already at least 17 locations in the UK registered with BOINGO. BT recently launched a public WLAN service in the UK called BT Openzone. Plans are to have 400 sites operational by 2004, increasing to as many as 4000 by 2005.

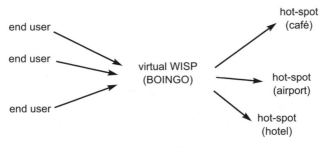

Fig 8.1   WISP set-up.

BOINGO collects fees from subscribers and pays fees to the hot-spot operators. BOINGO has three payment options:

- BOINGO As-You-Go is US$7.95 per Connect Day;

- BOINGO Pro is US$24.95 per month and includes 10 Connect Days;

- BOINGO Unlimited is US$39.95 per month and gives unlimited usage.

BOINGO also offers a variety of corporate pricing plans to suit small, medium and large organisations.

The BOINGO network can grow at little or no cost to BOINGO and the bigger the network becomes, the more benefits membership of BOINGO will offer.

8.4.2   3G

If high-speed data is the 'killer application' of 3G, then WLAN is a serious threat because:

- the current crop of 3G devices are not really suitable for data-rich services;

- WLAN can offer unmetered usage (but this can also be a disadvantage if the WLAN operator cannot charge for or manage usage);

- WLAN can offer faster data speeds than 3G (but WLANs do not work when users are moving at high speed, e.g. in a car).

3G devices are not suitable for services requiring broadband, because they are:

- bigger, as they have to be compatible with GSM and 3G;
- heavier than GSM telephones, as 3G electronics are heavier than GSM and the battery is also heavier (otherwise the usable battery life is limited).

It is not clear what sort of mobile devices people will prefer and the answer is probably quite a wide range. Some may be satisfied with a small mobile telephone that they use just for voice. Other may prefer an 'all singing, all dancing' mobile device that has a bigger, better screen and even a small keyboard. If the latter, these devices may work better with a WLAN that can offer more bandwidth than 3G.

Research commissioned by Compaq suggests that laptops are currently the preferred device for accessing data-rich services because users prefer a large screen and a keyboard to get the full benefit of the data-rich service. This was separately confirmed by Vodafone, who completed commercial testing of a new broadband wireless technology in New Zealand. They found that high-speed data service was more popular with laptop users than with handset users [12].

Today, WLAN cards often come factory installed in laptops, as part of the standard specification, e.g. Apple Powerbooks. If not, fitting a WLAN card is simple and inexpensive, and so there is little or no additional cost to equip the laptop to access a wireless network.

In addition, a WLAN can be free to use. Free-to-use networks are springing up in all sorts of places. Directories are being published. Warchalking is providing signs on the ground. Anyone equipped with a laptop and wireless network card can attempt to use these networks.

In public places, there may be no need for that, as a reseller called Wialess [13] partnered with 3Com recently in a scheme to provide free wireless Internet access for café and restaurant customers in the UK. The owner of the café or restaurant pays for the service (installation of ADSL and WLAN) because it increases the number of customers and/or increases the average revenue per customer.

WLAN also has a speed advantage. The maximum speed a single user can expect in a 3G cell (if stationary and alone in a small cell) is up to 2 Mbit/s (but the actual speed is commonly circa 384 kbit/s). With WLAN, in the same conditions, a single user can expect up to 11 Mbit/s (but the actual speed is commonly circa 5 Mbit/s). If the richness of the user experience is determined by the bandwidth, services delivered over WLANs will have the bandwidth to be richer than those delivered over 3G. Therefore, we think wireless device users may prefer WLANs to 3G for services that need high bandwidth. 3G faces the double whammy of WLAN being cheaper and better.

8.5   Capacity

Ad hoc networks often have greater capacity than fixed line Internet access. This gives them an edge, at least over short distances. For example, the minimum times it would take to send a 1.0 Gbyte file over network access in common use are given in Table 8.2.

Table 8.2   Minimum time to send a 1 Gbyte file[13].

Dial-up modem	ADSL	WLAN
56 kbit/s	512 kbit/s	5 Mbit/s
39.7 hours	4.3 hours*	26 minutes

* Or 8.7 hours, if restricted by the asymmetric rate of 256 kbit/s, e.g. for peer-to-peer.

This suggests the dial-up modem is impractical and ADSL not that much more usable. The WLAN can do the job in a reasonable time and has the advantage that the speed of wireless networking is increasing much faster than fixed line networking.

In the 2.4 GHz frequency band, IEEE802.11b networks typically offers speeds circa 5 Mbit/s (up to 11 Mbit/s).

In the 5 GHz frequency band, IEEE802.11a provides much higher speeds, up to 54 Mbit/s. Even higher speeds are in the offing. This sort of capacity will also give 3G tough competition.

Part of the business case for 3G relies on the take-up of new feature-rich services. But if similar feature-rich services can be provided more cheaply over WLANs to (other) wireless devices, then the case for expensive 3G mobile telephones and services will be undermined.

8.6   Summary

Initially, we believe *ad hoc* networks will be used primarily for access, but when people get used to living in 'wireless bubbles', they may well use the freedom 'tetherless technologies' give them in new and exciting ways — there will be no reason not to. Wireless devices are becoming smaller, lighter, cheaper and faster. If these devices perform useful functions, people will want to use them at home and on the move. Wireless *ad hoc* networks offer the connectivity these devices will need. The mobility comes at no extra cost.

Of course the mobility will be limited to the wireless bubble, but this need not be too restrictive. If the wireless technology is based on the international IEEE802.11 standards, mobile devices that work in one bubble will work in them all. The bubbles will not have 100% geographical coverage but then neither do mobile telephones or DSL. We expect wireless bubbles to develop where enough people want them and provide adequate coverage for most users. If so, this could have

---

[13] Assumes 'top of the range' transmission speeds over the fixed lines, although this is unlikely in practice.

interesting implications for fixed and wireless network operators serving this market.

References

1   LocustWorld — http://www.locustworld.com/

2   Engst, A. and Fleishman, G.: '*The Wireless Networking Starter Kit, a practical guide to WiFi networks for Windows and Macintosh*', Peachpit Press (2003).

3   Freedom to Tinker — http://www.freedom-to-tinker.com/

4   Flickenger, R.: '*Building wireless community networks*', pp 83-88, O'Reilly & Associates, Inc (2002).

5   Shapiro, C. and Varian, H. R.: '*Information rules: a strategic guide to the network economy*', Harvard Business School (1998).

6   Warchalking — http://www.warchalking.org/

7   E-xpedient — http://www.e-xpedient.com/

8   Got Toast — http://www.toaster.net/

9   Arwain — http://www.arwain.net/arwain.htm

10 Cambridge Ring North East — http://www.carnet.uk.net/

11 BOINGO^{TM} Wireless Inc — http://www.boingo.com/

12 '*The first real payback from third generation (3G) networks*', The Business Magazine Online (January 2003) — http://www. businessmag.co.uk/

13 Wireless Internet Access — http://www.wialess.com/

9

SECURING MOBILE AD HOC NETWORKS — A MOTIVATIONAL APPROACH

B Strulo, J Farr and A Smith

9.1   Introduction

If mobile *ad hoc* networks (MANETs) [1] are to go beyond their current, limited, use in military systems or where there is complete ownership by one party, then there will need to be security mechanisms in place. This chapter focuses on the security-related issues inherent in such networks, and in particular upon participation incentive schemes. A more general description of *ad hoc* network technologies can be found in Chapters 6 and 7.

Mobile *ad hoc* networks are formed from communications devices without any centralised control. Each device can act as a router to transmit its neighbours' packets as well as transmitting and receiving its own packets. Distant devices are reached by forming a chain of devices. This is illustrated in Fig 9.1 where device A wants to communicate with device E, but its transmission range only reaches device B.

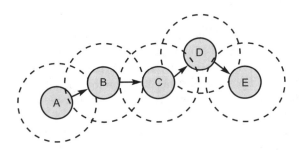

Fig 9.1   Communication chain in an *ad hoc* network.

Ad hoc networks have been used for many years by the military where they offer a high degree of resilience, and can be deployed rapidly and with no fixed infrastructure.

In a general, publicly owned *ad hoc* network, security issues arise because:

- devices are owned, operated and used by multiple distinct parties with diverse motivations and objectives;

- there are few, if any, centralised or well-trusted entities;

- devices have highly heterogeneous capabilities — in particular, very many of them have scarce resources (processing power, storage, bandwidth and especially battery power may be very limited).

Without security and a means of ensuring participation, a general, public *ad hoc* network is vulnerable to a number of attacks. There is a need to ensure that devices all co-operate in order to form a network since without this co-operation there is no network.

This then is the focus of this chapter — security techniques to maintain desired network function in the face of misbehaviour by one, or a small subset of devices. Our approach is to use an analysis of why devices may misbehave and then look at structures that can change these motivations. The emphasis is on addressing the question: 'Why should I forward packets for others?' The key here is to create appropriate participation incentive schemes to control both the use made of the network, and the resources contributed to the network.

The rest of this chapter is organised as follows. Security vulnerabilities are introduced in section 9.2. Section 9.3 analyses why devices might misbehave and how. Section 9.4 describes motivational techniques to overcome these concerns. Section 9.5 describes the components required and section 9.6 describes how these might be used in practice and describes some proposed systems. Finally conclusions are made in section 9.7.

9.2   Security Issues

There are two main security issues with *ad hoc* networks. The first is the need for privacy of communication in a network where communications transmission is being performed by devices owned by many different people/organisations. The other is that the network is vulnerable to a number of attacks (not necessarily deliberately) that can degrade the performance of the network or give unfair advantage to some of the participants.

To ensure privacy, the techniques used need not be sensitive to the nature of the network; we can simply import the techniques that are already used in the Internet. Authentication protocols, digital signatures, and encryption can all be used to achieve confidentiality, integrity, authentication, and non-repudiation of communication in *ad hoc* networks. However, these mechanisms are not sufficient by themselves. Zhou and Haas [2] present an approach for securing communications in an *ad hoc* network when some devices are trying to intercept communications.

The other main issue is that *ad hoc* networks provide a particularly hostile environment for another sort of security — protecting the functions of the network from the devices that make up that network.

In an *ad hoc* network, the functions of the network (packet forwarding, route discovery, etc) are provided by the user devices themselves rather than an independent network operator. The motivations of the people controlling these devices are inherently diverse, and often conflicting. There is no *a priori* reason for them to act co-operatively with the users of other devices. In the real world we cannot simply assume that devices will behave co-operatively because their manufacturer has programmed them to; the heterogeneity of the system implies that they will come from different manufacturers with different vested interests.

Furthermore, for the foreseeable future, it seems that users will always be able to find ways to re-program their own devices, as happens today with mobile telephone unlocking.

The anonymity of mobile devices to each other is another anti-cooperative factor. Since anonymity typically frees the sender of a message from fear of retaliation or confrontation, it can encourage dishonesty in communication [3].

These reasons mean that the functions of an *ad hoc* network are particularly vulnerable to disruption caused by the self-interest or active malicious intent of those user devices.

9.3   Device Misbehaviour

Device misbehaviour is any behaviour that acts against the co-operative requirements of an *ad hoc* network. Co-operation is key in order that the network can continue to function efficiently. This section discusses why devices might misbehave and the types of misbehaviour that could occur.

9.3.1   Why Devices Misbehave

Devices will misbehave if controlled or programmed to do so by their owners/users. The difficulty in adapting a device varies (in general-purpose programming environments this may be very easy, and in specialist-built hardware/software it may be much harder), but, if the benefit to be derived from the adaptation is sufficient, it will be attempted.

In order to help inform the design of security schemes, and in order to aid their comparison, it is convenient to classify different types of misbehaviour. We distinguish three distinct dimensions of misbehaviour:

- accidental/deliberate;
- selfish/malicious;
- individual/collusion.

Selfish behaviour can be passive or active and is normally for economic reasons, for example to save money. Malicious behaviour is usually directed and not done for economic reasons, but rather to disrupt a network as can be seen with computer viruses.

It is often difficult to distinguish between misbehaviour and proper behaviour in an *ad hoc* network. We can only observe the external behaviour of devices and this gives limited clues as to their motivation: devices can drop packets because of transmission errors. A device may be powered off maliciously or just because its battery is flat. This makes policing a hard task. Thus strictly accurate reporting of, and responses to, misbehaviours is somewhat pointless. Approximate solutions are likely to be as valuable and provide less of an overhead.

9.3.2   Types of Misbehaviour

There is a broad range of system functions that can be disrupted. These are listed below. Included are misbehaviours based on attacks on charging systems — such charging is itself one of the possible solutions to encouraging participation in *ad hoc* routing.

- Routing

  Attacks can be made during the route discovery phase, maintenance phase, or during routing itself. Types of attack include non-cooperation, false offers, pseudo-spoofing and malicious reporting of broken links. Pseudo-spoofing is a serious concern if there is any cost to send packets. A device could send packets using the identity of another device in order to avoid being charged for transmitting. Similarly, were destination charging to be introduced, a device could transmit to a device beyond its intended recipient and the real device would receive the data but pretend to be a forwarding device.

- Transmission

  Sending at the wrong times or the wrong power levels can cause disruption. Devices can do 'less work', and hence save battery power, by minimising their transmission power. A device doing this would appear to be behaving correctly when observed by neighbouring devices.

- Attacks on individual links

  These include eavesdropping, replay (to avoid any accounting mechanisms), modification or deleting of packets.

- Authentication mechanism

  A device could block another device's attempts to contact an authentication service and thus block them out of the network.

- Denial of service (DoS)

  DoS attacks can take many forms. A DoS attack can be made to overload part of the network or a particular device. Also, as related above, attacks on routing can deny service, for example offering routes but not routing.

- Attacks on incentive schemes

  For example, an incentive scheme, that financially rewards those that behave well, itself can be subject to attacks (e.g. to gain greater reward).

An important threat to overall co-operative behaviour — exacerbated by the high levels of dynamism and asymmetry in *ad hoc* networks — is that individual devices are able to manipulate their own responses subtly over varying time-scales. They can decide when, and under what circumstances (e.g. location, neighbours, traffic levels, remaining power) to behave well, and can trade these off in order to increase overall utility.

9.4   The Solution Space

This section will concentrate on those solutions that attempt to encourage participation, starting with general features of such systems and continuing with a discussion of the various dimensions these can take.

9.4.1   General Features

It is evident that some apparently desirable features make others difficult or impossible, for example anonymity makes complete accountability impossible. Thus particular systems will typically only encompass a subset of these. These motivational schemes are aimed, ultimately, at the users of the devices. Incentive schemes should:

- encourage participation in packet forwarding;
- discourage pointless use of network resources;
- impose a cost on devices for use of the network that is somewhat related to the amount of consumption;
- encourage correct routing behaviour;

and more subtly:

- if a punishment scheme is used, it must usually maintain the possibility of full-reintroduction into the community [4], since we must not seriously deplete our network over possibly short-term misdemeanours;

- monopoly situations, where some devices can charge excessively for some routes, and where the introduction of competition can only be achieved by physical infrastructure changes, for example moving a device about, must be avoided — such monopolies reduce global economic efficiency;

- as some communications are not strongly sensitive to the precise time at which they are made, for example e-mail, the incentive scheme should encourage devices to engage in such communications at a time that will minimise their demand on the resources of the network;

- it may be valuable for the network if devices sometimes do not offer to route (for example when there is over-capacity) because of routing overhead — but in this case we would need to motivate users to respond to demand as congestion increases or coverage drops (if devices have a stand-by mode, they might be in a position to offer to contribute if triggered to do so by other devices);

- availability of coverage should be rewarded, even if not actually exploited.

In addition, while stimulating devices to participate in routing is necessary, there are other good reasons to limit/manage the level of such participation. For example, discriminating over who you route for, and what QoS they receive, may prove to be a very effective sanction against those who do not route for others. Hence, perhaps, you should not be rewarded for routing for someone who has behaved anti-socially.

Most of the solutions examined attempt to select routes avoiding misbehaving devices or the symmetrical alternative of routing through the best behaving route.

In Fig 9.2 device C has been observed to be misbehaving. Consequently device A routes via devices X, Y and Z rather than C to get to D. This requires that misbehaving devices are excluded from the route discovery phase.

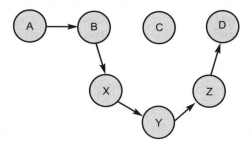

Fig 9.2   Device avoidance.

9.4.2   Rational versus Psychological Approaches

Solutions can be distinguished on the basis of whether they assume strictly rational behaviour, or whether they utilise psychological evidence. Indeed, there are a number of important psychological factors that might be exploited in order to develop practical solutions, the most important of which are described below.

- Rational choice

  People usually exercise rational choice only within broader, constraining, social contexts.

- Criticism and praise

  Most people are naturally motivated to criticise others more than praising them, though fear of reprisals can hugely moderate this. While the 'outrage' people feel at anti-social behaviour is often so strong as to motivate them to punish that person even at a cost to themselves [5], the fear of reprisals is so strong that it can nullify this behaviour.

  Hence the level of anonymity provided by the system is an important factor in determining whether they punish, and can determine the effectiveness of such punishment-based strategies.

  In contrast, people are rarely motivated to altruistically reward others, which can lead to under-reporting of good behaviour.

- Incentive/punishment motivation

  There are huge person-to-person variations in the actual utility offered by connectivity, battery power, applications, etc, and hence a wide variation in the effectiveness of specific rewards/punishments. In a given situation, people also have widely differing attitudes to risk — some being highly risk-averse, some risk-seeking, and others risk-indifferent. These effects combine to suggest that a rich set of incentive and punishment schemes may be needed to motivate large proportions of a diverse community.

  Market-managed solutions (see section 9.5.1) and reputation-based solutions (see section 9.5.2) fit especially well into the classification of 'rational versus psychological' approaches.
  While the former is based upon a clear theoretical foundation (assuming rationality), reputation-based systems usually rely upon strong psychological factors in order to work effectively. As an example, providing the feedback (especially negative feedback) after a transaction that reputation systems need is usually an altruistic act.

9.4.3   Reward versus Punishment

Solutions can also be classified on the basis of whether they seek to promote good behaviour, or aim to detect and punish misbehaviour (carrot versus stick). There are immediate consequences to this decision:

- if good behaviour is the norm, as we might expect it often will be, the process of individually rewarding good behaviour often requires far more overheads than may be necessary to police the misbehaviour;

- accurate detection of misbehaviour can be very difficult, and so there can be a high risk of injustices, which is especially important if there are high penalties.

Observed 'poor' behaviour (such as not forwarding a particular packet) may be attributable to overload, a genuine resource shortage, or true anti-social behaviour. Arguably, only the latter should be punished, and yet distinguishing between them may be difficult.

When a single act of misbehaviour might be reported or punished by multiple observers, co-ordinating an appropriate level of punishment might be difficult. Similarly, multiple acts of misbehaviour might only be measurable as an individual misdemeanour.

We can add to these two problems the fact that punishment mechanisms can themselves be maliciously abused.

Solutions differ too in the degree of coupling between rewarding contributions, and accounting for usage. Some systems adopt a strict reciprocity-based approach, where usage of the system is constrained by the degree to which that device has previously contributed to the community. For example, you need to make sufficient social contribution, say in terms of intermediary routing, so as to earn your right to use the network.

This can be supported by the use of either charging or reputation-based accounting systems; with a charging system you may need to send a form of internal currency with each packet in order for others to forward it or with a reputation system, your reputation may need to exceed a particular threshold.

Furthermore, some devices will act as gateways to the general Internet. This is a service to the *ad hoc* network that might need to be encouraged.

Finally, the degree to which QoS is supported, or is beneficial in the network, has an impact on the granularity with which contributions can be motivated. If users derive real, additional utility, i.e. monotonically increasing, rather than saturating utility, from improving network QoS, then they will be motivated to contribute as much as they are able.

In many cases it seems likely that the applications will be tolerant to, say, latency variations, and that 'good enough' performance will often suffice — devaluing its use as a reward mechanism.

9.4.4 Centralised versus Distributed Control

The final distinction we make is the extent to which a solution adopts a centralised or distributed approach to control.

An example in the context of a market is the manner in which prices are set. In a full, free market, individual players set their own prices for their services, and thus directly control the utilisation of their own resources (distributed control). An alternative approach is where a uniform price, or pricing policy is adopted (centralised solution). Equivalently, in reputation-based systems the reputation value may be aggregated or held centrally, or may be distributed among all participants. Factors in favour of a more centralised approach often include:

- efficiency (e.g. it may only require one-off acquisition of knowledge, for example, of price);
- fairness (avoids exploitation by devices in a favourable position);
- determinism (e.g. users of the network may prefer price certainty);

while those often favouring decentralisation are:

- scalability (fewer bottle-necks);
- robustness — to attacks, or faults;
- trust — no single authority is able to capture and thus, potentially, abuse user data.

Obviously the specific deployment context for a particular *ad hoc* network will hugely influence the trade-off of these factors.

9.5 Categories of Solution Components

There are two main ways of encouraging co-operation — to use either pricing or reputation. Whichever is chosen, there is a need to account for contribution to the good of the network.

9.5.1 Markets/Pricing

Co-operation could be engendered by rewarding devices that provide service to other devices. The purest and most basic approach to reward is to use real money exchange. The money a device receives for contribution is obviously straightforwardly exchangeable for consumption at another time but also comparable with money received for exchanges in other markets. Money as a valuation of contribution is a common underpinning for most such exchanges of value in the real world.

Accounting for money electronically relies on either a trusted third-party, such as a bank, or on trusted hardware 'wallets' on each device. Apart from this reliance on a centrally controlled infrastructure, all other aspects of trade of services could be decentralised to the devices themselves. Any pair-wise provision of service can be fairly counter-balanced by an exchange of money arranged by the devices concerned.

Where this free market does not motivate the desired equilibrium, various controls can be imposed. Prices may be partially or totally controlled in a non-local way.

The actual exchanges may be monitored and accounted for centrally by, for example, a service provider who somehow 'owns' the *ad hoc* network provision. Alternatively there may simply be standardised pricing and accounting processes that are mandated by separate social or technical mechanisms.

Systems may denominate their value in real currency. Alternatively they may use an internal currency that behaves like money but is confined to this market. Such currencies typically do have some external value but the use of an internal currency may allow greater central control, for example by adjusting the money supply. Some token systems can deviate from the usual properties of money, for example, by failing to be conserved in normal trade when rewards to contributors can exceed the cost to consumers.

9.5.2   Reputation

In real life and on-line, and despite their dependence upon irrational behaviour, reputation schemes often seem to provide a remarkably effective solution to the problem of stimulating co-operation. This has been shown in game theory experiments such as the iterated Prisoner's Dilemma [6].

A fundamental difficulty for reputation systems occurs when people are easily able to create new identities, and this problem is solved in different ways for different systems. Some reputation schemes allow for both positive and negative values of reputation to be expressed, and in such schemes new users begin with a reputation of zero and increase or decrease that value based upon their behaviour. Ordinarily in such a scheme, users with a negative reputation will be motivated to create a new identity for themselves, and so gain an improved reputation of zero [7]. To counter this, identity creation must be either controlled, for example it could be strongly tied to one's real-world identity, or discouraged, such as through the payment of a fee. Unfortunately, both of these solutions can have serious negative repercussions. The former strongly reduces one's level of anonymity — and, in an *ad hoc* network setting especially, this may mean an invasion of privacy, for example personal location tracking. The latter can mean that people are dissuaded from registering in the first place, which, in a system such as *ad hoc* networks is undesirable.

Alternative types of reputation scheme may choose to restrict the bounds of reputation to either positive, or negative values. Negative reputation schemes, which attempt to record and punish only misbehaviours, suffer the same 'identity creation' problem as described above, and must be solved in similar ways. Positive reputation schemes have been proposed [4], however, that employ an alternative solution: decrements to the reputation value are in proportion to the current value of the reputation. While this means that the same misbehaviour, performed by different people, is likely to result in differing degrees of punishment — it also ensures that people who have at some time made a positive contribution always have a better reputation than new users, and thus have little incentive to create new identities.

9.5.3   Accounting

Both markets and reputation are based on accounting for contribution. But data on contribution can be used more generically than for exchange of monies or for awarding reputation. One approach that we are pursuing within the MMAPPS 5th Framework project [8] is to account for and enforce co-operation through the use of more generic 'rules'. In the context of *ad hoc* networks, the rules would explicitly tie use of the network to measured contributions to it — for example, you might be required to show evidence that you have routed for X people the previous day before being able to authenticate, and thus use the network, yourself.

9.6   Solutions Design

We have described a solution space and some of the properties that can be achieved by different mechanisms. Here we describe some of the ways these mechanisms can be built into complete solutions along with their main advantages and disadvantages.

9.6.1   Free Market

A solution that is theoretically very appealing is simply to allow every device to charge every other device for any service provided — a complete and free market in services. Such a market, if achievable, will provide maximum efficiency (at least in an economic sense). Every device will only provide service if the requesting device is willing to pay enough to cover its costs; the requesting device will be willing to pay, if the value of the service to it (its utility) exceeds that price; competition will ensure that no-one can overcharge.

   In theory, such markets have many valuable properties. Opportunities for free-riding are minimal; everyone pays for every resource they use. Money flows to people providing service and encourages the socially optimum provision. The system can support anonymity because no long-term relationships between devices are required; once you have been paid you can forget about your history.

Such a market, however, can be extremely costly and difficult to implement. It is extremely difficult to arrange adequate competition; a requesting device needs to see many offers of service so it can be sure it is choosing a well priced one. In an *ad hoc* network this is extremely difficult and expensive. A device would need to receive pricing messages from many other devices before setting up any communications link. Furthermore, the actual physical location of devices can strongly limit a device's choice. In such monopoly situations a free market leads to inefficiency since the monopoly provider can charge so highly that usage is pushed down below the social optimum.

This is shown in Fig 9.3, where device M is in a monopoly position between two otherwise isolated sub-networks. In a completely free market device M could charge whatever it wanted.

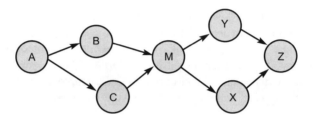

Fig 9.3   Device in monopoly position.

Even if these considerations can be tackled, the mechanisms needed to securely account for usage and transfer money are expensive, particularly in a highly decentralised environment.

The high levels of security needed to prevent fraud involving real money typically rely on multiple messages to third parties. This is particularly problematical in an *ad hoc* network where the sending of such messages is the commodity being traded.

There is also a human factors issue here — people wish to be involved in deciding whether to spend real money. If the transactions are frequent and of small value, then this is onerous. Yet people are reluctant to delegate such decisions to machines, particularly where prices are somewhat variable [9].

Some systems attempt to avoid these security and psychological issues by denominating payments in an internal currency that is somewhat decoupled from real money. However, it seems likely that such decoupling will always be only partial — if the system has real value to its users, then it will be extremely difficult to prevent users establishing an external market in the internal currency and hence giving it a value in real money. A further potential weakness of such systems is that, since the reward for contributions can only be spent on using the system, people might only be motivated to contribute to a degree aligned with their usage requirements — rather than to the maximum of their potential.

Finally, it should be noted that there are problems in defining and determining the quality of the service that will be provided for a given price. For example, *ad hoc* devices might expect to pay less for packets routed with long delays and pay nothing for packets that are dropped. However, if payment is to each forwarding device on a per-packet basis, then devices will pay more for long routes and will still have to pay even if the packet is dropped. If, instead, we attempt to define some sort of longer term session relationship, then we start to require the ability to judge whether devices are honouring the promises they have made at the start of such arrangements. This then begins to involve the ideas of reputation and social judgement, and becomes a hybrid system.

9.6.2   Controlled Market/Accounting

At least two solutions have been proposed based upon routing rewarded by the centrally controlled payments (rather than a complete free market). The 'Terminodes' project [10, 11] has examined charging models in which the payments are sent along with individual packets (funded by the source device), and also in which payments are exchanged for data packets (ultimately funded by the destination device). One design proposes the use of tamper-proof hardware modules on each device, a PKI-based authentication scheme, and a supporting accounting protocol to ensure that a device's use of the network is conditional upon its contribution to it. In particular, the tamper-proof modules enforce the constraint that devices are rewarded a single unit of currency (a 'nuglet') for each packet they correctly forward, and are charged a single unit for each network hop their own packets take to reach a destination.

This approach has many valuable features. It is largely decentralised and adds a fairly small overhead to communication. The incentives it gives are reasonably well aligned to resource contribution and consumption, though they do not encourage efficient routing and the charge for consumption is based on a routing estimate. The use of tamper-proof hardware is a dramatic, simplifying assumption that is certainly realistic for some contexts, though it does restrict applicability and add cost. It is not clear that this strict accounting for usage is totally desirable since it could leave users unable to transmit, even though they might be able, at a later date, to redeem an accounting debt incurred now. If this happens too often, it could strongly deter participation.

The '*Ad Hoc* Participation Economy' proposal [12] attempted to create a somewhat less-constrained market by merely imposing a uniform pricing policy on each device. The pricing policy was designed to ensure that the rewards for participating increased in proportion to the load on the network (a very desirable feature — and one that so far remains unique to this solution). In this design, the accounts themselves are held in a single 'bank', although the interactions with this centralised-entity were drastically reduced by the use of a micro-credit control protocol between individual peers. The bank devices are responsible for verifying payments are correctly made, and for ensuring the appropriate application of the pricing algorithm. The key problems that remained with this proposal appear to be how, in practice, to use the micro-credit scheme efficiently (extending very little credit to untrustworthy devices, and greater credit to trustworthy ones), and what sanctions the banks might use to punish misbehaviours beyond the crude punishment of exclusion from the system.

9.6.3   Reputation Based

Several reputation-based solutions have also been proposed. These include the PathRater [13] system that attempts to rate different routes through the network so that an originating device can select a route that avoids misbehaving devices. The ratings are performed by adjacent devices, sent back to the originating device, and aggregated. This particular system, however, fails to actually discourage the misbehaving devices — indeed the system indirectly rewards them by increasing the network capacity in their vicinity.

Another system, CONFIDANT [14], is inspired strongly by the principles of reciprocal altruism as described in Dawkins' *The Selfish Gene* [15]. A device detects malicious behaviour by other devices via observation or reports, then routes around misbehaving devices, and isolates them from the network. Devices have a monitor to record observations, reputation records for first-hand and trusted second-hand observations, trust records to control trust given to received warnings, and a path manager for devices to adapt their behaviour according to reputation.

This system relies on all (non-cheating) devices to act in the same rational manner. It is reciprocity based on good behaviour being rewarded by the ability to achieve your own communication. The protocol has a threshold of detected behaviours before a device is branded as malicious, thus it has an in-built mechanism to protect from false-positives (a low number of bad reports is not enough to brand a device as malicious). Removing malicious devices from the network can improve throughput, but is more sensitive to mobility because of the likelihood of coming into contact with a previously unknown malicious device. Also, it is vulnerable to colluding devices to attack another device and deny it service. It is a negative reputation scheme with all the problems of starting with new identities that this entails. Finally, it should be noted that this scheme only overcomes maliciousness rather than encouraging participation — for example, it does nothing to encourage users to turn their devices on and provide service to others.

9.7   Summary

This chapter has analysed the problem of securing mobile *ad hoc* networks in an environment where the devices are owned by a number of parties. It has outlined a range of possible approaches, and reviewed several of the more concrete proposals that have been put forward to date. It is plain that this is an important issue and that the design of solutions is at an early stage. All of the proposals have particular strong and weak points that are more or less important in different specific usage scenarios. No current proposed solution or combination of solutions can meet the needs of all such scenarios.

References

1   IETF MANET working group — http://www.ietf.org/html.charters/manet-charter.html

2   Zhou, L. and Haas, Z. J.: '*Securing Ad Hoc Networks*', IEEE Network, **13**(6), pp 24-30 (November-December 2002).

3   Levmore, S.: '*The Anonymity Tool*', University of Pennsylvania Law Review, **144**(5), pp 2191-2236 (1996).

4   Zacharia, G., Moukas, A. and Maes, P.: '*Collaborative Reputation Mechanisms*', in Electronic Marketplaces, Proceedings of the 32nd Hawaii International Conference on System Sciences (1999).

5   Fehr, E. and Gachter, S.: '*Altruistic Punishment in Humans*', Nature, **415**, pp 137-140 (January 2002).

6   Axelrod, R.: '*The Complexity of Cooperation: Agent-Based Models of Competition and Collaboration*', Princeton University Press (1997).

7   Friedman, E. and Resnick, P.: '*The Social Cost of Cheap Pseudonyms*', Journal of Economics and Management Strategy, **10**(2), pp 173-199 (2001).

8   MMAPPS project — http://www.mmapps.org/

9   Shirky, C.: '*The Case Against Micropayments*', — http://www.openp2p. com/pub/a/p2p/ 2000/12/19/micropayments.html

10   Terminodes — http://www.terminodes.org/

11   Buttyan, L. and Hubaux, J-P.: '*Stimulating Cooperation in Self-Organising Mobile Ad Hoc Networks*', to appear in ACM/Kluwer Mobile Networks and Applications (MONET), **8**(5) (October 2003).

12 Barreto, D. et al.: '*Ad Hoc Participation Economy*', — http://www.stanford.edu/~barretod/

13 Marti, S., Giuli, T. J., Lai, K. and Baker, M.: '*Mitigating Routing Misbehaviour in Mobile Ad Hoc Networks*', in Proceedings of Mobicom 2000, Boston (August 2000).

14 Buchegger, S. and Le Boudec, J-Y.: '*Devices Bearing Grudges: Towards Routing Security, Fairness, and Robustness in Mobile Ad Hoc Networks*', in Proceedings of the Tenth Euromicro Workshop on Parallel, Distributed and Network-based Processing, pp 403-410, Canary Islands (2002).

15 Dawkins, R.: '*The Selfish Gene*', Oxford University Press (1976).

10

THE USE OF SATELLITE FOR MULTIMEDIA COMMUNICATIONS

M Fitch

10.1   Introduction

A transmitting device on board a satellite has one big advantage over one on the ground — it has a very high vantage point. This means it can broadcast radio signals to a very large area, typically an entire country or continent. The wide-area coverage means that satellite systems are naturally suited to multicast and broadcast services, but they can also provide services to individual users where it is not possible or not economical to provide terrestrial connections. Satellite systems are also viable for fast, asymmetric or temporary connections. These attributes make satellite services suitable for the following customer groups and services:

- those in rural or remote areas, e.g. for broadband Internet connection and teleworking;

- trunk voice and data communications to remote countries;

- maritime and aeronautical, e.g. for communications, distress and emergency;

- those who require fast or temporary connection, e.g. for conferences, oil-field exploration, building sites, disaster recovery, etc;

- those who require asymmetrical or one-way connections, e.g. Internet Service Providers, business TV and push advertising applications.

There are several classes of satellite, e.g. transparent and processing, and there are several options for orbit type. However, the scope of this discussion is limited to the most common class (transparent) and to the most common orbit type (geostationary), in order to keep the number of variables in the disucssion managable. Transparent means that the satellite does not demodulate signals but re-transmits them back to earth at a different radio frequency. Geostationary means that they have an orbital period of one day and therefore stay in the same position in the sky. This corresponds to an orbit radius of approximately 37 000 km.

10.2   Radio Frequency Bands and Services

Satellites are operated at microwave radio frequencies in various bands, which are allocated by the International Telecommunications Union [1]. The allocations of frequency bands for satellite communications are scattered over the spectrum and some are regional and some global, but Table 10.1 gives an indication of the main frequency bands used and the kinds of services that run over them.

Table 10.1   Frequency bands and services.

Frequency band (GHz) (approx)	Name	Type of service (typical)	Example systems
1.5-1.6	L-band	Mobile voice and data, distress (Down-link)	Inmarsat, Thurya
2.4-2.6	S-band	Mobile voice and data	ICO / Globalstar
4-6	C-band	International dialled and leased circuits. Broadcast TV	Intelsat, PanAmSat, Dish
7-8	X -band	Military services	Skynet 4 & 5, NATO
11-14	Ku-band	TV broadcast, business systems, broadband Internet to homes and offices	SES-Astra, Eutelsat Intelsat
20-30	Ka band	Broadband to homes and offices	Hughes Spaceway
40-50	V band	Non real-time fixed services	Hughes Galaxy

For a particular service, both up-links and down-links are usually in the same frequency band, with the up-link towards the top end and the corresponding down-link towards the bottom. An exception is mobile services where the satellite operates cross-band with the mobile up- and down-link in L-band or S-band and the gateway up- and down-link in another band (e.g. C-band).

The L-band and S-band satellites are used for low-rate voice and data since they have limited spectrum available, these rates are usually 9.6-64 kbit/s. The comparable terrestrial systems GSM, GPRS and 3G can provide data rates of 9.6, 28.8 (typical) and 284 kbit/s respectively to each of a limited number of terminals from each base station.

The remainder of the frequency bands can support data services up to many 10s of Mbit/s, usually to fixed locations although there are sporadic mobile allocations in these bands that would support less. Satellites operating in these bands divide the band up into channels usually either 36 or 72 MHz wide and transmit each channel through separate amplifier chains called transponders. The maximum bit rates that can be transmitted through a transponder depends upon the following factors:

- the modulation scheme — the scheme consists of $m = 2^n$ different symbols where n is the number of bits mapped to each symbol and 'n' is typically between 1 and 4 (the number 'm' is the order of the modulation scheme);

- the use of a forward-error correction coding scheme — typical code rates are between ¾ and ½, which adds between 33% and 100% redundancy to the user data;

- filtering of the signal before transmission that adds a further 20-40% 'excess' to the bandwidth.

10.3   Bandwidth Efficiency

The following is a simple 'ball-park' calculation of bandwidth efficiency for the purpose of comparison of typical satellite and terrestrial mobile services. Let:

user bit-rate = 64 kbit/s;
modulation scheme = QPSK (quarternary phase shift keying), so $m = 4$ and $n = 2$;
coding rate = ½;
filter = 40% excess bandwidth.

The occupied bandwidth of this signal is:

(user bit rate/code rate) × (1/bits per symbol) × (1 + excess bandwidth).

So in this case, the occupied bandwidth = 89.6 kHz. The bandwidth efficiency of this scheme is 64/89.6 = 0.71 bit/s/Hz.

Another example is DTH (direct to home) Internet or TV from a satellite where a 34 Mbit/s multiplex is transmitted from a 36 MHz transponder, giving an efficiency of around 1 bit/s/Hz.

To compare this with some terrestrial systems:

- GSM has 8 TDM (time division multiplex) channels occupying a total of 200 kHz spectrum, each capable of 9.6 kbit/s data rate — the spectrum efficiency is thus 9.6 × 8/200 = 0.4 bit/s/Hz;

- IS-95 (CDMA) users share a throughput of 1.23 Mbit/s in a bandwidth of 1.25 MHz, giving an efficiency of 1.23/1.25 = 1 bit/s/Hz, assuming that the bandwidth is efficiently packed.

Cable TV broadcast is standardised by DOCSIS (Data Over Cable Service Interface Specification), the range of modulation schemes is from QPSK to

128QAM with excess bandwidth filtering of 1.25, resulting in a rough calculation of bandwidth efficiency of between 2 and 6 [2]. The various bandwidth efficiencies are shown in Table 10.2.

Table 10.2   Typical bandwidth efficiency for satellite and terrestrial systems.

Service	Bandwidth efficiency (bit/s/Hz)
Satellite mobile	0.7
Satellite DTH TV	1
GSM data	0.4
IS-95	1
Cable TV	2-6

From Table 10.2, it can be seen that satellite mobile services give a comparable or better bandwidth efficiency than cellular mobile services, in terms of bit/s/Hz for individual users.

Terrestrial cellular systems appear to be much more efficient because they support a much higher number of users in a given area. They achieve this by re-using the same frequencies many times in an area equivalent to a satellite footprint, whereas the satellite uses the frequency once or only a few times. As an unlikely example, but one that serves to compare satellite with terrestrial mobile systems, in the satellite L-band allocation which is around 100 MHz wide, with 25 kHz channel spacing, 4000 simultaneous voice channels could be supported. This is the limit to the simultaneous calls within the satellite antenna footprint if the whole of the band was used for voice from the same type of terminal and if frequency re-use is not employed. The size of the footprint on current generation L band satellites is roughly $^1/_3$ of the earth's surface, so that the user density supported is quite low. In contrast, terrestrial cellular systems reuse the spectrum many 10s of times in the UK and therefore can support many 10s of lots of users. Next generation L-band satellites will improve this comparison, since they have spot beams that will enable the frequency spectrum to be re-used many times in the overall footprint. It is unlikely that mobile satellite systems will ever support the same user population as cellular systems, but again it is stressed that the main benefit of satellite mobile is that access is possible in the air, at sea and in land areas where there is no cellular coverage.

Mobile user terminals, whether satellite or cellular, are fitted with antennas that have little or no antenna directivity and this means that the spectrum can only be re-used in a limited way. However, with fixed satellite services that use higher frequencies, the terminals use highly directional antennas that pick out the desired satellite. This means that the use of spectrum is much more efficient and every satellite location can re-use it. In fact, there is lots of room for new satellites in most parts of the geostationary orbit. For example, in the case of Internet DTH services, it is estimated that 300 suitable[1] Ku-band transponders can currently 'see' the UK. The capacity represented at 34 Mbit/s per transponder is 300 × 34 Mbit/s = 10.2 Gbit/s. The number of Ku-band transponders that can possibly be fitted into the sky above the UK within elevation angle limits of 10 degrees is around 2000, assuming a satellite is positioned every 3 degrees along the geostationary arc. Therefore there is certainly room for expansion — only 300/2000 = 15% of the total possible satellite capacity is in use.

10.4   Latency and Availability

Latency is another characteristic of satellite communications, since it takes radio signals about 240-280 ms to travel to a geostationary satellite and back (i.e. one 'hop'). This latency is not noticeable with most services but it can become noticeable with highly interactive services like voice and games. Latency on voice has tended to give satellites a bad reputation, but this is now remedied by technology development that allows satellites to offer a large range of voice and data services. Latency is different between one-way and two-way satellite architectures (see section 10.5).

Concerning availability, for frequencies of Ku-band and above, the availability of a satellite link is dominated by weather conditions, such as the presence of rain and clouds and other atmospheric effects. Link margins are calculated to give the required availability in different rainfall regions and ITU models are often used for these computations [3]. Satellite Internet services are typically dimensioned to provide 99.9% availability.

To some extent power control and other fade mitigation techniques can be used to overcome weather effects, but at Ka band and above, it becomes impractical to use power control to overcome all the fades. This is because many magnitudes of power adjustment would be required and power levels could become dangerously high with associated safety and interference issues. The solution is to tolerate the fades, whereby outages of several minutes per year have to be built into service level agreements and a slight degradation of services is expected. A discussion of the tolerance of TCP (transport control protocol) to fades is given later in this chapter (section 10.8.3).

Satellite links are also sometimes affected when the sun is behind the satellite when viewed from an earth station. The sun is a powerful source of radio noise, therefore any earth station pointed to the satellite eclipsed by the sun will receive this noise. Small earth stations (with antenna size below a few metres) are not disturbed by noise from the sun because their gain is not particularly high. However, larger earth stations suffer deep, but short-lived, outages because their beam-width is narrow and their gain is high. Such outages are predicted and, if appropriate, BT transfers critical traffic to alternative earth stations during these eclipses.

---

[1] Can support small dish applications.

10.5   Basic Architecture Options

10.5.1   One-Way and Two-Way

Satellite systems can be broadly classed as one-way (to user) or two-way. A one-way system is where traffic is carried to the user by satellite. Push TV is an example of a one-way system. Any return traffic associated with a one-way system needs to travel along a terrestrial route using PSTN, ISDN, etc. Such systems are sometimes called 'hybrid' because two different media are used for the different directions. A two-way system is where traffic is carried in both directions by the satellite. The round-trip time of a hybrid system is generally less than that of a two-way system. Figure 10.1 shows one-way and two-way satellite system architectures where the satellite is in the access network for Internet and intranet access.

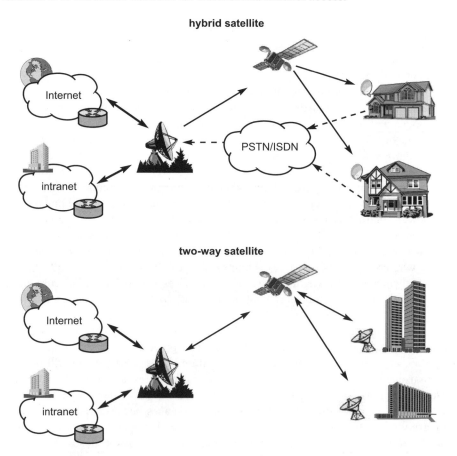

Fig 10.1   One-way (hybrid) and two-way satellite system architectures.

To give an idea of size of earth station equipment, an off-the-shelf one-way (receive-only) satellite terminal with 45-65 cm antenna, suitable for a TV or Internet feed to a house, will enable several Mbit/s to be received at Ku-band. The down-link multiplex of such services is around 38 Mbit/s gross bit rate. The receiver must receive at this rate in order to extract the wanted channel.

If the system is two-way, a return bandwidth upwards of several hundred kbit/s is available, but two-way terminals are much more expensive (about 10 times the cost of one-way). A low-cost 45 cm one-way satellite antenna is shown in Fig 10.2 and a VSAT (very small aperture terminal) two-way terminal 90 cm in diameter is shown in Fig 10.3. Both examples operate in Ku band.

Fig 10.2   Low-cost one-way satellite antenna (receive only).

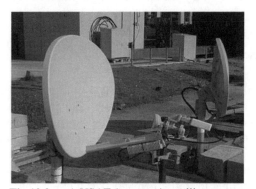

Fig 10.3   A VSAT (two-way) satellite antenna.

The transmitter power amplifier can be seen in Fig 10.3, at the back of the feed on the end of the arm.

An example of a two-way satellite service is the business satellite service offered by BT Openworld. This has two variants, the business satellite 500/1 designed for single users and the business satellite 500/4 designed for four users on a small network. These services are aimed at business users and marketed as having the following attributes:

- Internet access without tying up telephone lines;
- always on connection not requiring dial-up time;
- e-mail addresses;
- Web site space.

The latest products including bandwidths and costs associated with these systems can be found on the BT Openworld Web site [4].

Two-way satellite broadband access systems like this are 'always on' in the sense that there is no dial-up procedure, in this respect they are the same as ADSL.

10.5.2   Star and Mesh

The BT Openworld system is an example of a 'star' VSAT system with a central hub large earth station and many users. With star solutions, users can communicate only with the hub, therefore any traffic between remote users (remote A to remote B) must go via the hub and this means a double two-way satellite hop, which is not usually viable. The BT Openworld system is not intended for user-to-user communications.

A mesh configured satellite system, which is only an option for 2-way systems, permits direct communications between users' terminals, thus limiting the latency to just one two-way satellite hop. With a mesh system there is still a hub station, but this station only looks after the management of the connections and does not handle the user traffic.

10.5.3   Other Issues

With star VSAT solutions, the outbound signal from the hub is a multiplex that all VSATs receive. The inbound signals to the hub from each VSAT have to share the satellite capacity and this requires management using a scheduler. There are several techniques for achieving such sharing, an example being multiple frequency time-division multiple access (MF-TDMA), where several VSATs can transmit in each of several channels and each channel is divided into timeslots that are allocated via a control channel according to traffic demand. The hub has control of the scheduling of capacity allocations to the terminals. On these systems, the satellite capacity on the link from the terminals to the hub is not designed to be sufficient to cater for all of the terminals to transmit at once. The ratio between the number of terminals in the system and the number that can transmit at once is called the contention ratio. This ratio is typically between 20 and 50 and is chosen to balance the cost of the satellite link and yet provide acceptable service. The ratio is a critical system parameter and is set by the hub.

The regulatory issues are more complicated for a two-way terminal; it is necessary to obtain a licence to transmit, which in turn requires site clearance (an assessment of interference) to be carried out. To fast track the process, BT has worked with the UK Radiocommunications Agency, part of the Department of Trade and Industry, to develop streamlined procedures for domestic and small company users, where clearance can be done on-line by entering the post-code and house number [5].

Apart from the cost and regulatory issues, the other issue with two-way satellite is the double-hop round trip time (RTT) of around 560 ms plus terrestrial delays. A one-way system would give an RTT of about 280 ms plus terrestrial delays. The impact of this delay is discussed in section 10.8, but here it is stressed that satellite path latency has not proved to be a problem with broadcast or Internet/intranet access applications.

10.6 DVB

This is a standard for broadcasting digital TV over several kinds of transmission medium, written by the Digital Video Broadcast forum (DVB); the satellite variant is DVB-S [6].

The greatest use of satellite in the access network is TV broadcast (in terms of numbers of users) and the mass production of user terminals to the DVB-S standard for TV reception has resulted in them being very low cost. Data (IP) services over satellite are provided on the back of this development. In order to provide interactive TV and data services by two-way satellite, the DVB forum has recently published a standard for DVB-RCS (return channel by satellite) [7].

In order to carry IP packets over DVB, they are first encapsulated within a DSM-CC (digital storage media command and control) session. This session is then split into DVB transport packets. Figure 10.4 shows the encapsulation process.

DSM-CC was chosen as a mechanism for IP encapsulation because it offers MAC layer addressing and was already in place for control of data. The DSM-CC protocol is one-way and runs between the IP/DVB gateway and the user client. It is not optimised for IP encapsulation because it adds redundant fields. The loss in capacity caused by this inefficiency is between 10-20% depending on packet size, but there is work under way to improve this, e.g. by the European Space Agency [8].

Another issue is that whereas TV is one-to-many, data is mainly point-to-point at present with TCP as the most common transport layer protocol. As the use of IP multicast grows on the Internet, satellites will become an even more attractive delivery technology because the type of service is a closer match to the type of coverage.

Simultaneous delivery of data to any number of users is much more efficiently performed over a single node aboard a satellite than over a terrestrial network,

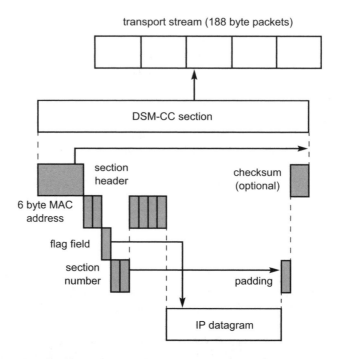

Fig 10.4   Encapsulation of IP packets into DVB.

where multicast routing algorithms and packet duplication processes place heavy demands on routers.

10.7   Interface to the Terrestrial Network

A satellite link is usually connected between two (or more) terrestrial networks as shown in Figs 10.5 and 10.6. The left-hand network could be run by a large network operator and the right-hand one can be as simple as a user PC or could be a home/

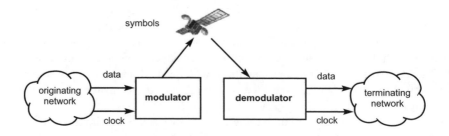

Fig 10.5   Terrestrial interfaces where the user network is small.

enterprise network with its own routers. Some of the difficulty in connecting different networks is due to timing and synchronisation problems caused by each having a different clock.

Figure 10.5 shows connection of the demodulator to a small user network that does not have its own reference clock. Figure 10.6 shows the connection of the demodulator to a larger network where the clock is used to time the output of the Doppler buffer fitted to the demodulator.

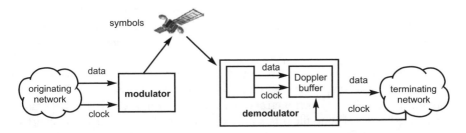

Fig 10.6   Terrestrial network connection where the user network has its own clock.

Working from left to right of Figs 10.5 and 10.6, the data that needs to go over the satellite is groomed from a core transport technology like SDH or PDH (synchronous or pleisochronous digital hierarchy). The data then has to go over a 'backhaul' to the satellite earth station. The cost of the backhaul can be minimised by installing the earth station close to the point where the grooming is done, usually called a point-of-presence (PoP). The originating network is usually a large one, synchronised across several nodes by a clock. The terminating network might not have a clock if it is located in one place (Fig 10.5), but it will have its own clock if distributed across several nodes (Fig 10.6). If there are two different clocks running at the ends of the satellite link as in Fig 10.6, they will run at slightly different speeds and a way must be found of coping with this difference. To make the situation more complicated, the satellite itself moves around slightly and hence causes shifts in timing. Even modern geostationary satellites with good station-keeping move a few degrees over a 24 hour period. Because of this movement, a Doppler shift happens which causes the speed of symbol arrival (and hence data-rate) at the demodulator to vary slightly. The timing errors are usually smoothed out by a buffer in the demodulator. The satellite modem will usually time its transmission of symbols on the air interface to the originating network clock and the buffer in the demodulator is used to take out the variations in data rate caused by satellite movement; such a buffer is called a 'Doppler buffer'. Data is clocked out of this buffer by the receiving network clock, so that the buffer is also used to mitigate the slight difference in network clocks. In a two-way system, this procedure is used in both directions.

The satellite link is characterised by a ceiling on the data rate. The networks on either end may not have a data rate that matches that of the satellite link, which is usually dimensioned quite carefully to keep costs to a minimum. For instance, a satellite link may be 9 Mbit/s, but the PDH/SDH network may only be capable of grooming the traffic at either 8 or 16 Mbit/s. The satellite modem may also be connected to an Ethernet LAN capable of 10 Mbit/s; it is generally unlikely that a satellite link data rate will be matched to the data rate of the networks that connect to it.

So in the general case, the bit rate has to be managed down to travel over the satellite. In the above example, the 16 Mbit/s stream has to be 'managed' down to 9 Mbit/s. Now this rate adaptation can be done at the link or network layer, but it is better done at the network layer, for the following reason. At the link layer, the cell or frame rate can be reduced to the wanted rate by discarding cells or frames and re-clocking. By doing this, every packet at the network layer will contain errors. If however the rate reduction is done at the network layer using a process like Committed Access Rate (CAR), some packets are discarded but those that end up being transmitted do not contain errors. In either case, a new clock at 9 MHz has to be generated and locked to the originating network clock. Rate adaptation interface boxes can be bought that perform these operations at the link or network layer and they generate the necessary phase-locked clocks [9].

10.8   Transport Control Protocol

10.8.1   TCP Performance

TCP is widely used over the Internet to support application protocols like FTP (File Transfer Protocol) and HTTP. Also some routing protocols like BGP (Border Gateway Protocol) and some network management protocols use TCP.

TCP is a point-to-point protocol and so is not a particularly good match to services suited to satellite, which are of a broadcast nature. Despite this, the performance of TCP over satellite is of interest because satellite is used for point-to-point services, such as Internet access for individual users where ADSL is not available.

Over any link, the TCP throughput is upper-bounded by the product:

$$\text{receive window size} \times 1/\text{RTT}$$

Windows 95 and 98 both have a default receive window size of 8 kbyte, whereas Windows 2000 has a default receive window size of 64 kbyte with window scaling enabled.

So, for example, if the RTT is 500 ms and the receive window size is 64 kbyte, the maximum throughput is 128 kbyte/s (= 1024 kbit/s) per TCP session. Users and applications can invoke multiple sessions to achieve higher rates, for example

HTTP v1.1 does this for multiple objects. The receive window size is advertised by the receiver, but there is another window used by the sender, whereby the sender can transmit TCP segments up to the end of the window without having received acknowledgements.

The sender's window starts small; it begins at 1 segment and grows towards the advertised receive window size. The growth rate of the sender's window depends on two main factors:

- the slow-start threshold (ssthresh) — the sender's window grows exponentially, doubling at each interval, up to ssthresh; after the ssthresh has been passed, the growth abruptly slows down so that only one segment is added per interval (the interval is determined from the RTT and is usually set to just greater than the RTT);

- the RTT itself.

The size of the segments can be varied, the optimum size being just large enough to be encapsulated into IP packets, without the need to fragment them. The default size of an IP datagram is 576 byte, so that the most common TCP segment size is 536 byte, equal to 576 byte less the length of the IP and TCP headers (20 byte each). Both ends must agree on the maximum segment size to be used; however, determination of the optimum size 'S' is not easy on a network path through several nodes and most implementations do not attempt to calculate 'S', but use the default.

The RTT on a geostationary satellite link is typically 500 ms, which causes the sender window growth rate to be lower than that for terrestrial links where the RTT is typically (but not always) less.

TCP assumes that, if acknowledgements are not received in a certain time, the cause is due to congestion on a network. TCP reacts to this by reducing the sender window size exponentially (by half each time an acknowledgement timer expires) and by halving the slow-start threshold, so that the window does not grow again as quickly.

Figure 10.7 shows the way in which the sender window grows at the start of a TCP session and then reacts to an acknowledgement time-out. This is not usually the case for satellite links where, depending on the frequency range, such an event is more likely to be caused by a rain-fade.

As the sender window is adjusted to avoid congestion, it is sometimes called the 'congestion window' (cwnd).

The assumption that time-outs are due to congestion may be true for networks like the Internet, but the same is not generally true for satellite links because these are usually dimensioned more carefully. Depending on the frequency band, the time-out can be caused by an outage caused by a rain-fade. Therefore, satellite links are sometimes fitted with performance enhancing devices (described in the next section) in order to fool the sender into believing that the RTT is much less than it really is, to speed up the window growth.

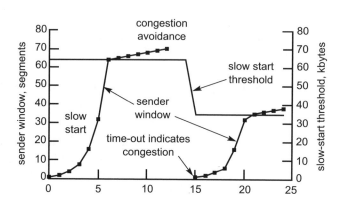

Fig 10.7   TCP sender window and slow-start threshold management.

The sender will stop transmitting segments at the end of the sender window if no acknowledgements have been received. If there is a throughput constriction on the flow of acknowledgements from the receiver back to the sender, this can effectively 'choke' the forward flow of TCP segments, where the system settles to sending a forward segment each time an acknowledgement is received. In this situation, the sender is right up against the window limit and a larger window will not improve the throughput.

The dependency of throughput on latency and on the free flow of acknowledgements in the return direction means that the data rate is not necessarily limited by the physical data rate of the satellite channel. When calculating the required physical satellite link bandwidth, the skill is in summing up the actual throughputs that the transport layer will achieve and then deciding on a contention ratio.

10.8.2   Performance-Enhancing Proxies

There are performance-enhancing proxies (PEPs) that can be bought and installed at satellite gateways to increase throughput with certain types of TCP-based services over a two-way satellite link. PEPs can only be used with two-way satellite systems: it is very tricky technically to install one on one-way (hybrid) systems and would be of limited benefit. Even on two-way systems, the PEP is only of benefit in a few situations.

There are various types of PEPs, but one common type works by manipulating the ACK segments in some way, e.g. by sending ACKs back to the sender to keep its measurement of the RTT falsely low. This increases the data rate from the sender in two ways:

- by reducing the slow-start time and hence increasing sender window growth rate;
- by increasing the rate of window-fulls of data sent.

The PEP buffers the data and sends it on to the receiver on a channel best-effort basis (e.g. using UDP), tracking ACKs coming from the receiver and taking care of any retransmissions required. Figure 10.8 shows a PEP arrangement where the TCP sessions are locally terminated to keep the senders under the impression that the RTT is falsely low as discussed above.

The transport protocol that runs over the satellite, the 'blank' boxes in Fig 10.8, will be optimised to run over the particular type of link. An example of this protocol has been observed to be a modified form of LAP-B. Such a protocol could have reduced headers, larger segments and some form of compression to optimise the throughput.

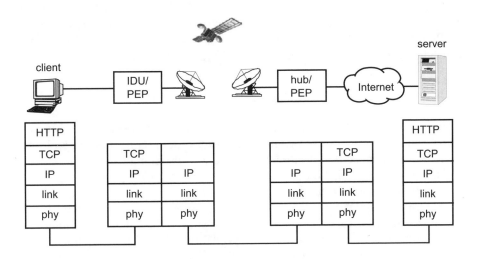

Fig 10.8 A PEP arrangement.

For the PEP to work, it has to detect when a TCP session is in progress and break into the TCP segment header; however, this is not possible where the TCP segments are encapsulated in an encrypted IP packet such as with IPSec tunnel mode. This problem crops up where IPSec encryption is used end-to-end, for example when a VPN is tunnelled through the Internet. For this reason, when IPSec VPNs are carried by satellite, they are not usually at an enhanced rate. Solutions to this problem could be to send the TCP header unencrypted (as in IPSEC transport mode) or to use application based encryption.

Table 10.3 contains results from an actual two-way satellite system showing transfer rates of an FTP session with and without encryption, from an earlier publication [10]. Three downloads of 5 Mbyte files were made in each case. Note that the encrypted data cannot be enhanced by the PEP and so the results given are equivalent to those with and without enhancement.

Table 10.3   Measurements of throughput.

Unencrypted	Transfer rate (kbyte/s)	Transfer rate (kbit/s)
Mean download speed	97	774
Maximum download speed	101	808

Encrypted	Transfer rate (kbyte/s)	Transfer rate (kbit/s)
Mean download speed	65	521
Maximum download speed	74	595

The PEP is enhancing the throughput by about 50% in the unencrypted case. The receive window size was set to 64 kbyte (Windows 2000). As stated earlier, the use of PEPs is not feasible with one-way satellite systems and are only useful in certain situations with two-way systems.

10.8.3   Tolerance of TCP to Fades

Fades are usually not a problem when the satellite link frequency is Ku band and below, since adequate power margins and power control on the link are built in to achieve availabilities of typically 99.9%. However, the rain attenuation at Ka band is more severe and availabilities are limited to typically 99.7% [11].

TCP has a session timer that aborts the session if the link is broken for a certain amount of time, typically 1 min. The timer is reset when ACKs are received at the sender. If the flow of ACKs stops due to a problem on the link such as a rain-fade, the sender carries on sending segments up to the end of the current window. After a time that is a bit longer than the RTT, TCP will retransmit the first segment that is not acknowledged. While no ACK is received, each retransmission interval is increased typically by a factor of two and this will continue up to the point where no ACKs have been received by the session timer, then the session will be aborted. More details of this mechanism can be found in Comer [12].

In order to make some measurements of TCP performance during fades, a test network was constructed as shown in Fig 10.9.

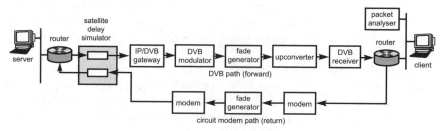

Fig 10.9   TCP performance test network.

The network consists of a server and client connected by DVB in the 'to-client' direction and a straightforward circuit in the 'to-server' direction. Satellite delay is added in both directions by a simulator. FTP 'get' requests were made from the client to the server which then sent test files of known size to the client.

To give an idea of the file transfer performance over satellite without a PEP, several test file transfers were made using the network in Fig 10.9 and it took an average 63 sec to transfer a 1 Mbyte file using an FTP 'get' command with RTT = 530 ms, a throughput of about 127 kbit/s. The receive window size was 8 kbyte (Windows 98), which is why the throughput is lower than the previous test results given in Table 10.3. It helps slightly that FTP opens up at least two TCP sessions, one for the FTP control and another for each file to be transferred. As stated earlier, HTTP opens several TCP sessions in parallel, usually one for each object on each page, so that the effect of latency is less noticeable on Internet access.

The fade generators were constructed to cause accurate gaps to appear in the signal, that could be timed from a fraction of a second to many minutes.

The circuit modem and DVB receiver lock-up times have to be taken into account and these were measured in isolation to be 11-16 sec and 1-3 sec respectively.

Regarding fade duration, the only existing model is a draft ITU Recommendation 'Prediction methods of fade dynamics on Earth-space paths [13]. This shows, as expected, that the number of fades longer than a given duration increases as the duration decreases. However, it is believed that all fades will last longer than 1 sec, but it seems that fades of a few seconds duration are of interest. The test network was therefore used to introduce fades between a few seconds up to several 10s of seconds.

As an example of what happens to TCP throughput during fades, Fig 10.10 is a plot taken from the packet analyser in Fig 10.9. It shows the throughput of two sessions that have been interrupted by a 5 sec fade and a 10 sec fade. Surprisingly, both fades result in the same effect on the TCP session.

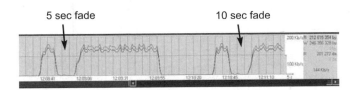

Fig 10.10   A plot showing the effect of a 5 sec and 10 sec fade.

The reason is that TCP retransmits the first missing segment after 2, 4, 8 and 16 sec. The retransmission at 8 sec was missed in the case of the 5 sec fade because of the time taken for the modems to re-lock. Therefore the TCP session was re-established at 16 sec for both cases. If the fade is longer than about 35 sec with a RTT of 530 ms, TCP never recovers, because the retry interval is beyond the TCP session time-out. A solution to this problem will require the endorsement of the IETF for wide implementation.

10.9   So What Does It All Mean?

This chapter will now describe some new opportunities for satellites in the overall mobile and wireless communications arena.

In terms of multimedia services, it has been established that satellites should be used in mobile and broadcast situations, where they have natural advantages.

One opportunity is public transport, where satellites can be used to communicate with trains, coaches or boats, which in turn may have a wireless LAN fitted. The transport vehicle could have several access link options, of which only one is satellite, with automatice selection according to location and the demand from the applications. One such vision for a train is shown in Fig 10.11.

Figure 10.11 gives a vision of a system where information is prepared for the user and managed using a system of agents. Developments for trains are taking place in Italy and a commercial service is now being offered by Linx in Sweden [14].

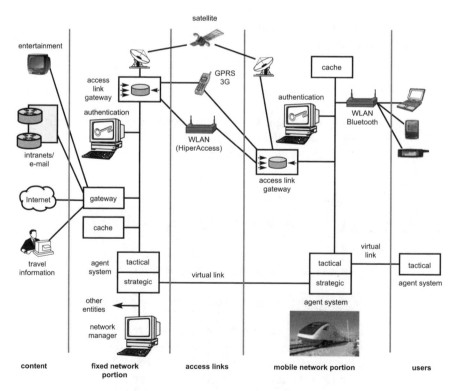

Fig 10.11   A possible architecture for a train.

The theme of satellite links with WLAN is a powerful one that is not limited to use on public transport. Remote communities who live in places that will not be served with ADSL can use this configuration. A satellite link can connect a satellite gateway to a WLAN base station that is mounted on a building and then houses within 50 m of the base-station can share the available bandwidth. Whether this is worth doing depends on how many houses are situated within the WLAN coverage area and what local arrangements can be set up to share the bandwidth and the cost fairly.

Yet another use of this idea is to place a satellite terminal in a vehicle, with the idea of parking the vehicle outside a hotel, exhibition hall or similar venue. WLAN base stations can then be wired to the vehicle to provide fast temporary coverage. Figure 10.12 gives the general idea.

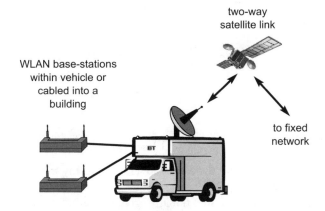

Fig 10.12   A satellite terminal on a vehicle can provide temporary coverage when combined with WLAN.

Yet another opportunity being explored is for overlay systems to provide broadcast-type data to 3G handsets and WLAN base stations provides another opportunity for satellite. This theme is being developed in an EU 6th framework project MAESTRO, where satellite gateways will be integrated with the 3G core network, with automated routing decisions over satellite.

The final opportunity mentioned in this chapter is broadcast content delivery, where content is distributed to the edges of a network, especially for applications like video-on-demand and Internet services. This idea makes use of the decreasing cost of data storage, by keeping caches updated with popular films and web-sites. Transmissions can be made regularly, e.g. every night, which makes the system less susceptible to rain fades, hence higher frequencies like V band[2] can be used. Figure 10.13 shows that the architecture is fairly simple, however there are significant problems to address, especially in the area of security.

---

[2] The definition of V band varies. Here it is assumed to mean around 40-50 Ghz (see Table 10.1).

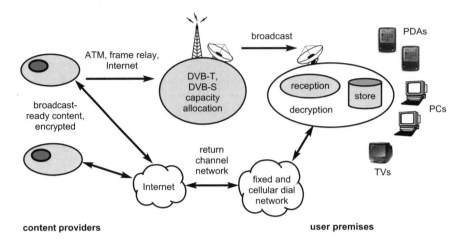

Fig 10.13   Broadcast content delivery.

In Fig 10.13, content can be requested via dial-up networks and then delivered to user premises where it is kept in storage until used.

10.10   Summary

Satellites are good at providing wide area coverage to many users, giving fast and flexible access. They are capable of carrying a wide range of services and the propagation delay, which has traditionally given satellite a bad name, is not generally noticeable except perhaps with highly interactive services like gaming and voice.

Satellites are already being used for TV broadcast and in the access network where ADSL is not viable. There are several exciting new opportunities for satellite in the near future, including links to vehicles, community WLANs, 3G overlay and broadcast content delivery.

The techniques for interfacing satellite links to terrestrial networks is well developed, including ways to overcome timing problems.

TCP is used over satellite for reliable point-to-point links. The performance of TCP is improving with increased window sizes becoming available. Fades are characteristic of satellite channels at frequencies of Ka band and above, however these are not yet much used. Even so, TCP is tolerant to fades up to several 10s of seconds. Performance-enhancing proxies (PEPs) can improve the speed of TCP sessions in some situations but cannot be used with one-way services or when IPSec is used, and we would question the need for PEPs even with two-way services.

Future technology and standards work will be aimed at reducing the cost of satellite capacity, for example by increasing the efficiency of IP over DVB/DVB-

RCS and developing protocol enhancements that enable higher frequency bands to be used more effectively.

References

1   ITU Radio Regulations: '*Frequency allocations*', Volume 1 Articles, Article 5 (2001).

2   Data Over Cable Service Interface Specifications (DOCSIS) Radio Frequency Interface Specification, SP-RFIv2.0-103-021218, table 6-6 (p 77) (1999).

3   ITU-R: '*Propagation Data and Prediction Methods required for the Design of Earth-Space Telecommunications Systems*', Recommendation P.618-7, ITU-R Propagation Series (2001).

4   BT Openworld — http://www.btopenworld.com

5   UK Radiocommunications Agency — https://www.e-licensing.radio.gov.uk/

6   ETSI EN300-421: '*Digital Video Broadcasting (DVB) Framing Structure, Channel Coding and Modulation for 11/12 GHz Satellite Services*', (1994).

7   ETSI EN301-790: '*Digital Video Broadcasting (DVB) Interaction Channel for Satellite Distribution Systems*', (2003).

8   European Space Agency ITT AO4372: '*Standards support of enhanced IETF IP encapsulation techniques for DVB-S*', (2003).

9   Metrodata — http://www.metrodata.co.uk/solutions/satellite/index.htm

10 Woodward, P., Hernandez, G. and Pell, T.: '*Performance enhancing proxies — are they as beneficial as they seem*', BT paper to IEE Conference on IP over Satellite (June 2003).

11 Satellite Today magazine archives — http://www.telecomweb.com/satellite/viasatellite/ask/archive/1017834517.htm

12 Comer, D. E.: '*InterNetworking with TCP/IP*', pp 191-212, Prentice-Hall (1988).

13 Draft Recommendation ITU 3/90-E.: '*Prediction method of fade dynamics on Earth-space paths*', (May 2002).

14 Linx — http://www.linx.se/templates/newspage.aspx?id=2203

11

EVOLVING SYSTEMS BEYOND 3G — THE IST BRAIN AND MIND PROJECTS

D Wisely and E Mitjana

11.1   Introduction

11.1.1   The Mobile Generation Game

In the beginning there were analogue mobile telephones; these were bulky, insecure and could only be used in a single country — analogue mobile systems are often termed first generation (1G). Next came 2G digital systems, such as GSM, providing secure voice services and, with international standardisation, roaming to over 150 countries today. 2G mobile services are, however, restricted to voice, short messages and the ill-fated dial-up Internet access WAP (wireless application protocol) system. GSM has recently been upgraded to offer low bit rate (10 to 64 kbit/s) data services via GPRS (general packet radio service) — the great advances of GPRS being that it can offer per-packet, as opposed to connection-time-based, charging as well as an 'always-on' capability for data. Being 'always-on' means that users are contactable in the data world, for example receiving e-mails and instant messages immediately.

The term third generation (3G) mobile system refers to a family of new air interfaces and access/core networks [1] — examples being the European-backed UMTS and North American-backed CDMA2000 systems. These networks are beginning to offer users higher data rates, up to 384 kbit/s, better quality of service support for data, including real-time applications, e.g. video, and new applications, such as location-based services [2]. Within the industry and research communities, there is considerable debate about what form fourth generation (4G) mobile systems will take. In Japan [3] and the ITU [4], 4G has been proposed as a new air interface, probably based on OFDM (orthogonal frequency division multiplexing) and offering data rates up to 100 Mbit/s with deployment in 2012 or later (see Fig 11.1).

At the same time as the debate about 4G systems has been taking place, wireless LANs (WLANs) have taken off across the world, being used for broadband (512 kbit/s+) hot-spot access to the Internet and corporate VPNs (virtual private networks).

WLANs have grown rapidly because they are low cost, operate in licence-exempt spectrum and offer high connection speeds. In the UK alone BT plans to deploy 4000 hot-spots within the next 2 years. In addition to WLAN deployment, incumbent operators across Europe are rolling out broadband access through a variety of DSL (digital subscriber line) systems — in Germany there are already over 4 million DSL connections and BT have announced a target of 5 million DSL connections by 2005.

Thus, through a combination of broadband access technologies, a significant number of European consumers have experienced broadband and helped drive the development of new services, such as presence and video/audio download and streaming.

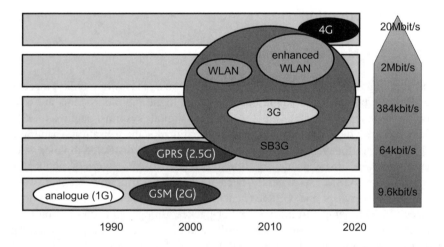

Fig 11.1 Mobile generation.

Some commentators believe that these developments will have a major impact on mobile systems long before 4G systems, with their new air interfaces and flexible networks, can be developed and deployed. The term 'systems beyond 3G' (SB3G) has been coined for the 'unification' of 2.5/3G wireless LANs and DSL systems, as well other access technologies, where users are able to access common services using TCP/IP running over them [5].

Figure 11.2 depicts the concept of systems beyond 3G with several access technologies, a common IP backbone, and service creation and intelligence pushed to the edge of the network.

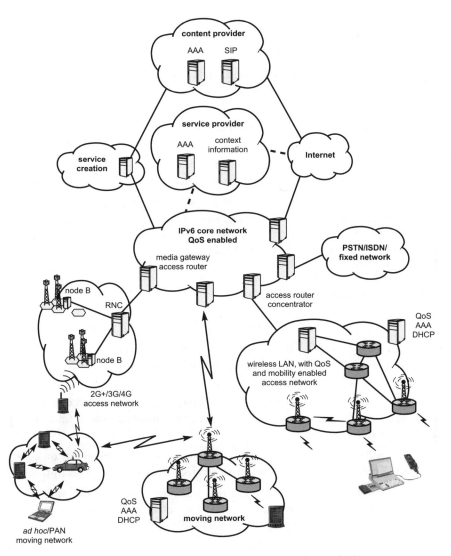

Fig 11.2   Typical architecture for a 'system beyond 3G'.

11.1.2   The BRAIN and MIND Projects

In order to investigate and research some of the key technical challenges that systems beyond 3G represent, a consortium of some of the world's leading operators, manufacturers and academia formed the BRAIN (Broadband Radio Access for IP-based Networks) project [6]. BRAIN has been succeeded by a follow-on project called MIND. The Mobile IP-based Network Developments [7] project was partially funded by the European Commission within the framework of the Information Society Technologies (IST) programme. The project partners of MIND were drawn from three key areas:

- manufacturers — Ericsson AB (Sweden), Nokia Corporation (Finland), Siemens AG (Germany), Sony International (Europe) GmbH (Germany) and Infineon Technologies (Germany);

- network operators — British Telecommunications plc (UK), France Télécom SA (France), NTT DoCoMo Inc (Japan), and T-Systems Nova GmbH (Germany);

- small enterprise and academia domain — Agora Systems SA (Spain), Universidad Politécnica de Madrid (Spain), and King's College London (UK).

MIND ran from June 2001 to November 2002. The overall aim was to facilitate the rapid spread and deployment of broadband multimedia services and applications that are fully supported and customised when accessed by future mobile users from a wide range of wireless access technologies.

11.1.3   Systems Beyond 3G

Systems beyond 3G could be said to have the following characteristics:

- IP connectivity — IP is the network layer 'unifying' protocol that runs, end-to-end, over a number of access technologies, including WLAN and 2.5/3G cellular mobile;

- enhanced WLANs — to provide low-cost, high-bandwidth coverage over campus-sized networks with support for the real-time hand-over of time-sensitive services such as voice and video;

- a potentially open business model — whereby users are billed customers of a service provider and can gain access to services from any access networkp;

- a flexible and dynamic IP service-creation platform — to allow the rapid creation of services, potentially by third parties, for delivery, suitably adapted, to an authenticated user with any terminal on any IP-connected network;

- vertical hand-over support — allowing users to move from, for example, WLAN to 3G and back again without re-starting a session and having the underlying service respond to the change of bandwidth and delay, etc;

- support for new network topologies — including *ad hoc* fringe networks (such as community networks) and moving networks (in trains, boats and planes, etc).

The BRAIN and MIND projects have tackled a number of key business and technical barriers to the development and deployment of SB3G. Initially the project has produced user scenarios — relating the technical developments to specific business opportunities and value chains — an example of which is described in section 11.2. Also described, in section 11.3, are the results of techno-economic modelling of SB3G, done in conjunction with the IST project TONIC [8]. Section 11.4 describes our architecture for a flexible and dynamic service creation platform based on session initiation protocol (SIP), and section 11.5 looks at middleware components and end-to-end protocols that have been developed within the project to run on top of SIP to allow the very rapid adaptation of services to changing QoS environments. This rapid adaptation reduces the QoS requirements for a given end-user experience and dramatically increases overall system efficiency. In section 11.6 we discuss seamless access — in the context of 3G-WLAN interworking — describing the various levels of seamlessness that are being developed. Sections 11.7 and 11.8 present an outline architecture for a SB3G network — meaning the underlying QoS, mobility and, to a lesser extent, security enhancements that are needed to current, fixed, IP networks. The final section examines how WLANs (especially HIPERLAN/2) can be enhanced to efficiently carry IP with a guaranteed QoS and how the performance of basic WLAN systems can be increased in range, functionality and efficiency.

11.2   SB3G Scenarios and Business Models

11.2.1   An Example Scenario

Based on the work of the original BRAIN project, MIND has defined three scenarios called 'leisure time', 'nomadic worker' and 'medical care'. These scenarios were used to establish the business model for systems beyond 3G and, subsequently, to derive the requirements on the network and service creation platform. The project rapidly realised that there were a very great number of possible scenarios and these three example scenarios were used to narrow down the focus of the project.

A section from the 'nomadic worker' scenario follows [9]:

'Stephanie Jones is a member of a large multinational corporation based in Frankfurt and therefore constantly on the move from one place to the next, often having to cross international borders. She has a Personal Wireless Assistant (PWA), which allows her to remain connected to the Internet, no matter where in the world she decides to go next.

In the morning, Stephanie checks her PWA for any new messages. She notes that the agenda for the morning has been modified. The scheduled meeting for preparing the review of the MIND project has been moved to Munich. She decides to go there by train. Stephanie knows some other people involved in the Munich meeting will take the same train at intermediary stations, and proposes that they organise an 'on the road' session in the train, in order to prepare for the meeting.

The first thing that happens when Stephanie enters the train is that her PWA informs her the train service provider is running a high performance mobile communication network. Stephanie logs into the train network by entering her secure pre-paid account number and sets up a videoconference to the MIND project leader to sort out some minor details. When the other members of the MIND project gradually join Stephanie in the train they attach their terminals *ad hoc* to Stephanie's because some sections of the review have to be worked over by the project partners.

The network created in the train is secure because Stephanie has sent a temporary session key in the invitation to the train session beforehand. Because some of the partners' terminals only feature short-range radio technology (e.g. Bluetooth), they use Stephanie's terminal to access the corporate server and ...'

11.2.2   Existing Mobile Business Models

The 2G value chain and business model is quite simple — basically consisting of service providers and network providers. Service providers sell handsets and billing packages provided by network operators (e.g. mmO$_2$ and T-Motion) who actually operate the networks. It was originally thought that service providers would offer billing and value added services, but in practice they have been reduced to little more than shops; the billing, service provision and corporate functions have been firmly controlled by the operators. Since 2G content — voice and SMS text messages — are all generated by the users themselves, the 2G value chain is effectively just about users buying handsets and billing packages from the operators through retail outlets. One of the reasons for this has been the 2G security systems — in particular the GSM SIM cards that are issued, authenticated and invoiced by network operators. In this case it might be argued that the current GSM operators are, almost, acting as both network and service providers. In the case of pre-pay, for example, this is fairly obvious.

For SB3G networks it is often suggested that the value chain will get more complicated — Fig 11.3 shows one possible version (see also Chapter 16).

Fig 11.3   SB3G value chain.

The suggested roles of the players are as follows:

- network operator (NO) — owns the spectrum and runs the network;
- service provider (SP) — buys wholesale airtime from the network operator and issues bills;

- mobile virtual network operator (MVNO) — owns more infrastructure than SPs (perhaps some switching or routing capacity) and issue their own SIM cards, but do not own spectrum or base-stations;

- mobile Internet service provider (M-ISP) — provides users with IP addresses and access to wider IP networks;

- portal (context) provider (PP) — provides a 'homepage' and hence access to a range of services that are managed in association with the portal provider;

- application provider (AP) — supplies products (e.g. software) that is downloaded or used on-line;

- content provider (CP) — owns music or Web pages, etc.

The SB3G business model might be likened to the current dial-up Internet access model. Suppose that I want to buy my wife a rare book for her birthday and decide that the Internet is the best place to look. On my computer (an Apple Mac) I have a useful function called 'location' that allows me to configure the machine to dial in to a number of ISP accounts. Perhaps I choose Freeserve today and initiate the connection. Underlying this is my PSTN telephone line — it belongs to BT but I have signed up with another company (CheapTel) who installed a dialler box and send me, what they claim at least, is a lower bill than BT would for the same service that they buy wholesale from BT. My access to the resources (the goods and services) of the PSTN is secured by the lock on my front door. When I reach the PPP login screen to access Freeserve, I have to enter a user name and password — this service is free but some ISPs still charge for access. I have my browser set to start Yahoo, which offers search facilities and easy links to common services — including books. From the Yahoo portal link I contact Amazon and locate the requested tome — I duly order it using my credit card as payment and relying on the authenticity of the digital certificate presented by the Amazon site and the encryption facilities of my browser to have a security association between my terminal and the Amazon secure server.

11.2.2.1   BRAIN Business Model

The BRAIN business model is derived from the basic SB3G model outlined above (Fig 11.4) and, as adapted to the specific BRAIN architecture (see Fig 11.11 in section 11.7). The BRAIN network architecture is simple in principle — it consists of an IP-based radio access network. The access network terminates at a BRAIN mobility gateway (BMG) — that looks to the core IP network simply as a normal IP gateway router (it might run BGP, for example) and has a service level agreement with the IP core network provider. The access network is owned by a single organisation — the access network provider in the BRAIN business model [10]. The security model for access to the network follows the usual IP AAA (authentication, authorisation and accounting) model. A request for access to the network is sent to the local AAA server (AAA-L) this identifies the user's home domain and, if an agreement exists between the two organisations, the AAA request is relayed by (for example) a proxy RADIUS client to the home AAA server for authentication (AAA-H).

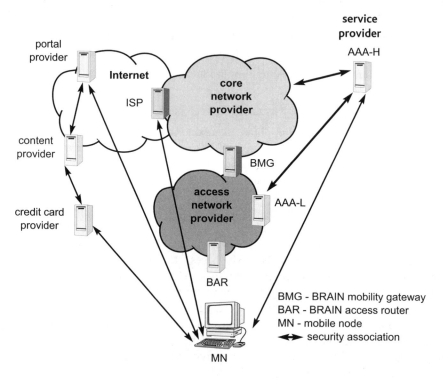

Fig 11.4   BRAIN business model.

In the BRAIN business model [11] the AAA-H belongs to a service provider who bills for the services used. This could be a UMTS operator, or at least an issuer of UMTS SIM cards, and the normal U-SIM challenge response and authentication mechanism can be carried over the IP-RAN. Indeed the integration of WLAN and UMTS is being studied by ETSI BRAN [12] — with the intention of allowing 3G users to connect via either WLAN or UMTS base-stations.

Once authenticated there is envisaged to be some policy, associated with the user relating to the QoS that the users are entitled to request from the network. The BRAIN network uses a bandwidth broker for admission control and access to network resources and this would obtain the policy from the service provider.

As far as the network is concerned that is the end of the story. In the BRAIN architecture, there is an enhanced service interface (ESI) that provides a QoS-enabled transport service to the higher layers. The role of the network is as a simple IP packet delivery service (with the complications of hand-over, address allocation, control of layer 2 and so forth). Services and applications are built on this transport interface but are insulated from the network details. When QoS violations are detected, or signalled from the network, the end terminals adapt, using session and application layer solutions.

The business can be complicated, with portal and ISP providers and distributed content providers all needing payment, but the key point is that that the same goes for a fixed IP network — the BRAIN architecture has hidden the mobile nature of the network from the applications and services.

11.2.2.2   MIND Business Model

Within the MIND project the key goals of the network layer research were to consider how *ad hoc* nodes can attach to an existing IP access network, how these *ad hoc* nodes can conspire to deliver IP packets on anything other than a best-effort basis, and how users can obtain a secure, end-to-end service under such circumstances.

The MIND network (see Fig 11.11 in section 11.7) extends the original BRAIN all-IP access network by the addition of:

- access network wireless routers (ANWRs) which are owned by the access network provider, mains-powered and portable, but change position only slowly — they lack the full capabilities of normal access routers and have only wireless interfaces;

- MIND mobile routers (MMRs), that act as routers through which mobile nodes may be connected to the network — they may be stationary but are considered lightweight and owned by end users meaning that they can also be mobile.

The MIND project is not concerned with stand-alone *ad hoc* networks; the main interest is in extensions to the BRAIN network to extend coverage, increase efficiency or lower cost, for example. If the extension is made by the same organisation that owns the BRAIN access network (BAN), then nothing has changed — the mobiles trust the new base-stations (whether they are mobile or just wireless). Analogously UMTS security now includes network authentication. When I sign on to a visited network it refers my access attempt to my HLR and gains a sequence number for its challenge-response to my SIM — the SIM checks this challenge response sequence number to ensure the visited network is trusted by my home network (or SIM issuer/service provider). It is very unlikely that users of MIND networks will want to sign on to mobile and wireless routers that are 'impersonating' BRAIN access routers — they either are proper BRAIN access routers or not.

The next level of complexity in MIND networks is probably the case where a single router (mobile or otherwise) is extending the network by one hop to a group of users (Fig 11.5).

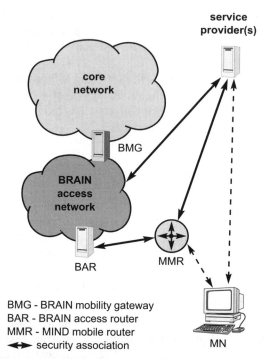

Fig 11.5   Single-hop BRAIN network extension.

The extender — which we can call the MMR — is not part of the BAN and is 'owned' by somebody other than the BAN owner. The MMR is also a standard BRAIN mobile node — since it is desirable not to modify the BRAIN access router (BAR) — and it must authenticate itself with the BAR in order to connect to the BAN. If the MMR and the mobile nodes (MNs) are 'friends' — i.e. they have a trust relationship — then they could rely entirely on the MMR to provide access to the network resources. This is similar to installing a wireless LAN in my house (e.g. the Macintosh Airport product) that allows all the family members to (simultaneously) share a single dial-up or ADSL Internet connection via a NAT (network address translator) box. In this case the business model is enhanced only by a trust relationship between the MMR and MN — If my daughter wants to buy a book I pay for the PSTN call and she pays for the book with her own credit card. The trust relationship might be established by private key exchange (e.g. using proximity (IrDA)) or pre-existing configuration (like file sharing today). In this case the MMR might be called a proxy access network provider.

In the case that the MMR is not trusted by the MN, then it certainly cannot pretend to be a base-station (as discussed above); therefore it must, effectively, act as a second network access provider. It is convenient to divide the relationship between the MN and the MMR into 4 classes (summarised in Table 11.1).

Table 11.1   Classes of auxiliary access network providers.

	MN-BAN relationship	MN-MMR relationship	MMR-BAN relationship
Class 0 Proxy	Not needed	Local (direct) relationship	Required
Class 1 Reseller	Not needed	Full relationship via Service Provider	Required
Class 2 Relay	Required	Full relationship via Service Provider	Required
Class 3 Repeater	Required	Not needed	Required

- Class 0

  Friends: as described above — a proxy access network provider.

- Class 1

  Trusted at the service provider level: the MMR is a full peer to the MN home service provider and is fully authenticated by the home network. The MN only sees the MMR offered service and not the BAN — an access network reseller.

- Class 2

  Trusted at the service provider level: the MMR is a full peer to the MN home service provider and is fully authenticated by the home network. The MN, however, also has a full relationship with the BAN (obtaining IP addresses, etc) — an access network relay service provider.

- Class 3

  No trust: the MMR might have a digital certificate and belong to a 'well-known' organisation or just be any old laptop — an access network repeater service provider.

An example of a class 2 access network provider might be a local authority that chooses to extend the range of a commercial WLAN service to areas that the network provider might consider uneconomic — perhaps a housing estate or the local library.

Class 3 providers might be normal BRAIN mobile nodes, e.g. a laptop, that people are happy can be used as range extenders in the knowledge that (hopefully) they might be able to make use of other people's range extensions in the future. In class 3 it is hard to see how the MNs could avoid having a full relationship with the BAN (i.e. sign on, authenticate and acquire IP address).

A much more complicated scenario is the case of a MANET (mobile *ad hoc* network) attached to the BAN (Fig 11.6). Here the MIND envisages that all the nodes of the MANET run a specific routing protocol. Also, because it is not a stand-alone *ad hoc* network, there will be some common relationship between the members of the MANET, e.g. in the example scenario described earlier (section 11.2.1), Stephanie and her colleagues form an *ad hoc* network. Another example might be that all nodes belong to students on a campus and to get access to the wireless LAN (and intranet) they must run certain software — including the MIND *ad hoc* routing protocol. In that case there would be a relationship between the *ad hoc* nodes, and, additionally, the active nodes would all have a full relationship with the BAR access network provider (since they would have to have MIND terminal software). The relationship between the nodes might include a new player — the co-operation service provider (e.g. a university server), that ensured cheats (that never forwarded packets) were excluded and (perhaps) performed load balancing and QoS functions within the MANET. Pure MANETs would not have the advantage of always having a central service, such as this, available.

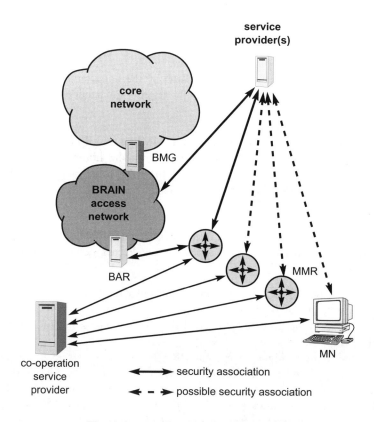

Fig 11.6   MIND MANET attachment.

In conclusion then, we can say that the MIND business model introduces some new players — in particular the auxiliary access service providers (proxy, reseller, relay and repeater) and the co-operation service provider. Because of the BRAIN/MIND architecture, the impact on players higher up the chain is largely through less reliable QoS and reachability issues. The fundamental business model is not drastically altered, however.

11.3   Techno-Economic Analysis of SB3G

Following co-operation with the IST project TONIC, a techno-economic analysis on the provision of IP services over SB3G, consisting of a combination of UMTS and an enhanced WLAN, has been performed [8].

Using the BRAIN architecture (explained in detail in section 11.7) for the WLAN infrastructure, the modelling assumed that a single operator deployed both UMTS and WLAN from as common a network, billing and service platform as was possible. The gateway node of the 3G network was used as the connection point to the BRAIN mobile gateway (BMG) and estimates were made of the cost of WLAN deployment and backhaul. The modelling estimated the costs and average revenue per user (ARPU) for the case of large and small countries deploying either UMTS alone or a SB3G of UMTS plus WLAN. Figure 11.7 shows the cash flow and cash balance curves in the large country case with (LY) and without (LN) WLAN deployment.

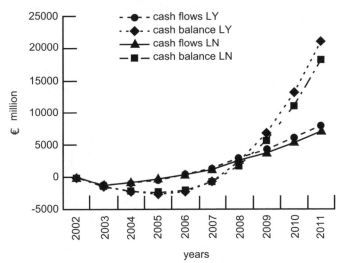

Fig 11.7   Example cash flows of an operator deploying UMTS with (LY) and without (LN) a wireless LAN complementary system.

The costs of WLAN deployments were, not surprisingly, much lower than those for UMTS, but the WLAN was able to grow overall revenues, by offering higher bandwidth services and attracting customers who were unwilling to pay for 3G, as well as cutting the cost of UMTS deployments by moving traffic on to the cheaper WLAN. It should be noted that the global discounted investment for WLAN turned out to represent only 2% of the total, while global discounted revenues resulting from it accounted for 8% of the total. The techno-economic study concluded that adding WLANs to the roll-out of UMTS, even where licences were not paid for, can be done at a relatively low cost. Recent moves from established network providers towards complementing their 2.5G and 3G offerings with WLAN in Europe (Telenor, Sonera), USA (T-Mobile, Nextel) and Japan (NTT DoCoMo), seem to corroborate the attractiveness of this combination.

11.4   A Flexible and Dynamic Service Creation Platform

A flexible and dynamic service architecture must be capable of supporting the BRAIN/MIND SB3G business models and, in particular, allowing the full range of value chain players, from content providers to users themselves, to participate in the service creation process. In traditional telecommunications networks the role of service creation has fallen, by and large, on to the network operator. Services such as divert when busy, voicemail and free-phone services involved deep changes, often at the exchange level, to the network. Operators were not prepared to open up the necessary interfaces or yield up the potential revenue to third parties, restricting service creation opportunities. With the IP revolution users can have, as we have seen in the business model example, a copper wire provided by a telco, Web access by an ISP, and buy goods in an on-line bookshop with a credit card. But on-line shopping is just one service and is typical of the client/server type services that are available today. If we are to move beyond these, to services, that include presence and real-time multimedia as well as offering users integrated services, then a new architecture for dynamic and flexible service provision is required. As an example a user might register that he is looking for entertainment tonight and his preferences include war movies and historical films. A cinema might be having trouble filling the screen for its latest blockbuster war movie and so advertises 50 cheap seats. A really useful service would put the two in touch and initiate (perhaps) a transaction.

In the past, service creation has been anything but dynamic with provision of services taking up to several days. In the Internet, multimedia future services will be created almost in real time. If I am at my local team's football match I may have a camera in my PDA and could use WLAN hot-spot to relay the game to the local network. This would probably involve a multicast, various rate adaptors and QoS requests to the network, as well as endless renegotiations to deal with changed loads, etc. All this requires a very dynamic architecture that can respond in real time to service requests and changes. Figure 11.8 shows an example studied in the projects of how SIP and Parlay (a generic network interface for third party applications — see Chapter 14 for more details) could be used to create a dynamic and flexible service platform in the case of the cinema example cited above.

11.5   BRAIN End Terminal Architecture (BRENTA)

End-to-end QoS is very difficult to deliver across different IP domains. User's data may have to traverse an *ad hoc* fringe, be subject to DiffServ per hop behaviour in

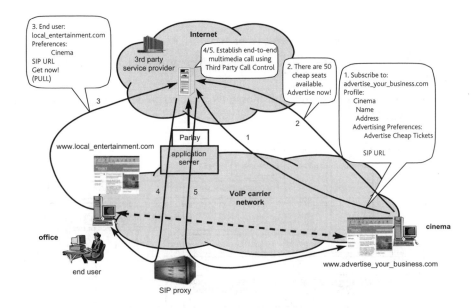

Fig 11.8 SIP used for service creation.

the BRAIN access network and, perhaps, little QoS support in a distant access network. Mobility, with sudden deteriorations of link quality and handover, makes any guarantees that are given liable to be broken several times during a session. Only in really homogeneous systems, where end domains are similar, services tightly defined, and radio resource continuously allocated, as it is in GSM, can mobile QoS be delivered by the network alone.

In the BRAIN/MIND architecture the network provides IP transport with a target QoS that is not a hard guarantee even over the access network. If applications were directly exposed to this kind of rapidly changing network QoS during a session, then the user-perceived QoS could be very poor indeed. What is needed is a way for nodes to measure and adapt to the QoS that the network is providing. Given the rapid nature of QoS fluctuation, some QoS information and an adaptation strategy will typically need to be negotiated and agreed by the end nodes of a multimedia session in advance.

There are other QoS considerations that must be taken into account at the end nodes as well as QoS adaptation. Firstly, it is no good launching a video application with a computationally intensive codec if the processor does not have spare capacity. Secondly, all the parties to the session must negotiate, in advance, details about the session such as codecs, data rates and data formats.

Figure 11.9 shows the complete BRENTA architecture which is capable of varying levels of these functions to different application types:

- legacy (type A);

- those that utilise session protocols (type B);

- those that can make use of a component API providing frame grabbers and packetisers (type C);

- those that can make use of a full-blown QoS broker to deal with all connection issues (type D).

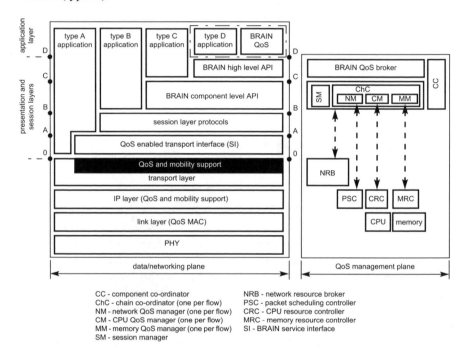

Fig 11.9   BRAIN end terminal architecture (BRENTA).

Full details of all the functions and components of the BRENTA architecture can be found in Kassler et al [13].

In order to create an implementation of the most important aspects of the BRENTA architecture, MIND has developed an end-to-end negotiation protocol that is transported by SIP. Currently SIP is the IETF protocol for session initiation, negotiation and termination. The session description protocol (SDP) is normally used to describe the type of session and the details, such as codec type and bit rate. In a mobile, heterogeneous environment, using SIP to renegotiate each time a QoS violation took place would be too slow and generate too many signalling messages. The project has designed a model for the end-to-end negotiation of terminal capabilities and a model for the end-to-end negotiation of QoS. Figure 11.10 shows an implementation of this functionality.

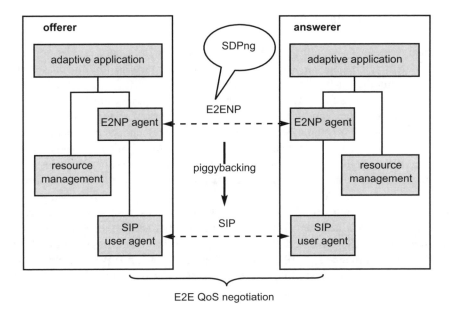

Fig 11.10   Implementation of QoS adaptation.

Central to the implementation is a new protocol, the end-to-end negotiation protocol (E2ENP) [14]. E2ENP allows end peers reaching an early agreement on what capabilities to use and which QoS levels to enforce. End terminals use the protocol to agree an adaptation path (AP) defining in advance which of the alternative negotiated paths to choose, when QoS violations occur. The enhanced service interface is then responsible for actually enforcing the QoS.

E2ENP uses next generation SDP (SDPng) and extends it to take account of its use to describe terminal capabilities, the adaptation path, and so on. E2ENP itself is carried within SIP messages and a SIP user agent is present at both end terminals.

When handover takes place between different networks, such as from WLAN to UMTS, then a major renegotiation and adaptation to QoS is required. The MIND/BRAIN architecture tackles this at several levels:

- terminal middleware — interprets the user's preferences and QoS requirements into underlying reservation and performs adaptation (e.g. using the end-to-end negotiation protocol) when QoS changes abruptly;

- IP layer — with end-to-end IP QoS signalling it is possible to set up common reservations across different underlying networks;

- IP$_2$W — the IP$_2$W interface is able to support IP QoS classes and frameworks directly (see section 11.7);

- multihoming — terminals can be connected to multiple access routers, multiple access networks or multiple ISPs.

11.6   Seamless Access

Typical users today might carry a number of mobile devices, such as a lap-top, PDA and mobile telephone. In addition they might also access multimedia services through a number of fixed terminals such as TVs, fax machines, PCs and so on. At present there is no simple way to maintain an identity, pay a common bill or allow users to access the same services from any network. Existing solutions, for a weather service for example, are different for mobile telephones, Internet PCs, etc.

Today seamless access is in its infancy — with GSM, mobile users can now roam to over 150 counties and get voice services, but even the (comparatively simple) intelligent services, such as dialling 901/902 for voice message retrieval, is only just becoming available. Corporate users can often download their e-mail over a variety of access technologies (dial-up, GPRS and so forth) — with most e-mail clients setting a limit on attachment size. To switch to another access technology, e.g. from dial-up to GPRS, involves activating the GPRS connection, with a SIM authentication process, possibly further authentication to an ISP, establishment of a new VPN tunnel and restarting the e-mail session; e-mail, however, is a simple application, compared to the presence and real-time services that are being developed.

Seamless access in the context of SB3G has a number of meanings in different contexts — it can mean common billing across a number of Internet access technologies, or it can mean a single identity (e.g. dave.wisely@bt.umts) and an associated authentication mechanism (e.g. a smart card/SIM) for all access technologies. Going further, all that users really want to be able to do is access services from any terminal wherever they are currently located (even if it belongs to somebody else), and seamless access can mean the adaptation and delivery of the required service to the user wherever they are located.

Table 11.2 provides a description of different 'levels' of seamless access.

Table 11.2   Seamless access integration levels.

Seamless access classification	Description
Common billing	Users are able to connect via a variety of access technologies and receive a common bill
Common authorisation	Users are authorised in the same way (e.g. by a smart card) across all access networks and their authorised services described in a common way
Common authentication	Individual users are authenticated to the network — even if they are connecting via a device owned and authenticated separately
Common QoS	Quality of service classes have at least a common definition in all access networks (even if they are implemented in very different ways)
Common services	Common services are adapted depending on the QoS, bandwidth, cost, etc, of the connection and the capabilities of the terminal
Seamless handover	Users can maintain sessions during handover between heterogeneous access technologies (e.g. WLAN and GPRS)

One of the most important aspects of seamless access, for SB3G, is WLAN — 3G interworking. In the future, users will want universal wireless access to services, via both WLANs and 2/3G networks. What is the best way of operating them together, i.e. so that they complement each other?

There are three broad ways of interworking wireless LANs and 3G, as outlined in Table 11.3. These reflect different trade-offs between, on the one hand, the required degree of modifications to standards and ease of development, and, on the other hand, the seamlessness of the interworking and amount of infrastructure commonality.

Table 11.3   General characteristics of different approaches to WLAN — 3G interworking.

Interworking	Outline	Advantages	Disadvantages
No coupling	WLAN and 3G networks are completely independent, from the perspective of users and networks	1. always possible 2. very rapid introduction	1. No common services 2. No load balancing = more expensive infrastructure 3. No handover[1]
Loose coupling	The two networks have common authentication and separate data paths	Supports a single operator offering a common service portfolio (including billing and security). Interworking with other public networks. The existence of two networks is largely hidden from the user. UMTS costs can be reduced by moving some traffic on to WLANs	1. Handover can be seamless if planned but might involve a few seconds' lost packets.
Tight coupling	WLAN network integrated into UMTS network, e.g. attaching to the GGSN as an extension to the PS domain[2]	1. WLAN looks like 3G systems — mobility and QoS support 2. Full integration of services	1. Lock in to 'legacy' 3G developments. 2. Considerable development needed to modify WLANs. 3. Not compatible with loose or no coupled operators.

[1] Although a terminal could maintain parallel connections on each network to mitigate this (multihoming).

[2] Or tight coupling would require WLAN operators to run at least a 3G IMS.

No coupling is available today and some degree of seamless handover is possible using mobile IP to achieve slow handover without breaking the session continuity. Within the IEEE there is work on the loose coupling approach — looking at defining extensible authentication protocol (EAP) methods for both SIM and non-SIM authentication for WLANs — and hence allowing mobile users to authenticate themselves to WLAN hot-spots.

3GPP, the UMTS standards body, is working on a tight-coupling solution where WLAN QoS classes are based on those of 3G data, and user data is routed back to the mobile user's home network, if they roam to a WLAN, for metering and service creation.

11.7   Network Architecture for SB3G

11.7.1   What Does All-IP Access Network Mean?

The term 'all-IP' is often used in conjunction with mobile networks and has come to mean a number of different things to different people. At its weakest, the phrase really only implies that all the traffic is encapsulated, but not necessarily transported, within IP packets in the network core as in UMTS Release 5, where IP packets from the mobile are only reconstructed at the SGSN and then tunnelled to the Internet proper.

In contrast in BRAIN and MIND we have taken a number of principles — mostly based on well-known Internet architectural concepts — to design an all-IP access network that transports IP packets natively and provides QoS and mobility support. Our first design principle is network transparency — the network's primary role is to deliver packets and all other intelligence functions are provided at the edge of the network.

This is an extension of the end-to-end principle [14] that really states that as much functionality as possible should be provided within the terminals, as opposed to within the network, since the end terminal is best placed to determine exactly what service is required.

Secondly we wanted to design a network that enabled and encouraged further evolution implying that the components were independent. To this end we tried to follow a strict layering model, e.g. only the IP layer processes had access to the link layer.

Our final precept was to only solve the special problems of mobile wireless access. We adopted existing and emerging solutions from evolving standards, such as the IETF, and contributed to developing these to take mobility, as well as QoS aspects, into account.

11.7.2   Basic Network Architecture

Figure 11.11 shows our basic network architecture based on the above principles. We have assumed an IP core network, where IP packets are highly aggregated and QoS is provided only for a few classes of traffic. The interface to the BRAIN access network is through a number of BRAIN mobile gateways (BMGs). These act as normal IP gateways running exterior routing protocols, firewalls and so forth. More than one BMG is recommended for resilience. At the other end of the network are BRAIN access routers having a radio link to mobile nodes and providing IP-level mobility, security and QoS functions.

The BAN hides the fact that the nodes are mobile, and all details of the particular radio access technologies, from the core. Mobile access as seen by the IP backbone,

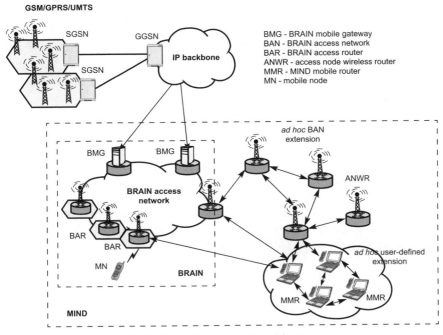

Fig 11.11   Basic network architecture.

at the network level, is very similar to access through DSL or UMTS (packet-switched domain).

The BAN is responsible for allocating one or more globally addressable IP addresses to a visiting terminal and has the basic role of delivering packets to that terminal as it moves within the access network. This gives rise to three fundamental functions that are required within the network — mobility management, QoS and security.

The basic architecture can be further enhanced by networks of self-organising unmanaged routing nodes. These are categorised as either ANWRs, which could be portable or rooftop base-stations, or MMRs that are envisaged as end-user devices — practical examples of these routers have already been given in section 11.2.2. This part of the network we have called the *ad hoc* fringe. Typical examples of the *ad hoc* fringe are, for the former, an operator wishing to provide temporary mobile coverage for an exhibition, and, for the latter, a University campus where WLAN coverage is patchy and extended by routing through terminals that are within range of a fixed base-station.

The unique features of this architecture are:

- simplicity — only a few new elements are introduced for mobility and QoS support, since fundamentally the network just delivers packets within the QoS requirement;

- robustness — the network is designed to be robust against a single point of failure;

- expandability — the use of strict layering and IP design principles means that the network is highly extensible;

- scalability — modelling has shown that the mobility management and QoS frameworks within the BAN can scale to cellular network sizes.

The IETF defines a very loose IP architecture and largely concentrates on protocol standardisation where one protocol is developed for each function that must be supported. However, the design of network elements, and interfaces between network elements, is considered an implementation issue. In 3GPP, by contrast, a complete, fully integrated, architecture has been defined with clear network elements, well-defined interfaces between them for equipment interworking purposes and tight integration between the layers. The BRAIN/MIND architecture identifies the key network elements and some of the protocols that provide support for mobility, QoS and security.

11.7.3   Interfaces

In fixed IP networks, the interface between layers within the stack (e.g. IP to link layer) contains very little functionality. For mobile networks, where QoS, handover and other advanced capability is required, much more functionality and co-ordination between the layer 2 and layer 3 is needed. The projects have designed an IP to wireless interface (IP_2W) [15] (see Fig 11.12), which provides a way for the IP layer to interface to a number of different wireless link-layer technologies. The IP_2W interface is only present in the mobiles and the access routers because the wireless link layer needs to provide much greater functionality than typical wired links such as Ethernet. In particular QoS must be addressed at the link layer, a MAC is required and link-layer addresses must be mapped to IP layer addresses.

The IP_2W interface supports discovery of link layer capabilities. Many functions, such as handover, can be entirely handled at layer 3. However, if layer 2 information, such as signal strength, is available, then better performance is possible. Layer 2 QoS must also be co-ordinated with layer 3 QoS to avoid inefficiency or instability, e.g. at which layer queuing and QoS differentiation takes place.

In order to implement the functionality of the IP_2W interface a convergence layer is necessary for WLAN technologies to build on the underlying functionality. Within the project we have specified a convergence layer for HIPERLAN/2 [16, 17] which is described in more detail in section 11.9.

In addition to the IP_2W interface the projects have also defined the enhanced service interface (see Fig 11.12). The ESI takes the Unix or Windows network socket concept further by including QoS support for applications or middleware allowing them to access the underlying QoS facilities provided by the network layer. Actual QoS might in effect be achieved through, for example, DiffServ or overprovision without the applications being aware of how this is achieved.

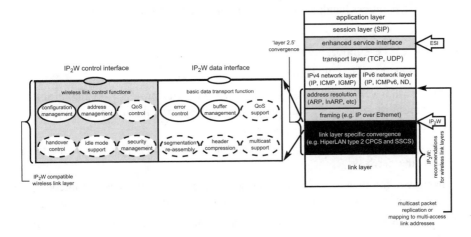

Fig 11.12 IP$_2$W and ESI interfaces.

11.8 Detailed IP Network Design

11.8.1 Mobility Management

Mobile IP has been developed over a number of years as an IP mobility solution. Users have a home agent and packets sent to this agent are tunnelled to a local 'foreign agent'. As terminals move they acquire new care-of-addresses and the tunnel end-point is moved. Mobile IP, however, does not provide seamless handover and has other disadvantages such as triangular routing and makes QoS support difficult [1]. Within the project we have taken the view that mobile IP might be used for IP macromobility (i.e. between domains) but that an IP micromobility protocol should run within the mobile access network. This is not only to overcome some of the listed disadvantages of mobile IP, but also because our business models suggest that the mobile has to be authenticated by the access network (although this may be with a remote operator) and then receives a globally routable address for the time that it is active within the network. This approach allows mobility-related issues to be confined within the access network. When a terminal receives an IP address from the access network it can then register this with a mobile IP home agent, but it could also use this address as part of a SIP registration process [18].

The role of the micromobility management function is to support terminal mobility allowing sessions to be continued as terminals move between access routers as well as supporting signalling of incoming sessions (e.g. paging). Some of the requirements on the micromobility management solution are:

- minimise mobility signalling traffic;
- provide seamless handovers (with minimum delay and without loss of packets);
- be scalable;
- be robust, i.e. support multiple routes or rapid rerouting;
- be compatible with other Internet protocols.

There are a very large number of IP micromobility protocols available for this type of mobility support. In order to classify and evaluate them, we decomposed the micromobility function into four parts:

- mobile node to access router protocol — how the mobile signals to, and interacts with, the network;
- handover framework — describing how the old and new access routers buffer and transfer packets to achieve seamless handover (e.g. by temporary tunnel);
- path updates — how the routers alter their routing tables to take account of the mobile handover;
- paging — such that nodes can report their location with a lower degree of accuracy than that of the cell.

From a full analysis and evaluation [19] it was found that no existing protocol was able to provide all the requirements for a micromobility protocol. Using concepts from existing protocols the BRAIN candidate mobility protocol (BCMP) [20] was developed to specifically provide such a solution. Essentially the BCMP is an IP micromobility protocol that allows users to connect to a wireless network, be allocated a globally routable address and then move around the mobile network without changing IP address, together with providing full support for real-time handover as well as paging for idle nodes. In this protocol users are authenticated either locally or remotely by an anchor point (ANP). If accredited the mobile is allocated a globally routable address from a range pointing at the ANP. The ANP tunnels received packets to the local access router and reverse tunnels packets from the mobile in the opposite direction (see Fig 11.13).

The BCMP has the advantages that it will run over existing routing protocols such as OSPF, offers good QoS control since the packets are tunnelled and reverse tunnelled from access router to anchor point, and supports a fast handover framework (planned and unplanned handover) and paging.

Within the *ad hoc* fringe there are really two very different types of network extension. The wireless routers (the ANWRs introduced earlier) might well be owned and controlled by the network operator and used, as an example, to provide temporary coverage at an exhibition. In this case the ANWRs must perform exactly like normal access routers in terms of authentication, handover and so forth. The key issue is how the ANWRs are securely added to an existing network without in any way compromising the existing system. We have explored several architectures for this attachment, the most promising being a gateway ANWR that attaches directly to a BRAIN access router.

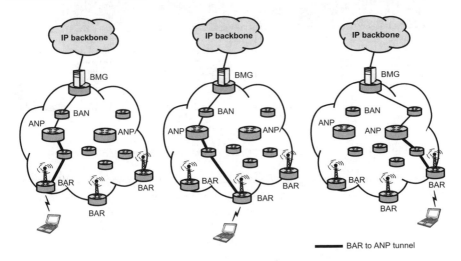

Fig 11.13   BRAIN candidate mobility protocol (BCMP).

A campus or large office, where WLAN coverage is patchy, typifies the other type of *ad hoc* fringe network extension we have investigated. Peripheral users can only reach the access routers via one or more relaying hops through other user's terminals. Under these circumstances the intermediate nodes should not impersonate access routers and so the end node must establish a direct security association with one of the access routers. In one architecture that has been advanced from within the project, called the virtual radio link (VRL), IP data and control packets are transported across the *ad hoc* fringe by an *ad hoc* routing function (see Fig 11.14). The end terminal is authenticated by the access network and obtains a globally routable IP address from the Anchor Point. However, it also obtains another IP address (possibly self-configured) for the *ad hoc* routing function. IP packets from the end node to the BAR are either tunnelled, IP in IP, or the access routers maintain an IP address mapping table to alter the headers appropriately. The mobile node to BAR protocol runs over the VRL as shown in Fig 11.14.

11.8.2   Quality of Service

In our QoS architecture (see Fig 11.15), the enhanced socket interface provides an applications programming interface (API) to RSVP for applications to signal QoS requirements to both the access network and the correspondent node. IntServ, per-

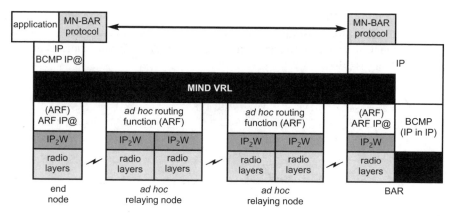

Fig 11.14   Virtual radio layer (VRL) in the user-defined *ad hoc* fringe.

flow, reservations are made over the wireless hop and then mapped to DiffServ per hop behaviours (PHBs) at the ingress to the access network. The access network, therefore, holds state only at the edge but can also be enhanced, for example, to

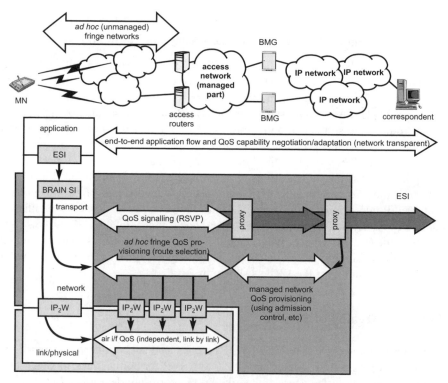

Fig 11.15   MIND QoS architecture.

provide better support for real-time services with the bounded delay PHB; this gives a hard guarantee on the maximum delay a packet will experience at each router. Another advantage of this architecture is that, even if end-to-end QoS is not available, localised QoS within the access network is still supported. Finally, nodes that only support DiffServ packet markings can obtain a better than best-effort service.

Within the user-defined *ad hoc* fringe, QoS routing [21], in which the routing protocol takes the requested QoS and computes a route that can deliver that level of QoS given the current state of the network, has been studied. QoS routing avoids the need for the access network to try and organise QoS within a part of the network that it does not control. It is likely that only DiffServ classes would be supported in the *ad hoc* fringe, but RSVP would still be used to signal application QoS requirements.

Another important aspect of QoS is radio resource management (RRM). In traditional cellular networks this refers to such functions as layer 2 admission control, congestion control, handover management and power control. An IP access network must provide similar functionality and it is clear that some of the IP RRM functions reside below the IP_2W interface and some, such as those involving more than one router (handover, non-local congestion, etc), must reside above the interface.

11.9   2G Wireless LANS

Current WLANs offer simple Internet or corporate VPN access from isolated access points. Within both the MIND and the BRAIN projects we have attempted to greatly enhance the performance of WLANs — producing what we have dubbed '2G' WLANs — with the following:

- support for real-time data services such as voice and video — including a powerful interface from the IP layer to the wireless link layer (IP_2W);

- support for campus-sized WLAN networks with real-time handover support;

- loose coupled solutions for handover to 3G;

- greatly enhanced range and efficiency for WLAN PHY and MAC layers;

- *ad hoc* network support.

Within the projects we have taken HIPERLAN/2 as a candidate second generation WLAN. HIPERLAN/2 operates within the 5 GHz band where it is expected significant spectrum will soon be globally available for WLAN operation. HIPERLAN/2 also offers a high level of functionality when compared to IEEE802.11b, for example, in the areas of QoS, connection management and dynamic frequency selection.

The project has designed and specified a convergence layer to enable HIPERLAN/2 to efficiently transport IP packets with defined QoS (see Fig 11.16), adding to the convergence layers already available for Ethernet and ATM over HIPERLAN/2.

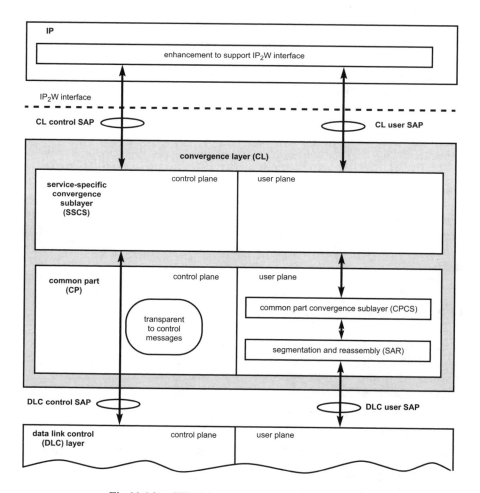

Fig 11.16   HIPERLAN/2 convergence layer for IP.

The convergence layer specification, that has been submitted to ETSI BRAN for standardisation [16, 17], provides support for:

- IPv4 and IPv6;
- IntServ and DiffServ;
- detailed address resolution schemes;
- the IP_2W interface;
- QoS mapping and scheduling;

- paging complying with IETF drafts;

- unicast, anycast and multicast.

Another important aspect of WLANs is that they cover only a relatively small area, typically 50 m indoors and 100 m outdoors. This is limited by the permitted transmit power. Obviously, if this can be extended then costs can be reduced and new scenarios covered. Also the spectrum can become crowded. The 2.4 GHz ISM band is unlicensed and also used by Bluetooth short range radio links and, at a typical hot-spot, it is not hard to imagine that the capacity will be greatly reduced by non-operator-owned WLAN and Bluetooth links. In addition to increasing capacity, mitigating interference, and extending range, is the need to support higher mobile speeds as the *ad hoc* fringes will sometimes include moving vehicles (see Fig 11.17).

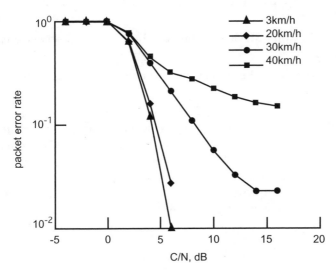

Fig 11.17   HIPERLAN/2 degradation with terminal speed.

Within the project physical layer enhancements to HIPERLAN/2 for extended range and greater efficiency have been studied, including multiple antennas, turbo coding and adaptive modulation. Efficient MAC strategies for vehicle-to-vehicle communications and multipoint-to-multipoint were studied as well [22].

11.10   Summary

The concept of a 'system beyond 3G' is based on the combination of several access technologies like 3G for cellular coverage, WLAN for hot-spot coverage and DSL for broadband fixed access. These are unified by running IP at the network layer and are interconnected by an IP backbone.

By offering users a choice of bandwidth and price, as well as common services, it is expected that take-up and use of multimedia services will be greater than if the various access networks exist only in isolation. There are many technical challenges that need to be solved before advanced multimedia services can be created and delivered seamlessly across such a system — in particular:

- introducing QoS within IP networks to create an end-to-end solution;

- increasing WLAN functionality and efficiency — in particular support for Mac-layer QoS, greater range and better integration with IP layer functionality;

- support for more flexible network topologies — including *ad hoc* fringe networks and moving networks;

- mechanisms for the rapid adaptation to QoS changes, e.g. from vertical handover, such rapid adaptation being much more efficient than inflexible bandwidth reservations;

- dynamic and flexible service creation platforms for rapid service creation must be developed;

- business models must be developed to support the business case for systems beyond 3G.

Within this chapter we have tried to outline the progress that the IST projects BRAIN and MIND have made towards solving some parts of these challenges. We believe that, with the roll-out of 3G, WLANs and DSL the world over, there will be powerful economic reasons, for either operators or third parties to unify the customer experience (e.g. billing, services, personalisation) through the commonality of IP. We believe it will be systems beyond 3G that will come to dominate the telecommunications arena over the next decade.

References

1   Wisely, D. R.: '*IP Mobility*' , Chapter 5 in Wisely, D. R., Eardley, P. and Burness, L. (Eds): '*IP for 3G: Networking Technologies for Mobile Communications*', John Wiley (2002).

2   Ralph, D. and Searby, S. (Eds): '*Location and Personalisation: Delivering Online and Mobility Services*', The Institution of Electrical Engineers, London (2004).

3   Nakajima, N. and Yamao, Y.: '*Development of 4th generation mobile communication*', Wireless Communications and Mobile Computing, **1**(1) (January 2001).

4   ITU report on systems beyond IMT 2000, GSC-8/GTSC-1/GRSC-1, Ottawa, Canada (May 2003).

5   Wisely, D. R.: '*The challenges of an all IP fixed and mobile telecommunications network*', Proceedings of PIMRC 2000 London (2000).

6   IST Project — http://www.ist-brain.org/

7   IST Project — http://www.ist-mind.org/

8   Mitjana, E., Wisely, D. R., Canu, S. and Loizillon, F.: '*Seamless IP service provision: techno-economic study by the MIND and TONIC projects*', Proceedings of the IST Mobile and Wireless Telecommunications Summit 2002, Thessaloniki (June 2002).

9   Robles, T., Mitjana, E. and Ruiz, P.: '*Usage scenarios and business opportunities for systems beyond 3G*', Proceedings of the IST Mobile and Wireless Telecommunications Summit 2002, Thessaloniki (June 2002).

10   Hancock, R. and Eardley, P.: '*Modular IP architectures for wireless mobile access*', Proceedings IST BRAIN International Workshop, London (November 2000) — http://www.ist-brain.org/workshop.htm

11   '*Service scenarios and business model in BRAIN project*', Proceedings 1st BRAIN International Workshop, London (November 2000) — http://www.ist-brain.org/

12   ETSI BRAN — http://www.etsi.org/bran/

13   Kassler, A., Burness, L., Khengar, P., Kovacs, E., Mandato, D., Manner, J., Neureiter, G., Robles, T. and Velayos, H.: '*BRENTA — Supporting mobility and quality of service for adaptable multimedia communication*', Proceedings of the IST Mobile Communications Summit 2000, Galway, Ireland, pp 403-408 (October 2000) — http://www.ist-brain.org/

14   IETF RFC 3234: '*Middleboxes: Taxonomy and Issues*', — http://www.ietf.org/ and http://www.postel.org/mailman/listinfo/end2end-interest/

15   Manner, J., Kojo, M., Laukkanen, A., Liljeberg, M., Suihko, T. and Raatikainen, K.: '*Exploitation of link layer. QoS mechanisms in IP QoS architectures*', ITCom 2001, Denver, Colorado, in: '*Quality of service over next-generation data networks*', Proceedings of SPIE, 4524, pp 273—283 (August 2001).

16   Bonjour, S., Hischke, S., Lappeteläinen, A., Liljeberg, M. and Lott, M.: '*IP Convergence Layer with QoS Support for HIPERLAN/2*', IST Global Summit 2001 Barcelona (September 2001) — http://www.ist-mind.org/

17   '*Functional specification of the IP service specific convergence sublayer for HIPERLAN/2*', IST MIND contribution to ETSI BRAN#28 (April 2002).

18   Wisely, D. R.: '*SIP and conversational Internet applications*', BT Technol J, **19**(2), pp 107-118 (April 2001).

19   Eardley, P., Mihailovic, A. and Suihko, T.: '*A framework for the evaluation of IP mobility protocols*', Proceedings of PIMRC 2000, London (September 2000).

20   Keszei, C., Georganopoulos, N., Turanyi, Z. and Valko, A.: '*Evaluation of the BRAIN candidate mobility management protocol*', IST Global Summit 2001 Barcelona (September 2001) — http://www.ist-mind.org/

21 Lin, C. R. and Liu, C. C.: '*An on-demand QoS routing protocol for mobile ad hoc networks*', Proceedings IEEE International Conference on Networks (ICON2000), Singapore, pp 160—164 (September 2000).

22 Dohler, M. et al: '*HIPERLAN/2 for vehicle to vehicle communication*', Proceedings of PIMRC 2002, Lisbon (September 2002).

12

ECONOMIC TUSSLES IN THE PUBLIC MOBILE ACCESS MARKET

G Corliano and K Khan

12.1   Introduction

Existing access systems to communications services have traditionally been designed according to the classical vertical communications model in which a system provides a set of services to users in an optimised manner. In particular, third generation (3G) mobile communications services specifically attempted to provide every kind of application through a single radio interface.

In their original vision, the capabilities of different wireless access systems should have been supported in all radio environments by that single interface. During its definition and standardisation phase, it turned out that there is no single radio interface that can be optimised for all applications, but rather a variety of heterogeneous access systems is required, each system supporting specific sets of applications (see Fig 12.1).

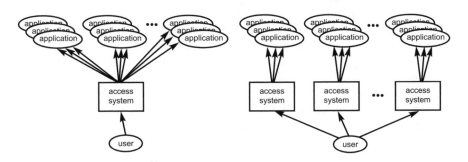

Fig 12.1   Multiple access systems.

At the moment, these access systems are mainly characterised by a horizontal communications model, where different systems — such as cellular, nomadic, cordless and fixed systems — converge into a common, flexible and seamless platform (see Fig 12.2) to meet different service requirements and radio environments [1].

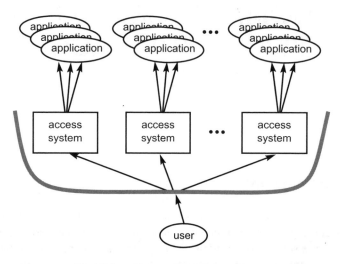

Fig 12.2   Converging access systems.

This convergence has accelerated the confluence of the conflicting economic interests of players in traditionally separate market sectors — thus the convergence of these sectors. However, whereas technology systems have predictable behaviour and evolution, the players in the converging market sectors have not. Clark et al [2] point out how, historically, the essence of successful societies is the dynamic management of evolving and conflicting interests. Such societies are structured around 'controlled tussle' — regulated by mechanisms such as laws, judges, societal opinion, and shared values. These mechanisms ensure that the tussle is not too strong, because that creates paralysis, and that it is not too weak, otherwise the size of the whole market would be smaller than it could actually be.

Similarly, we argue that the future of communications networks and services will increasingly be defined by tussles that arise among the various involved parties with clashing interests. This motivates our exploration of the market for public mobile access (the mobile market), the systems specifying the business interactions between mobile users and access providers (the business interface) and its limitations.

The business interface describes how users and providers interact when access services are to be purchased. In particular, it specifies how providers make their offers and users select and accept them. We observed that the original design of the business interface of the mobile market did not consider the unpredictability of the outcome of economic tussles. The approach of the mobile industry was to encode in its designs the result of these contentions (e.g. use of roaming agreements for user mobility), without allowing for variation. Because the market currently does not capture the business opportunities created by the combined deployment of different access networks, the business interface's designs should now be reconsidered to accommodate the convergence of different access systems and then allow for evolution. Therefore, we propose a set of principles for the design of the business interface that we have used to identify specific structural limitations with the current business interface and, on the basis of our study, we conclude that the future business interface should allow:

- providers to disseminate their offers to customers (possibly not end users) in an automated way;

- customers to be able to select the offers in an automated and intelligent way and then dynamically establish *ad hoc* (both short and long-term) business relationships with providers;

- end users to hold, at the same time, more than one business relationship with different providers, without experiencing any switching barriers;

- providers to dynamically reconfigure their metering systems on visited networks without having to publish their novel offers across all their competitors.

In this chapter, firstly we describe a reference model for mobile access that defines the business interface and its basic components — business relationships and functions. We then go on to discuss the current interface, illustrating the reasons why it is not appropriate and why the mobile industry has not been — and will not be — able to evolve it. We then formulate a set of architectural principles. We illustrate the specific technological limitations of the current business interface emerging from the application of these principles. Finally, we present five related novel capabilities of the future business interface.

12.2   A Reference Model for Mobile Access

This section presents a brief description of our reference model for mobile access. We first introduce communications and access services. Then, we discuss the market model as a composition of relationships among players and functions implemented by players. Finally, we define the business interface between end users and retail providers as a specific section of the market model. The notion of business interface plays a key role throughout the chapter because it specifies a method to capture the dynamics of the mobile market and to understand how providers supply and customers use mobile access services.

12.2.1   Communications and Access Services

Communication services include the Internet, traditional telephony, broadcasts, messaging, and end-to-end application services in general. To access these services, mobile users need to deal with service providers, possibly through an intermediary broker. Service providers supply network-based services and may supply complementary services to network-based ones. Brokers supply complementary services and possibly resell network-based ones supplied by providers.

Network-based services can be of information (e.g. content, portals) or connectivity type; the latter can be in turn divided into access or backbone connectivity. Our focus is on access connectivity.

We distinguish three access services (see Fig 12.3) — wireless, fixed, and Internet access. Depending on the communications service, some or all of these access services are involved. Service providers can supply one or all of these access services, depending on their business model.

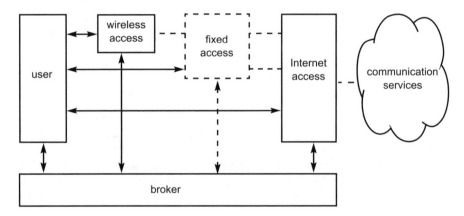

Fig 12.3   Access services.

12.2.2   Modelling Market Dynamics

The market model (or market structure) describes how market players behave and relate to each other. Key inputs to the market model are the business models of the existing players, the corresponding investment models, the imposed government regulation, and existing and emerging technology innovations (see Fig 12.4).

The output of the market model is its outcome, described by factors such as competition status, market size, and so on.

We decomposed the mobile market model into two components — relationships and functions. Relationships imply a common communications language and have different characterisations. Communications protocols, contracts and description languages are examples. Functions allow providers to deliver the service to

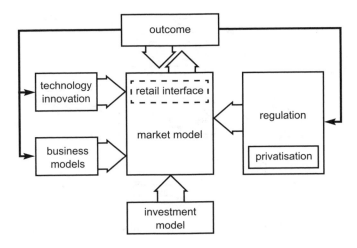

Fig 12.4   Market dynamics.

customers, run the business aspects of the service delivery, and support the establishment and management of relationships. The set of implemented functions determines the role of a player. Routing, handover and accounting are examples of functions.

As a result, a market model consists of an arrangement of a set of functions onto a set of players, capable of establishing appropriate relationships (see Fig 12.5).

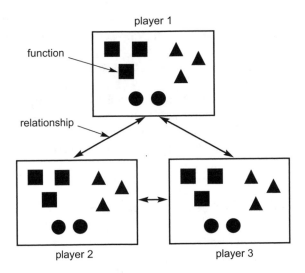

Fig 12.5   Functions and relationships.

In a market model, we distinguish three planes — the technical, business and support planes (see Fig 12.6).

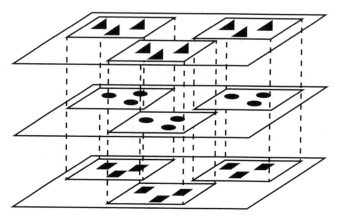

Fig 12.6   Technical, business and support planes.

At the technical plane, players communicate through technical relationships and implement technical functions. A technical relationship captures how services are delivered to users, in terms of protocols and mechanisms. Technical functions deliver the service to users and depend on the service to be provided (e.g. access or backbone connectivity, information); routing and handover are examples.

At the business plane, players communicate through business relationships and implement business functions. A business relationship defines the terms on which users are allowed to use access services. Business functions allow providers to run the business aspects of the service delivery and support the establishment and management of business relationships; they do not depend on the communications service.

The support plane involves the issues related to assisting users during and after the delivery of the service.

The business interface is that section of the market model's business plane, involving only retail players (end users, access providers, brokers), which defines how end users and providers interact when access services are to be purchased. In particular, it specifies how providers make their offers and users select and accept them.

12.3   The Current Business Interface

The current business interface has three essential characteristics: it is based on roaming agreements; the use of roaming agreements is encoded in its design; and providers compete on a nation-wide basis. We observed that such a business interface, relying on nation-wide competition only, is not appropriate for a mobile market where local and wide area networks co-exist. Moreover, because the obligatory use of roaming agreements is encoded in designs, the industry cannot make the business interface evolve — and therefore cope with the alterations of economic equilibria — except by rethinking current designs.

This section first illustrates the current business interface, focusing on roaming agreements and the reasons why they represent a barrier to user's mobility. Then, it highlights how, under circumstances other than the current ones, the business interface can have different instantiations for different outcomes of economic contentions (i.e. it can be altered to accommodate different commercial models). Finally, it outlines how, thanks to co-existing different access networks, and local and wide area competition, the business interface can provide more complex and richer service environments to mobile users.

12.3.1   A Business Interface Based on Roaming Agreements

The business models of cellular market players have been unchanged for some time. They involve mobile users gaining access to communications services by subscribing to a prior agreement with an access provider, the home provider, and maintaining a long-term business relationship with it. Users are therefore tied to that provider for the duration of the agreement and most of the time they do not have control over the service being provided. Most importantly, they maintain the business relationship with their home provider even when using access services (i.e. visiting access networks) of other providers.

The roaming mechanism consists of a set of inter-provider (roaming) agreements allowing customers on visited networks to get the (approximately) same service environment as their home ones. This mechanism specifies that the visited provider meters the user, but presents the charge to the home provider, which in turn presents the charge to the user.

Roaming agreements allow users to have technical mobility — visiting access networks of different providers; but they do not allow users to maintain contractual mobility — changing provider (i.e. offer) when visiting different networks, and having more than one business relationship at a time. Users are constrained to use alternative mechanisms — such as 'phone hire' in visited countries — which do not allow them to use their own environment (e.g. number, services, profiles). Roaming agreements, although valuable for facilitating the establishment of roaming for new providers, impede flexibility — which would offer improved user benefit — because they do not allow users to choose. Also administrative and billing arrangements, which are set up for roaming by providers, reduce flexibility and potentially lower pricing. In addition, the process to choose the access network, when roaming, depends on a variety of factors. Users may have some influence, but it is complex and not used to any great extent.

Roaming agreements represent a barrier to user's mobility, created by a perceived mutual self-interest between providers. Reducing or eliminating this barrier would make technology and business innovation more likely to increase user's demand — and more innovating providers would lead to an increased market size.

12.3.2   Business Interfaces as Synthesis of Components

We describe market models as the resynthesis (i.e. composition) of basic components. Accordingly, business interfaces consist of an arrangement of a set of business functions onto a set of retail players able to establish appropriate business relationships.

If the business interface is designed so that it can accommodate alterations of economic equilibria, then the business interface can take a whole range of instantiations and the industry can re-synthesise them as the commercial environment evolves.

We consider two scenarios in which the outcome of the economic tussles differs. On the one hand, we consider a business interface in which providers can capture mobile users as they walk into their coverage area. Users and providers are able to establish *ad hoc* business relationships, without the necessity of any pre-existing commercial agreement between providers. Users stay with a provider out of loyalty and not because of long-term subscriptions. On the other hand, we consider the current business interface, based on roaming agreements, where users visit other networks without changing their home provider and relying on existing inter-provider agreements. These two scenarios represent the two ends of a spectrum of instantiations of the business interface, where:

- each instantiation is the result of existing economic contentions;
- the design of the business interface accommodates the variation of the result of such contentions.

The approach of the mobile industry to the design of the current business interface has been to encode specific business models into designs and standards, without considering the unpredictability of the outcomes of economic tussles. The current standardised interface, based on the obligatory use of roaming agreements, is the result of such an approach — a circumstantial interface has been turned into the interface, in that the use of roaming agreements has been encoded in its designs. The business interface is therefore not flexible and open enough to accommodate the convergence of different access systems and allow for variation — we can have interfaces based on roaming agreements, but we cannot attempt to re-synthesise different ones.

The evidence is that the mobile industry is trying to accommodate the convergence of different access systems by extending to all access systems the use of the business interface based on roaming agreements. The standardisation process of mobile access services is accordingly following such a design approach. In addition, it appears that, in the recent past, the introduction into the market of many innovations did not change the tradition and did not imply any evolution. The major risk which market players will soon be facing is market saturation — and hence stagnation.

12.3.3 Business Interfaces of Increased Complexity

Evidence shows that the investments of current market players allow for a combined deployment of wide- and local-area access systems. However, the current mobile market does not capture the business opportunities created by this combined deployment. Market players still tend to adopt business models of the large-scale provider type, competing with each other on a nation-wide basis.

A business interface based solely on nation-wide competition is inappropriate for capturing the realities created by the deployment of local-area access networks, especially in the context of a market driven by total user mobility. The future business interface will be different in different places, and involving different players. We expect scenarios of increased complexity, based on small business investment and local competition, as well as on large-scale providers also competing on a wider basis [3].

For example, there may be a situation where small providers and private network owners provide on a wholesale basis mobile access to large-scale retail virtual providers. Or, it could be a situation where national network operators wholesale mobile access to local virtual providers. The first is a scenario where evidently both private network owners and virtual operators share common interests — the former get revenues by selling the service to the latter, which in turn get revenues by providing their customers with value-added services and seamless roaming facilities across multiple networks; in other words, public customers would make parasitic usage of private networks. The second scenario would allow small virtual providers to easily enter the market and generate revenues.

A business interface designed to accommodate these and other scenarios would open up the market, creating a much more complex and richer service environment for mobile users.

12.4 A New Business Interface

Following the investigation of the current business interface, we observe that its design does not accommodate variations of the outcome of economic contentions. This peculiarity becomes of critical relevance now that the business interface is inappropriate to seize the new business opportunities created by the convergence of heterogeneous access systems.

On the basis of this study, we formulated a set of principles for the design of the architecture of the business interface. The application of these principles revealed specific structural limitations of the current business interface. We isolated them and translated them into new capabilities of the future business interface.

12.4.1   Architectural Principles

As the architecture of the current business interface does not allow for variations of the outcome of economic contentions, we assert that the key architectural principle is to design for variation in outcome [2]. This implies that the business interface has to provide to its players the means to dynamically respond to changes.

In the following, we outline the design approaches motivated by this observation.

- Design for choice

  Players should be able to make choices (assessing several objects and selecting between them) and express preferences. To make complex choices, automated agents should assist players. To express dynamic preferences, player policy and decision mechanism need to be as separate as possible.

- Design for user-centricity

  The business interface should allow, where possible, for the control of its components from the edge of the network — possibly, where appropriate, from the user device.

- Design for orthogonality

  The business interface should provide orthogonal components, so that modifications of one do not affect the other ones. In particular, different business functions of different players need to be orthogonal, so that no third party dependencies are created.

- Design for openness

  The components of the business interface should be as open (to communication) as possible. Open components can be used as closed ones, but closed components cannot be used as open ones.

12.4.2   Limitations of the Current Business Interface

Following the application of the above-mentioned architectural principles, specific structural limitations of the current business interface emerged. In particular:

- the inability of providers to supply electronic service offers to customers (possibly not the end user);

- the inability and unwillingness of users to assess novel offers because of the potentially large range of choice (both in terms of offers and providers) and the lack of automated systems to assess new offers on behalf of customers;

- the inability of users to make complex selections at the retail level provides no incentive to make good competitive selections at the wholesale level;

- the inability of users to establish business relationships based on credentials and not on existing inter-provider commercial agreements;

- the inability of users to hold, at the same time, more than one business relationship, without experiencing unwanted switching barriers (such as SIM cards and phone locking in the cellular market);

- the inability of providers to preserve their pricing scheme innovation, when using a model based on roaming agreements.

These limitations create the illusion of a *status quo* being in everyone's interest, but in reality, they restrict user's benefits, service providers' competitiveness, and the market's capability to evolve.

12.4.3   Capabilities of the New Business Interface

The limitations of the current business interface, emerging from the application of our architectural principles, motivate five related novel capabilities of the business interface:

- automated service offer dissemination — providers should be able to disseminate electronic service offers to customers (possibly not end users);

- automated policy-based offer selection — customers should be able to make intelligent and complex decisions in an automated way;

- *ad hoc* business relationships — customers should be able to dynamically establish *ad hoc* business relationships with service providers on a per-session basis and independently from any existing inter-provider agreement;

- offer-driven reconfigurable metering systems — providers should be able to dynamically reconfigure their metering capabilities on visited networks' systems without having adjacent providers to necessarily support their charging capabilities;

- managing multiple business relationships — users should be able to hold, at the same time, more than one business relationship with different providers, without experiencing any switching barriers.

12.4.3.1   Automated Offer Dissemination

Access providers must inform customers of the services they offer so that potential buyers may purchase them. Currently, this takes place in many forms, but none of these allows for an automated dissemination of electronic service offers to customers (both end users and brokers).

Customers and providers should be able to exchange information about users' service requirements, available access services, and associated offers. They can do so in three different ways:

- providers push their service offers to end customers (push model);

- providers push their service offers to the edge of the network — then, customers get only those in which they are interested (hybrid model);

- customers contact providers in order to get the appropriate service offer (pull model).

The choice to design a dissemination infrastructure in one model rather than another depends on a number of factors:

- potential customer — the problem of disseminating offers changes depending on whether the customer is the end user or the broker, e.g. end users can communicate via wireless, whereas brokers would not;

- type of offers — when wireless users are the potential customers, wireless access providers occupy a privileged position (compared to fixed and Internet access providers), in that they are at the edge of the network and it is relatively easy to reach them;

- communication layer — an offer dissemination protocol can be designed at three different communication layers of the Internet protocol stack: at the network interface layer (where they would be conveyed as system information in case of wireless access), at the IP layer (if they are conveyed in packets), and at the application layer.

Our work extends the tariff distribution protocol [4] by investigating the technical issues related to the dissemination of offers in heterogeneous wireless environments.

12.4.3.2   Automated Policy-Based Offer Selection

Competition is the principle upon which every market is based and consists of that never-ending tension between supply and demand. Users' decisions are driven by the way services are offered, and reciprocally, the way services are offered is driven by users' needs and preferences.

When purchasing a service, users tend to make a careful, although manual, evaluation of a number of alternatives. How much freedom do providers have when formulating their offers? And then what are the means by which they can better match users' needs?

The current paradigm sees providers making the effort of matching the price of a service with the subjective value that users give to it (i.e. their overall willingness to pay). The better they are able to match users' needs and preferences, the more they are competitive. However, this approach causes a number of disadvantages to both users and providers. Users apply a decision mechanism that is essentially manual — they make decisions purely on the basis of their limited understanding of the offers and their limited capability of making proper matches in a multi-dimensional space. Providers, because of users' manual decisions, limit their competitiveness, in that they are not free to be innovative with their offers, seeking a better match with users' willingness to pay. The limited human understanding prevents providers from offering more complex offers to users, which would provide a better match, but would not be useful for the improvement of user's decisions. In conclusion, the *status quo* stops providers from innovating their offers.

These considerations lead us to propose a new approach that removes from providers the effort of matching service offers with the user's subjective service value, and which instead makes customers (end users, brokers, and possibly other providers) able to make automated decisions. The capability would provide customers with the automated selection of access service offers, based on a per-session decision mechanism selecting the provider whose offer minimises the difference between the service offer and user's subjective value of the service. Customers could make complex decisions and therefore providers would produce complex new offers, seeking the best match with a user's needs.

We advocate a selection mechanism provided through an intelligent software agent able to make decisions on behalf of the customer, or possibly with minimal initial user involvement. An agent can be seen as an encapsulated computer system, situated in some environment and capable of flexible, autonomous action in that environment in order to meet its design objectives [5]. Our agent assesses providers' offers against the user's needs, and preferences, expressed through a generic buying policy. It then selects the best service offer from the available alternatives.

The agent takes two logical inputs (the current set of available offers, and the user preferences), and has one single output — the best offer. Two main modules accomplish the selection task — the decision analyst elaborates the inputs and produces a set of complete user preferences (in the form of a policy); the decision maker uses these to apply the selection algorithm.

Future communications networks will be multiservice, in that some pieces of information will be given preferential treatment over others (i.e. quality of service). This creates a need for constant pricing innovation, because there will be an incentive to use pricing to regulate network usage and the conception of novel and more complex offers will be required to apply new pricing strategies. Therefore, the real effectiveness of the selection capability will strongly depend on two challenges:

- how good providers will be at engineering novel offers to better match users' needs;

- how good the research will be at formalising users' needs and per-task preferences, in the form of a generic buying policy.

The Problem of the Selecting End

A telecommunications service involves at least two parties communicating. It is interesting to explore the factors that determine the ends of the communication involved in the offer selection.

We suggested an automated selection mechanism, supported by user buying policies. Its task would be to select the provider whose offer minimises the difference between the service offer and the user's subjective service value. This implies a strong assumption: only one of the parties requesting the service is paying for it.

Because a communication involves at least two ends, we investigate whether a different re-apportionment of the service costs [6] between the involved parties might influence the choice for an offer, rather than for another. One approach could consist in allowing each party to compare the service offer with the overall value that all the parties assign to the service (i.e. the overall willingness to pay, expressed by all the ends).

The challenge is the computation of the overall value of a communication as the addition of the single values that each end assigned. The suggested decision mechanism would then be distributed between all the ends of the communication, introducing a further degree of flexibility into the system — providers would be enabled to seek the best match innovating their offer in a sense that takes into account different possible re-apportionments of the costs, by, for example, encouraging some of the ends to pay for the service, and not others.

It should be noted that, in order to find out whether a different apportionment of the costs is feasible, the automated agent needs to know the buying policies of all the involved parties. Each involved party would receive the offers of those providers in their coverage area; each automated agent will select, with respect to everyone's buying policy, the best offer on behalf of its requesting party, by matching each offer with the amalgamation of the subjective values of the services expressed in each party's policy.

An alternative to multiple users' selection is to let an intermediary broker deal with the selection, and thus deal with the specific re-apportionment of the costs at a higher-level.

The Broader Context of Automated Offer Negotiation

In a broader context, the problem of automated offer selection represents the counterpart of the problem of automated service offer generation. Selection and generation of service offers constitute the general problem of automated offer negotiation.

Automated negotiation is the process through which a group of parties come to a mutually acceptable agreement on some matter [5]. Every negotiation is regulated by its own rules, to be agreed by the involved parties, determining the actions that can be undertaken and those that cannot. Automated negotiation is achieved through the dynamic interaction of two or more automated intelligent agents.

Given its inherent importance in many contexts, negotiation theory makes use of many different approaches (ranging from artificial intelligence, social psychology, to game theory). However, despite their variety, automated negotiation research can be considered to deal with three broad topics. The negotiation protocol is the set of rules that govern the interaction. This covers the types of participants, the negotiation states, the events that cause negotiation states to change, and the valid actions of participants in particular states. The negotiation object represents the range of issues over which agreement must be reached. At one extreme the object may contain a single issue (e.g. a price), while on the other hand, it may cover hundreds of issues. Orthogonal to the agreement structure, and determined by the negotiation protocol, is the issue of the types of operation that can be performed on agreements.

12.4.3.3   Establishing Ad Hoc Business Relationships

The current model for public mobile access makes roaming agreements the obligatory way of providing mobility to end customers outside the home network coverage.

We suggest an approach where customers can dynamically establish and manage per-session business relationships with providers, without experiencing any barrier due to pre-existing agreements between providers.

We assert that a business relationship has five dimensions. To establish each one of them, it is necessary to support a number of business functions. Mobile users should be able to establish each dimension with an appropriate party they select. It appears necessary then to re-think the design of current business functions in order to support these capabilities (see Fig 12.7).

In this section we briefly outline the five dimensions of a business relationship and the business functions we believe are the basic components supporting its establishment.

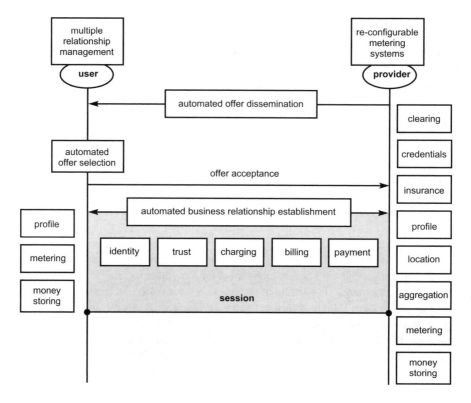

Fig 12.7   Establishment of *ad hoc* relationships.

Business Relationships

A business relationship consists of (sometimes explicit) statements specifying a set of expectations and a set of obligations. Control is required in establishment, release, and possibly transfer. We assume that the establishment of a business relationship is always necessary to purchase a service.

We have identified the following five dimensions of a business relationship.

- Identity

  An identity relationship is achieved through the mutual exchange — by the ends of the relationship — of information about one's identity. The establishment of identity relationships is naturally associated with user and provider authentication mechanisms.

- Trust

  A trust relationship consists of gaining the certainty that some party, owning a share of control, will not attempt to defraud the other ends of the relationship (e.g. by abusing control). The ends of the relationship ascertain the credentials of each other.

Trust relationships may be based on existing identity relationships. The establishment of a trust relationship is naturally associated with user and supplier authorisation mechanisms.

- Charging

  A charging relationship identifies the service provided to the user, along with the technology used to deliver it. It identifies the offer associated with the service and associates accounting and metering functions to be applied.

- Billing

  A billing relationship identifies billing and billed parties, along with the details regarding the billing model (e.g. the billing period) and the data format which will be used for the call data records.

- Payment

  A payment relationship is an agreement between the receiver of the payment and the party obliged to make the payment. It refers to the transfer of funds between these two parties. It specifies the form of the payment and payment instructions for any financial institutions involved.

Business Functions

Business functions allow providers to run the business aspects of the service delivery and support the establishment and management of business relationships. We have identified the following ones.

- Credentials provision

  Credentials provision consists of providing (typically selling) to users credentials in exchange for some form of guarantee (e.g. payment). Users spend these credentials (i.e. send these credentials to providers for verification) to purchase services. This function typically supports the establishment of trust relationships between users and providers.

- Money storing and management

  Money storing and management consists of storing the money paid by customers to get valid credentials to purchase services.

- Profile storing and management

  Profile storing and management consists of maintaining, in appropriate formats, user personal information, buying, security, and privacy preferences, and possibly credentials.

- Insurance

  Insurance functions ensure that, if some party does not behave properly (e.g. abuse control or try to defraud), the implications (e.g. money not paid) of such actions will be appropriately covered.

- Metering

  Metering refers to the measurement of resource usage by the users. The service offer will determine the type of metering system that is used.

- Accounting

  Accounting consists of collecting the information that describes a user's service utilisation and mapping a particular resource usage into technical values. The network then maintains this set of technical parameters in the so-called accounting record.

- Clearing

  Clearing consists of collecting the bills, produced by providers, that a user receives over a certain period of time, settling them, and then providing to the customer a single bill for final payment. The purpose is not only to make transparent to users different types of payments, but also to decrease the number of payment transactions to which customers would be subject.

- Aggregation

  Aggregation consists of a clearing function for inter-provider settlements, comprising three main processes — collecting billing information, delivering billing summaries to providers, and calculating the inter-provider payment obligations.

12.4.3.4 Reconfigurable Metering Systems

The convergence of Internet and mobile worlds is a process that needs to be supported by a widespread standardisation process. In particular, in terms of pricing infrastructures, these two worlds show a substantial difference — whereas the Internet is at the early stages of its path to build its pricing infrastructure, the mobile world already has a solid and deep-rooted one.

Unfortunately, a conflict arises when the principles of mobility are applied to Internet services and, in particular, when these two different pricing infrastructures come into contact. In accordance with the original Internet design principles — keeping the network as general-purpose as possible [7, 8] — the paradigm suggested for its pricing infrastructure provides for a scheme where prices can be determined, and charges accessed, locally to the edge of the provider's network where the user's packet enters, rather than computed in a distributed fashion along the entire path. This approach is called edge pricing [9] and its major advantage is that no uniform pricing standards need to be developed, since interconnection involves bilateral agreements that allow each provider to use their own pricing policy, to experiment with new ones, and very likely to exploit them as a competitive advantage. In brief, it allows providers to innovate their pricing schemes.

Extending the original Internet service with user mobility requires providers to support an adequate pricing model, through which they can supply the service to their customers and charge them for it. Unfortunately, the application of the current pricing model of the mobile world breaks the Internet model. In fact, the mobile pricing infrastructure provides for a mechanism — roaming agreements — supported by a paradigm that is incompatible with edge pricing.

The establishment of roaming agreements involves home providers implementing pricing schemes (i.e. charging capabilities) that are supported by visited providers' metering capabilities, and vice versa, which constrains the possible set of offers. This forced compatibility prevents the home provider from innovating its pricing scheme and its offers. It forces the provider to choose between either not making the new offers to roaming customers, or submitting the required metering capability for standardisation across all its competitors, destroying any advantage of the innovation. This happens because the majority of providers prefer to impose high switching barriers to customers as a self-protection mechanism. This prevents providers from innovating their own pricing schemes.

We conceived an innovation removing this limitation. It consists of the capability for providers to remotely re-configure visited providers' metering systems as novel offers are applied. When a novel pricing scheme is applied to roaming customers, they can re-configure the remote meter by sending a translated version of the new offer. This translation describes the low-level configuration information of the meter, necessary to support the new pricing scheme.

12.4.3.5   Multiple Business Relationship Management

Users wishing to express their preference for a service provider are limited in the current mobile access market. Once they establish a business relationship, they in fact start experiencing barriers in switching to an alternative provider. For example, in the case of cellular communications, providers will program the mobile device so that it only works with their networks, using subscriber identity module (SIM) card locking. Physical measures to stop users switching provider are currently applied to heavily subsidised pay-as-you-go devices; and new soft measures are being applied to annual contract devices, such as loyalty discounts. These soft measures allow the user to change access provider by inserting the SIM card of the provider they wish to use, but they require users to carry several cards and have long-term business relationships with the access providers they wish to use.

Each of the user relationships involves storing the details and profile information about the user at the provider end. If the user holds multiple accounts then this duplicates the information across multiple providers, and more critically, involves the user having to establish multiple one-time long-term business relationships. As a result, we explore a more efficient mechanism, where the user establishes *ad hoc* business relationships in an automated way and dynamically maintains several relationships at the same time.

The new mechanisms need to be user-centred, as the mobile device may contain private information about the user, which will be used to automatically join a new provider. This will require measures to build user confidence that devices will only distribute private information to the access provider they wish to use. These measures will require security mechanisms; this may also include third party insurance, which limits user's liability in the event of fraud.

12.5   Summary

This chapter argues that the original design of the business interface of the mobile market has not considered the unpredictability of the outcomes of economic tussles — the mobile industry encoded in its designs the result of such contentions, preventing further variation. Because the market currently does not capture the business opportunities created by the combined deployment of heterogeneous access networks, the business interface's designs should now be modified to accommodate the convergence of these access systems and allow for evolution.

We have formulated a set of design principles, whose application led us to identify specific architectural limitations of the current business interface. On the basis of this study, we concluded that the future business interface should allow:

- providers to disseminate their offers to customers (possibly not end users) in an automated way;

- customers to select the offers in an automated and intelligent way;

- customers then to dynamically establish *ad hoc* business relationships with providers;

- end users to be able to hold, at the same time, more than one business relationship with different providers, without experiencing any switching barriers;

- providers to be able to dynamically reconfigure their metering capabilities on visited networks' systems without having adjacent providers to support their charging capabilities.

Although the mobile industry is slowly becoming more conscious of the limitations of the current business interface, there is still little evidence of the development of capabilities allowing more flexible and open interfaces. We believe that the capabilities we suggest provide a more general model enabling the mobile industry to resynthesise different business interfaces and therefore to cope with the dynamic variations of the outcome of economic contentions.

References

1   Mohr, W. and Konhauser, W.: '*Access network evolution beyond third generation mobile communications*', IEEE Comms Mag (December 2000).

2   Clark, D. D., Sollins, K. R., Wroclawski, J. and Braden, R.: '*Tussles in cyberspace: defining tomorrow's Internet*', SIGCOMM '02 (August 2002).

3   Clark, D. D. and Wroclawski, J. T.: '*The Personal Router*', White paper, MIT Laboratory for Computer Science (March 2000).

4   Heckmann, O., Darlagiannis, V., Karsten, M., Steinmetz, R. and Briscoe, B.: '*Tariff Distribution Protocol (TDP)*', IETF, Internet Draft, draft-heckmann-tdp-oo.txt (March 2002).

5   Jennings, N. R., Faratin, P., Lomuscio, A. R., Parsons, S., Sierra, C. and Wooldridge, M.: '*Automated negotiation: prospects, methods and challenges*', Group Decision and Negotiation, **10**(2), pp 199-215 (2001).

6   Clark, D. D.: '*Combining sender and receiver payments in the Internet*', MIT Laboratory for Computer Science, Version 3.1.2, Presented at Telecommunications Research Policy Conference (October 1996).

7   Saltzer, J. H., Reed, D. P. and Clark, D. D.: '*End-to-end arguments in system design*', ACM Trans on Computer Systems, **2**(4), pp 195-206 (1984).

8   Clark, D. D. and Blumenthal, M. S.: '*Rethinking the design of the Internet: the end-to-end arguments versus the brave new world*', ACM Trans on Internet Technology, **1**(1), pp 70-109 (August 1996).

9   Shenker, S., Clark, D., Estrin, D. and Herzog, S.: '*Pricing in computer networks: reshaping the research agenda*', ACM Computer Review, **26**, pp 19-43 (April 1996).

13

ENABLING APPLICATIONS DEPLOYMENT ON MOBILE NETWORKS

M J Yates

13.1 What is a Network API?

The term 'network API' has a variety of interpretations. The focus of this chapter will be on interfaces designed for third party application developers to use capabilities in public scale networks. This is the commonly accepted term in the Parlay Group [1] or Open Service Architecture [2] (OSA) community. In this context a network API is a specification of a programmatic interface that can, when implemented, be used to access various network capabilities or data. The emphasis is on a specification that a programmer can use and understand. In the simplest case this might mean a Java or C++ language specification. However, abstract interface descriptions such as unified modelling language (UML) and interface definition language (IDL) are also used for the specification because they are independent from specific programming languages. The trade-off is that the interface description must be translated into each required programming language. Additionally, because network APIs are usually presumed to be distributed, the on-the-wire protocols must be treated in a consistent way between clients and servers if interoperability is to be achieved using the API specification. Example distribution protocols are Internet interoperability protocol [3] (IIOP) or simple object access protocol [4] (SOAP).

A network API describes a set of methods or procedures that can be called to perform a network action or retrieve network data. Typical examples from the OSA/ Parlay 4.0 specification are to manipulate call-legs of a circuit-switched call or to return the network determined status of a terminal. The term network API is not used here to describe APIs relating to bit transport, like Internet protocol (IP) sockets.

13.2   Why are Network APIs of Interest?

Clearly, systems such as intelligent networks (INs) already employ APIs for constructing services on a service control point (SCP). However, these APIs are vendor specific, even though their implementations ultimately rely on the capabilities of international standards such as IN application protocol (INAP), and transactional capabilities application part (TCAP). From the early 1990s onwards there was increasing realisation that the telecommunications industry was potentially stifling the creation of new content and telecommunications services that were integrated with enterprise IT systems. This was because the APIs in public networks were effectively inaccessible even if they were present to open standard — open in this context meaning a publicly visible specification rather than an accessible implementation on a network. World-wide collaborations such as TINA [5] and DAVIC [6] laid the foundations for a revision in thinking which brought operators to an acknowledgement that opening network functionality might create a growth in revenue-generating applications and network traffic, in a similar way to the growth of Internet servers.

There are now several trends that are encouraging fixed and mobile operators to start experimenting with open standard and accessible network APIs. A key motivation is the market saturation and projected revenue decline from voice-only services on fixed and mobile networks. With mobile penetration exceeding 70% in most European countries, and huge investment in 3G assets there is a compelling need to increase average revenue per user (ARPU). This must come from new, compelling voice and data services.

For mobile networks in Western Europe, data revenues are expected to grow to around $12-18bn by 2005, representing up to 20% of total revenue. These estimates are very sensitive to the commercial models the industry adopts to encourage third party providers to supply content and applications that attract users to increase their spend. It is on this latter point that the value of network APIs is seen, by creating an environment where:

- novel applications can be created by third parties and easily deployed inside the operator's network using standard APIs;

- operators can make those open standard APIs commercially accessible so that third party providers can operate their own applications.

Although the vision of a vibrant third party applications market is enticing, either on broadband or mobile networks, there are still commercial uncertainties that are setting the pace:

- robust implementations of the network APIs must be commercially available;

- skills must be available to use the APIs and integrate implementations with network elements and operational support systems (OSS);

- operators are wary of opening APIs that merely result in existing service propositions having duplicated/replaced implementations without commensurate reduction in the cost base or business agility;

- business cases must be carefully constructed to ensure that predictions of new revenues are genuine and not the result of new product lines simply substituting existing revenues without market growth or other benefits;

- uncertainty in the stance of national regulators to the access and pricing of network APIs, in what is currently an emerging and therefore risky market; operators want to be confident that the standards, implementations, and benefits will allow regulatory obligations to be met;

- third party application providers must have quantifiable commercial incentives, such as new or satisfied customers and shared content or traffic revenues; the latter has already been seen in the mobile short messaging service (SMS) and premium rate service market.

The operator's conundrum is that it is much easier to quantify the adverse impact on existing business from deploying immature network APIs than believe in cited speculative benefits, such as overall revenue growth. This is an innovator's dilemma.

13.3   Forums Specifying Network APIs/Services

There are a number of international, industrially and governmentally supported, organisations that have been active in specifying network APIs, particularly in the last ten years. Precursors to the currently active groups were collaborative projects such as TINA [5] and DAVIC [6], which have been influential in thought leadership if not specification.

Currently the largest groups are the Parlay Group [1], ETSI-OSA [2], 3GPP [7], IETF [8], and the OMA [9]. At the time of writing, the OMA has subsumed more specialised groups such as the Location Information Forum. This chapter will not consider in more detail forums such as W3C [4] and OMG [3] which have produced technology standards on which network API functionality can be deployed, for example SOAP Web Services and IIOP respectively. JAIN is a technology-specific group which is focusing on taking specifications from other forums and establishing interoperable Java language specifications.

It will be seen that these organisations are focusing on interfaces that access either data (e.g. terminal location or state) or transport functionality (e.g. call routing or messaging).

Heterogeneous networks have a variety of network elements from which information or capabilities can be derived. For example, call-routing functions might be offered via an IN-SCP on the switched telephony network and a session initiation protocol (SIP) server on an IP network. Herein lies the first of many compromises for API designers:

'Should an API specification that is agnostic to specific network technologies reflect the union or intersection of functionality found in the underlying networks?'

In reality there are no absolute rules used.

The aforementioned organisations are not entirely independent. The industry groups with open membership have no formal role in establishing standards, whereas ETSI [2] and ITU-T [10] have this specific objective. Organisations such as Parlay Group have therefore sought endorsement through submission to ETSI via the member companies (for more detailed information on Parlay, see Chapter 14 and Moerdijk and Klostermann [11]).

Table 13.1 explains some of the relationships. The list of inputs and sources is not exhaustive nor does it cover the entirety of each group's activities.

Table 13.1   Organisations and their responsibilities.

Industry Group	Comments
ETSI-OSA	Requirements and specifications into ETSI are submitted by members directly. Where relevant, these have been compiled in other groups such as 3GPP and Parlay, ETSI-SPAN being equivalent to Parlay 4.0 specifications. Independent publication.
3GPP	Requirements and specifications submitted into the ETSI-OSA process. 3GPP publications are taken from the ETSI/Parlay set. 3GPP releases have so far been a subset of these specifications (no additional functionality).
Parlay Group	Parlay specifications 1.0, 2.0 published independently. Parlay 3.0 and Parlay 4.0 submitted to ETSI via member companies and published by ETSI. Parlay X 1.0 Web Services published independently [11] (see also Chapter 14).
LIF [12]	In process of subsuming into OMA. Existing specifications remain.
OMA	Assumed responsibility for LIF and WAP. Currently developing new specifications.
PayCircle [13]	Input from Parlay 3.0 Content Based Charging. Independent publication. Expect convergence and agreement of APIs between Parlay and PayCircle.
JAIN [14]	Input from Parlay specifications used to produce the Java language mapping rules of Parlay 3.0.

The functional segmentation used by the ETSI SPAN/ Parlay 4.0 specifications group is described in Table 13.2.

Several of the functional areas, such as call control, are subdivided further to separate simple call control from complex multiparty conference capability with independent call-leg control.

Table 13.3 indicates the areas of functionality covered by the different organisations, and the typical network elements used to support the APIs.

Another interesting perspective on Table 13.3 is to look at the specification technologies that the different groups have adopted, shown in Table 13.4.

Table 13.2   ETSI SPAN/Parlay functional segmentation.

Functionality	Description
Call control	Real-time call leg routing for simple or multiparty calls
User interaction	Interactive voice prompt and response functionality
Mobility	User location determination
Terminal capabilities	Determination of terminal characteristics and state
Messaging	Messaging over a variety of media
Presence	Determination of disposition of a user as represented by agent
Accounting	Account transaction, balance and history determination
Content charging	Debiting and crediting accounts based on content
Data session	Create and control a data communications session
Connectivity Management	Establishment and control of QoS-based pipes between end-points on a provider's network
Policy management	General policy control functionality applicable to a variety of other services
Framework	Functions to authenticate, discover and request services

Table 13.3   Functionality covered by industry bodies.

Industry group	Functional area	Underlying network elements
OSA/ Parlay	Framework	None — used to find and access interfaces below
	Call-control	SCP/switch SIP/media gateway
	User interaction	SCP/switch Voice-XML server
	Terminal	Home location register (HLR) Visitor location register (VLR) SIP server
	Data session	Routers, circuit switches
	Messaging	SMS Centre (SMSC) Multimedia messaging service centre (MMSC) Instant messaging service centre (IMSC)
	Connectivity	Routers, switches
	Account management	OSS Business support system (BSS)
	Content-based charging	OSS/BSS
	Policy	OSS/BSS
	Presence and moblity	HLR/VLR presence server
LIF	Location information	HLR/VLR gateway location centre (GLC)
PayCircle	Payment and charging	OSS/BSS
W3C	Communication session initiation	SCP/switch or SIP/Media Gateway
JAIN	Various	All elements relating to the Java specification areas covered
SMPP Developers Forum	Messaging	Short message service centre (SMSC)

Table 13.4   Specification technologies.

Industry group	Specification technology
ETSI-OSA/3GPP	Normative UML OMG-IDL mapping Informative Web service description language (WSDL)
Parlay Group	UML WSDL
LIF	HTTP/XML
PayCircle	WSDL
W3C	XML/IETF text transport
JAIN	Java
SMPP Developers Forum	Byte level datagram

13.4   Types of Application

The range of applications exploiting atomic network API capabilities is only limited by the inventiveness of application developers. Applications in the mobile market have been well characterised, a typical segmentation being:

- voice telephony;
- video telephony;
- messaging;
- information browsing;
- commerce transactions;
- info-tainment;
- networking.

From the perspective of the network designer, the features that support the delivery of these services are a more relevant distinction. For example, both info-tainment and 'intelligent' real-time communications applications may make use of presence information — data about the called party such as:

'What is his location?'

'Is her mobile active?'

'What is his preferred call routing option at this specific time?'

'Is the user authenticated?'

Access to these types of function is needed for applications in a breadth of market segments. Presence applications may be built for consumers, communities of interest (e.g. supporters clubs) or for the enterprise employees to improve organisational efficiency or agility.

13.5　　Network APIs Deployment

In commercially viable deployments there is a much wider set of requirements that extend beyond the functionality embodied in an API.

Figure 13.1 shows a schematic deployment of network APIs in a tiered architecture. The minimal basic requirements for the API-surround are the account management of the applications that are accessing the APIs. At a basic level these management functions include authentication, authorisation, and accounting that in themselves imply customer relationship management (CRM) provisioning and accounting systems.

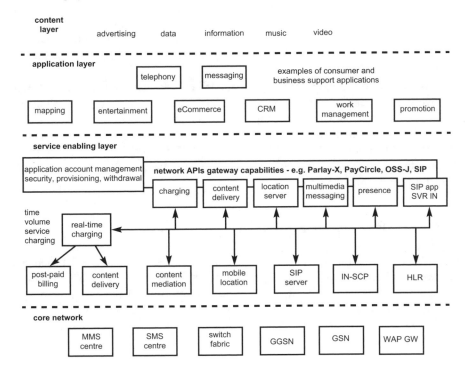

Fig 13.1　　Schematic of a tiered API architecture.

In addition to provisioning at the API-access level, there is the issue of fulfilment of API services that have specific charging schemes within the mobile network. An example is premium rate and receiver-pays messaging to consumers for info-tainment applications. Here, the action of an application sending such a message requires correlation of events including invocation of the API, use of the messaging sub-system, interaction with a post or pre-pay billing system, and network delivery confirmation. Similarly, use of presence-related capabilities by an application provider may incur charges that are offset against call or traffic charges paid by the end-user. In all these cases the application provider may be receiving a percentage of the traffic revenue collected from end-user charges and so the successful execution of an API request must be part of an overall revenue settlement system.

Network API capabilities can be exposed to fraudulent exploits already known from the IN world. For presence and location services there are legal data protection and privacy contracts to consider. For these reasons the API surround must be of commercial strength.

Part of the API surround is satisfied by the framework functions in Parlay/OSA that specify methods to enable an application to authenticate, use an API service directory, request an API, sign for its use, and monitor the status of the API gateway. From the gateway's perspective the framework helps achieve version control of interfaces and selective exposure of unique interface references. However, these latter capabilities can also be achieved with the other industry standard mechanisms such as the universal description, discovery and integration (UDDI) for Web Services.

It is also worth noting that, in all but trivial services, the use of run-time discovery of interfaces using semantic search criteria, followed by dynamic invocation is hard or infeasible for capabilities that incur charges or have non-trivial service-level agreements (SLAs). In such cases, signed contracts and understanding of terms and conditions may need to be in place before use of the API capabilities. At present, automation of this in conjunction with network API technology is too immature for commercial use.

13.6   Developer Forums

From the viewpoint of a developer creating an application for the provider there are various stages leading to the final operation of an application. These are illustrated in Fig 13.2, where it can be seen that a good proportion of the activities do not directly involve use of the network APIs. These activities may represent a significant fraction of the total deployment and maintenance cost, incurred by the time and skills of people.

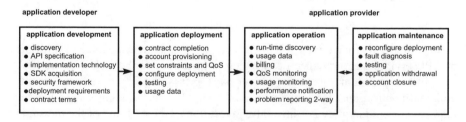

Fig 13.2   Application life cycle from a developer's perspective.

It should also be noted from Fig 13.2 that the people or enterprise involved at the development deployment phases of an application are not necessarily involved in the operation or maintenance. A typical example is where an independent software house constructs an application that is sold into several enterprises that will operate or maintain this application through their own in-house IT team.

What is also clear is that systems and processes for supporting a small number of fairly trusted, high-revenue applications are unlikely to be scalable for the support of large numbers of smaller-revenue, untrusted applications. The cost of engagement of these developers must constrain costs to within the price of opportunity for the application. For example a short-lived application to support a topical event or passing fad has a limited revenue opportunity, development and operation time.

This has an implication for the way that a network operator markets the overall business proposition built around the network API capabilities.

To attract custom the operator might promote its choice of interfaces, and provide developers with help and test facilities; on the other hand the operator might seek market pull from service providers contracted to features like billing, usage statistics, and API resilience.

Another alternative, seen in the area of SMS, is where the network operators distribute their capabilities via third party aggregators. In these cases the developer can also benefit from the aggregator's interconnect to multiple network operators. This is useful in cases where the application needs to access a variety of networks, and for which there is no operator interconnect. An example of this is mobile location where look-ups may be limited to the operator holding the subscriber record and MSISDN.

13.7 Platform Resilience

The ambition to support large numbers of untrusted applications introduces serious technical challenges for the API gateway technology and supporting frameworks.

Figure 13.3 illustrates some aspects of the in-depth security required to provide API capabilities to client applications that do not have assured behaviour.

There are benefits to both the network operator and application provider for having policing in-depth at all tiers in the network API infrastructure. Besides the prevention of malicious attacks, other motivations are:

- for the application provider:

  — minimising testing and validation cost and time;

  — preventing one application accidentally impairing another application's ability to access network APIs;

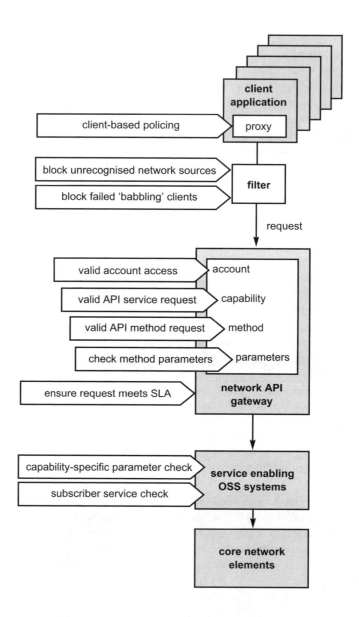

Fig 13.3 Schematic network API deployment.

— protecting the application provider from costs generated by their own application that over zealously consumes chargeable services;

— graceful or informative service refusal;

- for the network operator:
  - — ensuring that all invocations are associated with a valid account and commercial agreement;
  - — ensuring that an application is constrained to its service-level agreement;
  - — ensuring underlying network resources are not overloaded by too many requests, or unnecessarily loaded by invalid requests;
  - — verifying that subscribers are able to participate in the request (e.g. a location look-up charged to a pre-pay account).

The schematic in Fig 13.3 shows there are several stages to satisfying the policing needs of the operator and application provider. One of the objectives of the operator is efficient use of system resources. For several classes of network API the underlying core network resources have a restricted throughput that could be exceeded by aggregated use across a large number of applications, for example call servers or location centres. In these cases the policing is aimed at ensuring the resource is not troubled with requests it must reject or that statistical bursts by applications do not cause overloading.

It is worth noting that there are solutions that devolve the burden of some of the policing into the application domain. This is achieved through the use of managed proxies that reside alongside the application and are incorporated into the deployment by the application developer. This obviously has an impact on the developer's environment and may also, therefore, need consideration from a regulatory context. Nonetheless, this has an obvious advantage in that the processing demands of policing are not entirely concentrated on the interface gateway.

13.8   Summary

This chapter has given an overview of current network API initiatives and technologies. It aims to show that alongside the current generation of resilient network API gateways, there are important business processes that must be supported if the application market is to truly take off. The key emphasis should be on providing application developers and application providers with the tools and facilities needed to drive down and constrain cost and skills in the process of delivering applications to benefit users.

References

1   Parlay Group — http://www.parlay.org/

2   European Telecommunications Standards Institute — http://www.etsi.org/

3   Object Management Group — http://www.omg.org/

4   World Wide Web Consortium — http://www.w3c.org/

5   Ionue, Y., Lapierre, M. and Mosotto, C.: '*The TINA Book: a co-operative solution for a competitive world*', Prentice Hall (1999).

6   Digital Audio Visual Council (DAVIC) — http://www.davic.org/

7   3rd Generation Partnership Project (3GPP) — http://www.3gpp.org/

8   IETF — http://www.ietf.org/

9   Open Mobile Alliance — http://www.openmobilealliance.org/

10 ITU-T — http://www.itu.int/ITU-T/

11 Moerdijk, A-J. and Klostermann, L.: '*Opening the networks with Parlay/OSA APIs: standards and aspects behind the APIs*', IEEE Network Magazine, pp 58-64 (May/June 2003).

12 Location Interoperability Forum (LIF) — http://www.openmobile alliance.org/lif/

13 PayCircle — http://www.paycircle.org/

14 JAIN — http://java.sun.com/

14

THE PARLAY API — ALLOWING THIRD PARTY APPLICATION PROVIDERS SAFE AND SECURE ACCESS TO NETWORK CAPABILITIES

R Stretch

14.1 The Historic Lead-Up to Parlay

14.1.1 Next Generation Switch (NGS)

In 1996 BT and MCI embarked upon a joint initiative to design a new switching technique that would be able to integrate both narrow and broadband network topologies. This joint initiative was named 'The Next Generation Switch'. The concept was to produce a network switch that would be modularised, each module performing a different function, as shown in Fig 14.1.

A number of switches are interconnected across a network, with each switch represented by a vertical row of functions, and performing a different task. The terminology used in this chapter is as follows:

- service control (SC) — supplementary services;
- call control (CC) — basic call half;
- bearer control (BC) — bearer and signal services;
- resource proxy (RP) — a 'wrapper' for the resource complex.

This creates three levels of API definition:

- service control — call control;
- call control — bearer control;
- bearer control — resource proxy.

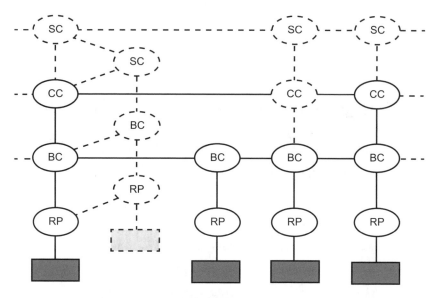

Fig 14.1   NGS architecture.

The meetings comprised representation from BT, MCI and a number of well-known equipment vendors from both the USA and Europe. The idea was to define each module in turn and then define the application programmable interfaces (APIs) that would provide the information flow between the modules. The success of the project foundered on the simplistic view that it would have been possible to build switches using modules from different vendors, with a guarantee that the 'plug and socket' method (API) would provide problem-free working.

There is no need to elaborate upon the reasons why this project failed — suffice it to say that all was not lost. BT salvaged the small amount of work that had been done on the call control to service control API and later utilised the concepts in the early Parlay initiatives.

Before moving on to the development of Parlay, we need to understand some work that was being undertaken (partly in parallel) within the Telecommunication Information Networking Architecture Consortium (TINA-C) [1], the concepts of which would also be utilised in the definition of the Parlay architecture and API.

14.1.2   TINA-C

The TINA consortium was set up to assess, within the telecommunications community, the common need for improving the way services are designed and the common opportunity for providing tomorrow's service offerings, based on increasing customer demands. It was discovered that similar studies on software architectures were being conducted in many parts of the world. A core team, consisting of engineers from the consortium's member companies, was set up in New Jersey, USA, to co-ordinate the efforts towards achieving this goal through the definition of a common architecture.

TINA-C was the 'Flag Ship' of distributed architecture design. Their work spanned a number of years and resulted in a lot of interesting development built on a solidly designed architecture. For an initiative such as TINA to succeed the results needed to be fed into open standards forums. Unfortunately this did not happen early enough, the result being that valuable work lies dormant in the archives of technology. The architecture designed by TINA is shown in Fig 14.2.

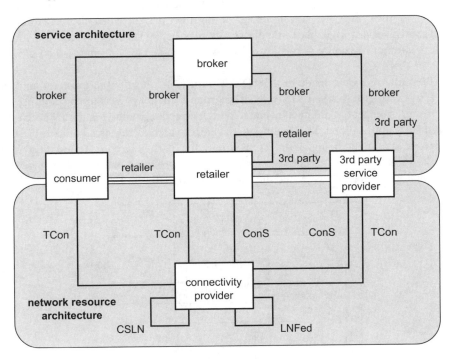

Fig 14.2   The TINA reference point architecture.

As can be seen from Fig 14.2, services are 'triggered' (launched) on the request of consumers who have first established an access session with a retailer. The retailer allows the consumer to choose from the services to which they have subscribed (explicitly or implicitly). Depending upon the purpose of the service, a communication session is established between the consumer and another consumer, or between the consumer and the service provider itself for actual service invocation. In turn, a communication session will be established through the services of the connectivity provider, who manages the end-to-end connections.

The interfaces shown here result in relationships between a number of actors (or 'stakeholders', as TINA puts it) in the model:

- one stakeholder playing a consumer role and one stakeholder playing a retailer role;
- one stakeholder playing a retailer role and one stakeholder playing a service provider role;
- one stakeholder playing a retailer or service provider role and one stakeholder playing a connectivity provider role.

At this point, it is worth mentioning some relevant work that was being undertaken within international standards bodies.

BT, having already started work on the Parlay initiative (see section 14.2), was trying to encourage the intelligent network (IN) standards groups in the ITU-T and ETSI, to look into the development of an interface between the SCF and a distributed service logic platform, classically outside the network domain. BT met a fair amount of resistance here and so for a while decided to concentrate their efforts within TINA-C.

Presentations were made to the TINA forum which resulted in the creation of a new working group called TINA-IN. The group's challenge was to consider how a TINA environment could be attached to an IN, effectively allowing the IN access to third party applications. The aim was to develop an architecture that could utilise the early development being undertaken within the Parlay Group. To do this they realised that some sort of gateway access must be provided that would manage the interaction between the SCF and the TINA platform (Fig 14.3).

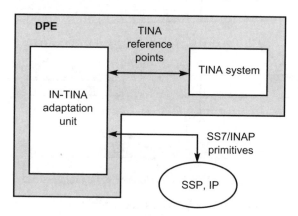

Fig 14.3   IN-TINA adaptation unit.

They called this gateway the IN-TINA adaptation unit, the fundamental purpose of which was to provide some conversion between INAP (the protocol understood by the IN SCF) and those protocols used by the TINA system.

The IN-TINA adaptation unit was made up of two separate gateways, as shown in Fig 14.4.

The first part would provide conversion between SS7/INAP primitives and CORBA, with the second gateway translating between CORBA and the TINA

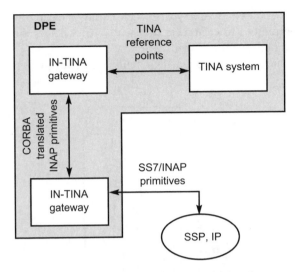

Fig 14.4   Expansion of the adaptation unit.

reference point architecture. The two gateways would effectively be physically separated, the INAP/IDL gateway being co-joined within the IN SCF.

Specification work within this project started to slow down as many of the member companies withdrew from TINA. This would appear to be mainly due to the difficulty TINA had in convincing the rest of the world to apply work that had been developed behind closed doors — a lesson for us all.

14.2   The Parlay Initiative

14.2.1   Introduction

The Parlay organisation, originally formed back in 1998 by BT and four other companies, continues to go from strength to strength. Today Parlay consists of some 65 telecommunications companies, including major operators, equipment vendors, application developers and third party service providers.

Parlay's vision, of providing an API allowing third-party application providers the opportunity to connect to existing network technologies, has not only been achieved technically, but has also proved itself practically.

14.2.2   Network Independence

One of the main strengths of Parlay has been its network independence. Service provision has always meant developing a service around a particular network technology. This means a high cost in providing overlay networks, with a finite lifetime, often around two to three years. With Parlay one can provide a myriad of application opportunities to customers irrespective of whether they originate their call from fixed, mobile, or IP networks.

Parlay has achieved this through their unique gateway (GW) technology (see Fig 14.5).

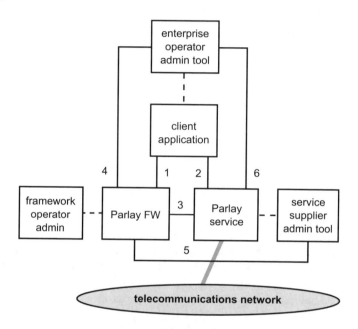

Fig 14.5   The Parlay GW architecture.

Gateways are usually seen as forming some sort of firewall between trusted and non-trusted networks. The Parlay gateway, sometimes known as the open service srchitecture (OSA), not only provides this capability, but also provides the unique opportunity for application providers to request a network to route, terminate, find and communicate to users.

As can be seen from Fig 14.5, the Parlay gateway consists of two parts:

- the service;

- the framework.

14.2.2.1   The Service Part

The service part is inextricably linked to the framework, which provides both the ability for the network operator to negotiate with the application provider (client application), and the initial point of contact to the client application to discover the services on offer from the network that can be utilised by the application. The services provided by Parlay are listed below.

- Call control services

  These services give the application the ability to request the network to route calls between two or more parties depending on the type of application being invoked. There are four call control services — generic call control (used for basic connection of two parties), multiparty call control, multimedia call control, and conference call control (each of these are explained in more detail in section 14.4.1).

- Mobility service

  This service basically allows applications to request the network to provide location information about the user. This information can then be utilised to send to users data that is specifically applicable to the locality where they reside. Examples of its use could be to direct the user to the nearest bookshop.

- Generic user interaction

  This gives the application the ability to request the network to play an announcement and collect information from the user. Examples of this could be to input a pin code.

- Terminal capabilities

  Some application implementations are dependent on the type of terminal connected to the network. So, in the case of a mobile telephone, the screen capabilities are rather diminished compared to that of a PC on an IP network. However, if a mobile is used in conjunction with a PC, then capabilities are enhanced. This service allows the application to poll the terminal to see what capabilities are available.

- Data session control

  Allows the application to set up and manage data sessions in an underlying network, typically for use within a GPRS network.

- Generic messaging

  This service has been developed to allow the application to manage message folders such as traditional e-mails. An example of its use is where the application is providing a call distribution service. In the event that an agent is unavailable, the application can invoke a mail message to be sent to an agent indicating the calling user's address/telephone number.

- Connectivity management

  This service allows the application to state the type of QoS that is required for certain applications. Although this facility was provided early on in the specification process, it no longer seems to be a favourite way of applying QoS, especially with protocols such as RSVP within the IP network.

- Content-based charging

  This is one of the newer and most useful services to be developed within Parlay. Previously, a certain charging profile would be defined within the generic call control service and this would remain active throughout the duration of the call. However, with the advent of content-based charging, charges can be defined for specific parts of a call. Money can be reserved from a credit card, debited, or credited — even the lifetime of credit reservation can be governed within the API.

- Account management

  The purpose of this is to provide a way in which network operators and application providers can manage the charges made/received between both parties during the interaction of information across the API and within the network.

- Policy management

  The policy management API defined an information model that can be applied to a range of policy domains. The information model is derived from the IETF policy core information model. The API allows definition of new policies and then management of events and statistics in the operation of these policies. Typical applications of this API could be for controlling access to QoS characteristics of a network or charging constraints on prepaid accounts.

- Presence and availability management (PAM)

  The API has been produced through liaison with the PAM Forum. The PAM API allows for the abstraction of the presence of a user and their availability, independent of the network that they use. The API also allows for the management of the identity and grouping of users. With the PAM API an application would be able to determine when a particular user switched on their cellular telephone and if they were willing for this information to be transmitted to a restricted set of users.

14.2.2.2   The Framework Part

This has been known, and is still referred to, as the jewel in the crown of the API. As was stated earlier, the framework is the initial point of contact for all application providers when they initiate contact with a network operator. The application provider is firstly authenticated, then given authorisation to use the API, and then allowed to discover which services, sometimes referred to as service capability servers (SCSs), are available for their use. It may be that the network operator has ten different SCSs available, but only allows application provider X access to eight. This decision may be based upon the application provider's ability to use certain SCSs or upon their trustworthiness.

The framework API contains interfaces between the application server and the framework, between the network service capability server (SCS) and the framework, and between the enterprise operator and the framework (these interfaces are represented by the circles in Fig 14.6). The description of the framework separates interfaces into these three distinct sets — framework to application interfaces, framework to enterprise operator interfaces, and framework to service interfaces.

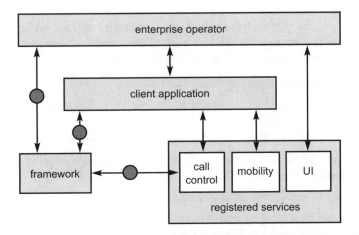

Fig 14.6   Framework interfaces.

Some of the mechanisms are applied only once (e.g. establishment of service agreement), while others are applied each time a user subscription is made to an application (e.g. enabling the call attempt event for a new user). The basic mechanisms between application and framework are listed below.

- Authentication

  Once an off-line service agreement exists, the application can access the authentication interface. The authentication model of OSA is a peer-to-peer model, but authentication does not have to be mutual. The application must be authenticated before it is allowed to use any other OSA interface. It is a policy decision for the application whether it must authenticate the framework or not. It is a policy decision for the framework whether it allows an application to authenticate it before it has completed its authentication of the application.

- Authorisation

  Authorisation is distinguished from authentication in that authorisation is the action of determining what a previously authenticated application is allowed to do. Authentication must precede authorisation. Once authenticated, an application is authorised to access certain service capability features.

- Discovery of framework and network service capability features

  After successful authentication, applications can obtain available framework interfaces and use the discovery interface to obtain information on authorised network service capability features. The discovery interface can be used at any time after successful authentication.

- Establishment of service agreement

  Before any application can interact with a network service capability feature, a service agreement must be established. A service agreement may consist of an off-line (e.g. by physically exchanging documents) and an on-line part. The application has to sign the on-line part of the service agreement before it is allowed to access any network service capability feature.

- Access to network service capability features

  The framework must provide access control functions to authorise the access to service capability features or service data for any API method from an application, with the specified security level, context, domain, etc.

The basic mechanisms between framework and service capability server, and between framework and enterprise operator are shown below.

- Registering of service capability features at the framework

  SCFs offered by a service capability server can be registered at the framework. In this way the framework can inform the applications upon request about available service capability features (discovery). For example, this mechanism is applied when installing or upgrading a service capability server.

- Enterprise operator's service subscription agreement

  This function represents a contractual agreement between the enterprise operator and the framework. In this subscription business model, the enterprise operators act in the role of subscriber/customer of services and the client applications act in the role of users or consumers of services. The framework itself acts in the role of retailer of services.

14.2.3   New Ventures within Parlay

Parlay has just completed Version 4.0 of the API and work is currently under way clearing up the small errors that have been found within the specifications (detailed above). Parlay has also recently completed a requirements capture for Release 5.0 of the specification and already has plans in hand for Release 6.0. These new requirements include upgrading the existing interfaces with more capabilities and capture of new interfaces providing the API with greater service offerings.

Over the past three years Parlay has been concerned to ensure that its specifications receive international recognition. This has been achieved by evangelising the API in the ITU-T, ETSI and 3GPP. Obviously a venture such as this needs management, otherwise there is a danger of three different specifications being released that satisfy the same requirements. To minimise this a 'Joint Group' consisting of members of 3GPP [2], ETSI [3] and Parlay [4] has been formed to capture the requirements from all, within one specification. This specification is produced under the management of ETSI as a European Standard. 3GPP takes a sub-set of this specification and publishes it under their specification range for mobile communications, thereby ensuring complete adherence to one interface specification. Management of this joint initiative was initially undertaken by BT and has been seen as a cornerstone of the development process.

14.3   Parlay X Web Services

The Parlay APIs are designed to enable creation of telephony applications as well as to 'telecom-enable' IT applications. IT developers, who develop and deploy applications outside the traditional telecommunications network space and business model, are viewed as crucial for creating a dramatic market growth in next generation applications, services and networks. The Parlay X Web Services are intended to stimulate the development of next generation network applications by IT developers who are not necessarily experts in telephony or telecommunications.

Parlay X Web Services are powerful yet simple, highly abstracted, imaginative building blocks of telecommunications capabilities that developers and the IT community can both quickly comprehend and use to generate new, innovative applications. These building blocks are defined to favour broad applicability rather than deep or specialised functionality, reflecting the 80/20 rule (i.e. 80% of applications can be implemented using 20% of the functionality available), in particular:

- each is abstracted from the set of telecommunications capabilities, focusing on simplicity over functionality;

- the interaction between an application incorporating a Parlay X Web Service and the server implementing the Parlay X Web Service will be done with an XML-based message exchange;

- Parlay X Web Services follow simple application semantics, allowing the developer to focus on access to the telecommunications capability using common Web Services programming techniques;

- Parlay X Web Services are not network equipment specific, nor network specific where a capability is relevant to more than one type of network;

- Parlay X Web Services are application interfaces and do not provide an implementation of AAA (authentication, authorisation and accounting), service level agreements, or other environment-specific capabilities — rather, they rely on proven and reliable solutions provided by the Web Services infrastructure.

Unless otherwise specified, in this chapter the term 'application' will refer to software that invokes a Parlay X Web Service, and the term Parlay X gateway is used to describe a server that implements one or more Parlay X Web Services. In telecommunications parlance, an implementation of a Parlay X Web Service on a Parlay X gateway would also be referred to as a 'service'.

14.3.1   Relationship of Parlay X Web Services to Other Parlay Activities

In addition to Parlay X Web Services, there are two other activities within the Parlay Group that address other areas of Web Services' use.

Firstly, the Parlay specifications include a WSDL realisation for the Parlay APIs as defined by the Parlay specifications.

Secondly, the Parlay Group has published a set of white papers on the use of Parlay and Web Services. These cover a variety of topics, including the infrastructure that will be present in a Parlay X Web Services environment.

Figure 14.7 shows the relationship between Parlay X Web Services and Parlay APIs. A Parlay gateway as depicted in Fig 14.7 typically implements the Parlay APIs. Applications can interact with the Parlay gateway over a network. Applications can communicate with the Parlay gateway using CORBA or Web Services transports. The application itself can, in principle, be written in any language as long as it can make the proper method invocations and correctly handle the corresponding responses for the API version that the Parlay gateway offers. Thus the application could be, for example, a Java program, a Visual Basic program, or an XML script.

In general, the interaction between the application and the Parlay gateway can be complex and can exploit the full richness of the Parlay API. For example, the application can create a call (i.e. a multiparty communications session over an underlying next generation network) and route each leg of the call independently.

Parlay X Web Services represent an abstraction and simplification of the Parlay APIs and thus is shown above them in Fig 14.7. It is anticipated that most Parlay X Web Services will be implemented by invoking functionality on a Parlay gateway; however, there are cases where the Parlay X server will implement the Web Service in another way, for example, via a direct connection to the network elements.

As mentioned, a Parlay X application can be written in any language as long as it can make the proper Web Service invocations. There are several development tools, for different programming languages, to create, deploy, and interact with Web Services. Further information on Parlay X Web Services can be obtained from the Web site [4].

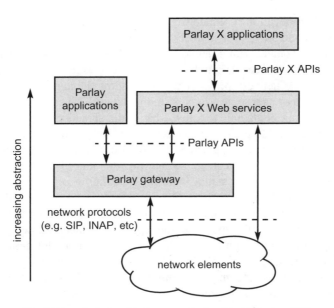

Fig 14.7 Relationship between Parlay X and Parlay APIs.

14.4 A Detailed Examination of Parlay's Call Control Interfaces

The basic concept of the Parlay API is to allow application developers/providers to gain access to network resources, so that they are able to sell their applications to users. It is also there to allow the network providers to make available a set of networks capabilities/resources to the third party application providers, thereby allowing their users to actually 'log on', as it were, to the applications provided by that third party.

To facilitate this, it is necessary that the application can request the network to control or route calls between users — this concept is known as call control. (One should realise here that the application does not have control of the network resources used in routing the calls, but requests the network to undertake these actions via a gateway or firewall resource at the edge of the network.)

Call control also allows events pertinent to the call to be sent to the application, thereby giving the application the ability to decide upon an appropriate action that may need to be taken in relation to that particular call.

The two players in this basic architecture are known as the 'client application' and the 'gateway'. The gateway effectively provides a firewall between the 'outside' or the untrusted domain and the network. The gateway is also responsible for deciding what to do with the requests that are sent from a client application towards the network and those received from the network aimed towards some particular client application. The gateway may also provide the appropriate mapping between the API and the underlying network protocol (e.g. Parlay to SIP).

When making contact with a network operator, application providers will carry out all the negotiations necessary to be able to connect, as it were, their applications to the network. The network provider will then allow the application providers to request the gateway to set some 'triggers' in the network. These initial triggers will result in the network being able to know what to do, should a user dial a particular number that relates to an application being provided by the client.

So when a call arrives at a particular network node it will 'set off' one of these triggers. This will, in turn, indicate to the gateway, that it needs to send a message to the client application provider stating: 'We have a caller who wishes to gain access to one of your applications.'

As one can see, the concepts are very basic but the resultant messages that flow across the interface do rely upon the application developer knowing something about routing and dealing with events related to calls within a telecommunications network environment.

It is interesting to note that the idea behind the concept known as 'Parlay X' is to remove this stigma, thus allowing application developers to design applications without the need to serve time in a telecommunications network environment — but that is another story!

14.5   Types of Call Control

In the development of call control, the Parlay organisation decided that it would be appropriate to group together aspects of call control under different headers. Some of this was historical and some of it was for ease of reference.

There are four variants of the Parlay call control:

- generic call control;

- multiparty call control;

- multimedia call control;

- conference call control.

14.5.1   Generic Call Control

Generic call control was thus named because it contained all the procedures necessary to take account of a basic two-party call, i.e. it deals with setting up, tearing down, dealing with network problems/events, etc, to do with one user communicating with either another user on the network via a third-party application, or communicating with some sort of resource within a network (see Fig 14.8).

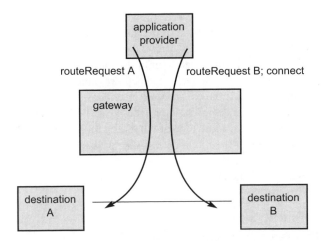

Fig 14.8   Generic call control.

The sorts of application to which this call control relates were initially seen to be 'the simple applications'. However, it was found that although generic call control is deemed to be a basic part of the API, it does not indicate that the applications themselves that use generic call control should be simple. For example a user may access an application that, among other things:

- requires the user's geographic location;

- uses charging mechanisms in the network that monitor the usage of different elements of the application usage;

- requests the type of terminal capability of the user.

Even though all of these other elements are relevant to the application, the application may only need to connect one user to another.

As can be seen from Fig 14.8 this is achieved by sending three simple messages to the gateway, one requesting to route a call to destination 'A' and the other to destination 'B'. At the end of this a message (connect) can be sent requesting the two parties be 'connected' together.

14.5.1.1   UML

To fully understand call control we need to explore UML (unified modelling language). The specifications have been written in UML. This modelling language was designed so that interface specifications such as that defined by Parlay could be written in a non-specific implementation manner. The beauty of this is that the resultant specification can, through an appropriate 'software program', be converted to implementations such as CORBA, IDL, Java, etc.

The generic call control interface (i.e. the set of operations known as methods that are sent from the client application to the gateway and vice versa) are separated into two halves. One half refers to those methods sent from the gateway to the client, and the other half, those sent from the client to the gateway.

As one would expect, the generic call control interface contains the least commands or 'methods' of all the call control interfaces.

The UML procedures are shown later (see Figs 14.10 and 14.11 in section 14.5.1.3) to allow you to gain some idea of what they look like. It is not the intention to go into all the specifics of the methods and IDL procedures.

As we have used object modelling concepts such as UML, it is worth understanding what this means. This is best explained by looking at a simple 'sequence diagram'.

14.5.1.2   Sequence Diagrams

These diagrams are produced to indicate the interactions between the client application and the gateway and also which methods from the generic call control interface (in this case) are used.

The number translation application does what it appears it should do — it translates one network destination number to another. An instance of its use is where an individual may be constantly on the move. He may sign up with an application provider and ask him to route all calls to one particular destination during the hours of 9.00 a.m. and 5.00 p.m., another destination in the evening (like his home number) and yet another for weekends.

This would mean that the application provider would need to set a trigger in the network (on request to the network provider). When the original directory number (1234) is received within the network, the trigger sends a network indication via the gateway to the application, which, by reference to the application logic, will provide a different destination number (4355) back to the network. This translated number is the one used by the network to onward route the call.

Notice from the sequence diagram in Fig 14.9 that there are five boxes on the top line. These boxes indicate which objects are activated by the application (remember we are using object modelling techniques here). The three on the left are those that reside within the client domain, hence they are preceded by the term IpApp. The two letters Ip refer to an abbreviation created by Parlay called 'Interface Parlay'. It is just a way of identifying those objects owned, as it were, by the Parlay organisation. The two boxes on the right hand side are those objects that reside in the gateway or 'service' side of the interface.

The extreme left object refers to the actual application logic itself. So for this to work the application must ensure that those triggers are set in the network. To do this the App creates an object IpAppCallControlManager. Its job is to manage all calls that access applications owned and run by this application provider:

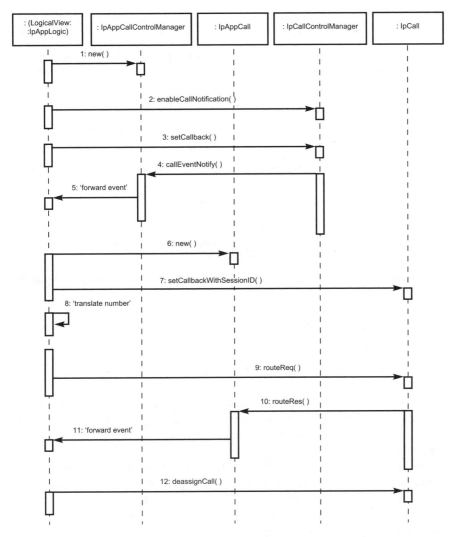

Fig 14.9   A simple number translation application sequence diagram.

- the logic then sends information to the other end of this call control manager interface (i.e. the other half of the manager) with message 2 — this provides all the information needed to set triggers such as the dialled number;

- message 3 sends information to that manager in the gateway to indicate the address of the object within the application domain, i.e. the object reference;

- message 4 is the indication (or method) that is sent to the call control manager advising that a caller has dialled that elusive number relating to your number translation application;

- message 5 is an internal message (i.e. not a method across the interface) sent to the application — this contains the dialled number (in this case for translation);

- the application logic prepares the interface by sending another internal message asking for an object to be created that will take care of routing this particular call, when the number gets translated (6 and 7);

- while this is happening the application translates the number (8);

- message 9 requests the network's call manager to route the call to a different destination number;

- when this is done (10), the result of whether the routing was successful or not is sent back to the call object;

- which, in turn, notifies the application with an internal message (11);

- the application, being satisfied with this result, tells the call object using message 12 to delete the call object relating to this call, which effectively deletes the call object on the application side.

Note that the call still continues in the network — it is just that the application requires no further interaction.

All of this may seem rather long-winded but it does explain the use of sequence diagrams and shows what creating objects is all about and what the objects do.

14.5.1.3   Generic Call Control Interfaces

Moving on from the modelling process, we now look, in more detail, at the generic call control interfaces. The generic call control service (gccs) consists of two parts or packages, one for the interfaces on the application or 'client' side and one for interfaces on the 'service' or gateway side.

The class diagrams in Figs 14.10 and 14.11 show the interfaces that make up the generic call control service package and the generic call control application package. Communication between these packages is indicated with the <<uses>> associations, e.g. the IpCallControlManager interface uses the IpAppCallControl Manager, by means of calling/callback methods.

Figure 14.10 class diagram shows the interfaces of the generic call control application package and their relations to the interfaces of the generic call control service package. The IPCallControlManager in Fig 14.10 is used by the application to:

- create call objects in the gateway;

- enable/disable notifications or triggers in the network and change the criteria used for the triggers if necessary;

- control/regulate the flow of information between the application and gateway;

- query the criteria used in the triggers that have been set.

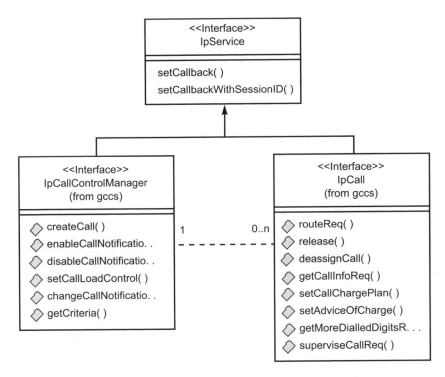

Fig 14.10 Service interfaces.

The IpCall object or interface can:

- route calls;
- release calls;
- destroy or de-assign the call object;
- get information such as charging, applicable to the call;
- set a particular charge rate or plan;
- notify the user about the charging costs at certain points in the call;
- request further information, such as dialled digits, needed to further the application;
- supervise the call, such as being notified occasionally about the duration of the call.

The other interfaces shown in Fig 14.11 are used by the gateway to return the results of the information asked for by the application and also to indicate if any errors have been received from the network.

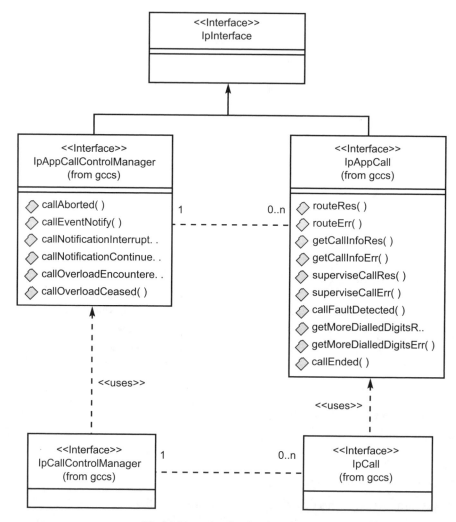

Fig 14.11   Application interfaces.

As has been shown, the generic call control interface methods allow client applications to have sufficient control over their calls within a network. Any information above and beyond this, such as managing more parties in a call, is taken care of by the other call control interfaces, described in the following sections.

14.5.2   Multiparty Call Control

The first thing to say about multiparty call control is that it is an extension of the generic call control service. Originally this service was developed with the intention of being used in conjunction with generic call control, thus providing the ability to manage control of individual call legs. However, this decision changed when the JAIN organisation [5] initiated contact with Parlay.

Parlay, ETSI, 3GPP and JAIN formed a 'Joint Group' responsible for defining the Parlay interfaces. This Joint Group defined the API with a number of underlying network topologies in mind, such as an INAP intelligent network, a mobile environment, H.323, SIP, etc. When the JAIN organisation joined, the group started looking not only at the underlying network topologies but also at an implementation-specific environment within the gateway and client domains.

JAIN was developing a Java-based implementation and sought the agreement of the whole group to make certain changes to the basic call control interfaces. Some of these changes were because of the problems encountered when mapping the Parlay call control on to that of JTAPI and the Java call control (JCC). The Joint Group agreed that the Java solution would not work successfully with generic call control, and therefore the clients implementing Java environment solutions would need to use multiparty call control as their basic call control solution. This meant that the existing multiparty call control had to be updated to include all of the capabilities of the generic call control.

This appeared to make generic call control redundant, but we had to take into account that some early market solutions had already implemented generic call control (3GPP Release 99 for example). So we could not just delete generic call control or even change the existing methods. This now means, that to the purists, we have a certain amount of redundancy, i.e. two interfaces that overlap.

14.5.2.1   Inheritance

The redundancy described above may better be explained by describing how inheritance works.

As mentioned before, the group uses object-oriented modelling techniques to describe the API, notably the UML. This form of description allows one to define separate packages of methods (interfaces) and for the programmer to inherit methods from other packages. This can be explained in a simple way.

Figure 14.12 shows the conventional inheritance that was first described. In object modelling this diagram shows programmers that, if they implement the multiparty call control (MPCC), they can inherit, with this, all of the capabilities that are available within the GCCS. In other words we do not have to add these to MPCC, they are already provided through inheritance. The same goes for multimedia call control (MMCC) and conference call control (CCC), they inherit from everything above them in the inheritance hierarchy. However, now that we had decided that the generic call control would not work properly with a Java environment and that 3GPP had already used GCC, we needed to split the inheritance hierarchy. Figure 14.13 shows the split.

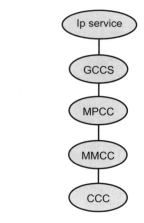

Fig 14.12   Conventional inheritance.

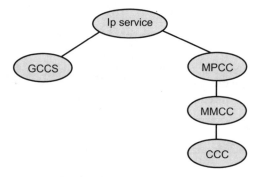

Fig 14.13   Split inheritance.

With this split we now see that MPCC does not inherit the capabilities of GCC; however, the rest of the inheritance underneath MPCC remains intact and GCCS remains part of the overall picture.

It was therefore necessary to enhance the MPCC with all the capabilities it once inherited from GCCs — hence the apparent redundancy.

14.5.2.2   MPCC capabilities

We are now able to describe the extra capabilities of MPCC. Figure 14.14 shows all the methods contained within MPCC. You will see many of the methods here mirror those in the GCC (for the reasons given above), but extra capabilities are added that allow the application provider to add more than two legs to a call and to control those particular legs.

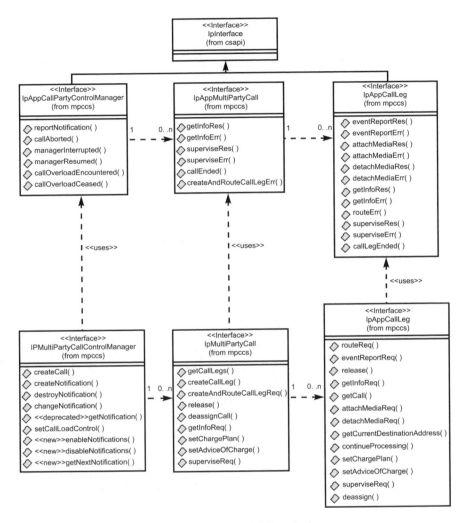

Fig 14.14 The MPCC methods.

You will see that added to the interfaces are methods applicable to call legs themselves. This will allow you to route individual legs (however many there may be) and to connect them together. It will allow you to receive events applicable to individual legs, so you know exactly what problems happen and where. One of the added capabilities is to include what are known as convenience methods. These methods combine two functions into one command or request, e.g. createAndRouteCallLeg — this combines create a call leg object and route that leg to a particular destination. These convenience methods are not always used, as some feel they not only create legs but also create problems.

The MPCC can therefore be split into three basic parts:

- the MPCC manager — which is responsible for managing the total set of calls from that client;

- the multiparty call interfaces — which are there to manage individual calls irrespective of how many legs are contained within a call;

- the call leg interfaces — which manage and manipulate individual legs in a call.

14.5.3   Multimedia Call Control

With the advent of new technologies and the availability of a myriad of network terminal types, it was essential that the telecommunications market went back to basics and came up with a satisfactory solution of how best to capture this new revenue-earning corner of the market. The use of the Internet and packet-based networks revolutionised the telecommunications scene and enabled the integration of different types of network technologies.

Parlay therefore had to come up with a solution for application providers, to not only route their calls but also to request the network to provide different types of media stream which would satisfy their end-to-end connection requirements.
Multimedia call control does what it implies and enhances the functionality of the multiparty call control service with multimedia capabilities.

To handle the multimedia aspects of a call the concept of media stream is introduced. A media stream is a bi-directional stream and is associated with a call leg. These media streams are usually negotiated between the terminals in the call. The multiparty call control service gives the application control over the media streams associated with the legs in a multimedia call in the following way:

- the application can be triggered, or told when a media stream is established, meeting the application-defined characteristics;

- the application can monitor the establishment (addition) or release (subtraction) of media streams of an ongoing call — in other words, during a call the end user may, for example, want to stop watching a video stream and just continue with voice;

- the application can allow or deny the establishment of media streams — this may happen because the application may not have the necessary terminal type, or the user's credit does not run to this sort of call;

- the application can explicitly subtract already established media streams;

- the application can request the media streams associated with a specific leg — in other words, ask what media streams have already been set up.

The multimedia call control service has four different aspects or interfaces:

- IpMultiMediaCallControlManager — this manages all media-related calls that exist between the client provider and the gateway;

- IpMultiMediaCall — this object controls and receives specific media-related events for one call;

- IpMultiMediaCallLeg — this object controls and receives events from one specific leg of a call involved with a media stream;

- IpMultiMediaStream — this currently only allows the application to subtract a media stream from a call, i.e. to remove a particular stream while leaving other streams (if they exist) up and running.

Figures 14.15 and 14.16 show each part of the multimedia call control interfaces and their relationship with the multiparty call control interfaces — it can be seen that they are inextricably linked.

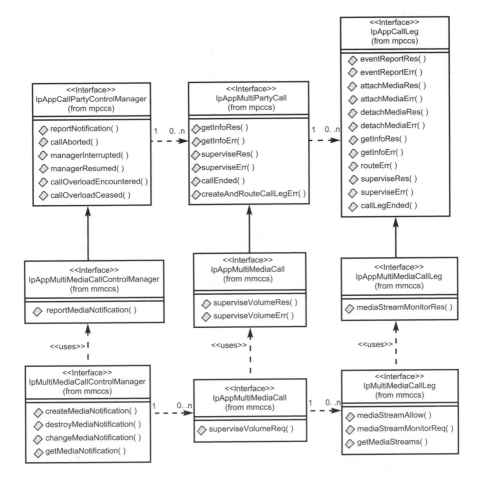

Fig 14.15 MMCC application interfaces.

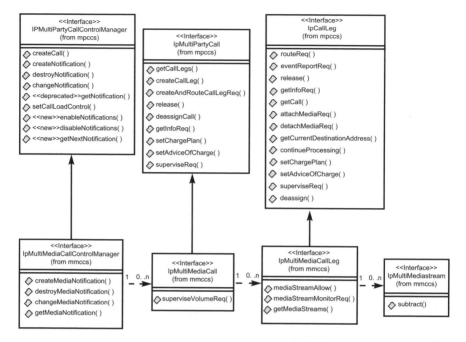

Fig 14.16   MMCC service interfaces.

14.5.4   Conference Call Control

Conference call control was defined to allow the application provider to manage and set up conference calls. The capabilities provided are the same as you would expect from any ordinary conference call in which you take part on a regular basis. In a normal scenario one would book a conference bridge and send the number plus password out to everyone you want on the bridge.

This can still effectively be done; however, the conference call control in Parlay allows for some further, more complicated, scenarios to take place. It is best described diagrammatically.

Figure 14.17 basically shows a simple scenario. The application is able to reserve a conference bridge in the network via the API. It sets a trigger in the network to be notified every time a caller joins the bridge. It is then able to allow that caller to join or be rejected based upon the originating number or some other process.

In this instance four members have joined the bridge and media streams have been connected between them — thus they are all conversing together.

Figure 14.18 shows another unique capability provided by the conference call control API. The application can, where necessary, move the callers to sub-conference bridges. This allows user A and B to converse and C and D to converse separately. In other words there is no media stream between AB and CD.

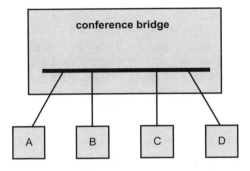

Fig 14.17   Simple conference call scenario.

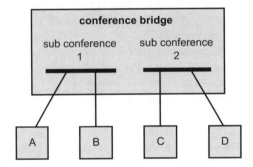

Fig 14.18   Sub-conference call scenario.

The application can then, if needs be, add user D to sub-conference bridge AB, thus leaving C in sub-conference 2, as shown in Fig 14.19.

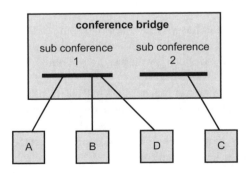

Fig 14.19   Alternative sub-conference call scenario.

All of these complex manoeuvres require one to use call leg control. The IpSubConferenceCall interface has a command 'moveCallLeg' which allows this to take place, thereby moving one leg of a bridge to another. If you cast your mind back to the inheritance hierarchy you will remember that CCC lies at the bottom of the tree and therefore inherits everything above it. This is important as it also relies on the multimedia call control interfaces.

Another possibility here is for the CCC to request from the network the possibility of reserving a bridge in advance. The network can send information back to the application stating whether or not this is possible. CCC can reserve the bridge and then forget about it until the first user joins. At that moment the application is notified about the creation of the bridge after which it then takes care of the process as shown in the examples above.

If a mistake has been made and the application does not want the bridge any more, it can free up the resources that have been reserved.

The CCC application also provides other conference capabilities, e.g. it is able to make one of the parties the chair of the call. The chair usually has overall control (and is often the one who is paying).

As we are using CCC in conjunction with multimedia call control there is the availability of providing video streams for all parties on the bridge or sub-conference bridges.

The CCC is able to elect a speaker, i.e. the one everyone else has to stop and listen to, and the video-stream of that speaker can then be transmitted to all other parties.

Before this takes place the chair alone can inspect the video-stream, to see if it is decent or not!

For the sake of completeness, reproduced in the Appendix (as Figs 14A.1 and 14A.2) are the CCC interfaces. They also show their relationship with multimedia call control and so look a little busy.

14.6   Summary

It is hoped that this chapter has served to remove some of the questions surrounding Parlay and the interfaces associated with the API. It has tried to show all of the call control interfaces, so far defined by Parlay and adopted by ETSI and (some) by 3GPP.

The work is still ongoing. At present Parlay Release 5.0 is being worked on and this may result in additional interfaces and possibly some enhancements to the existing ones.

In the future, there may even be new call control interfaces, but it is likely that four is all that will ever be required!

References

1   TINA-C — http://www.tinac.com/

2   3GPP — http://www.3gpp.org/

3   ETSI — http://www.etsi.org/

4   Parlay — http://www.parlay.org/

5   JAIN — http://java.sun.com/jain/

Appendix

The CCC Interfaces

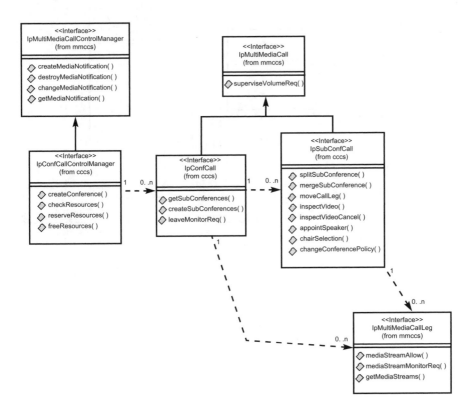

Fig 14A.1   CCC service interfaces.

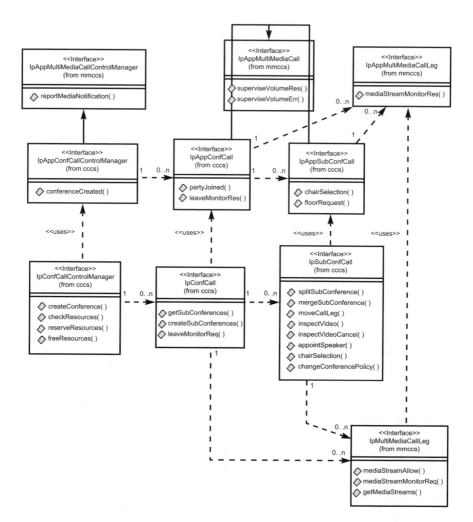

Fig 14A.2 CCC application interfaces.

15

RADIO SPECTRUM MANAGEMENT FOR TETHER-LESS COMMUNICATIONS

J S Dixon

15.1 Introduction

Consumer demand for tetherless and mobile communications devices is growing enormously, which is resulting in an increasing demand for radio frequencies. This requires that the available frequency bands are carefully managed, to ensure that the current users can continue to operate without receiving (or causing) harmful interference, while providing opportunities for new applications and systems to be deployed. Consequently, spectrum management is an important consideration for all radio uses.

15.2 Spectrum Management

15.2.1 Why Does the Spectrum Need to be Managed?

To many people, the need to manage the radio spectrum is an unnecessary burden, which is seen to restrict use of the frequency bands, and slow down innovation. However, spectrum management is necessary for very good reasons. The users of the radio spectrum are many, and varied, and so are their radio systems, which are used for much more than telecommunications.

15.2.2 The Many Uses of the Radio Spectrum

There are many uses of the radio spectrum and they can be categorised as follows:

- aeronautical uses for both mobile communications and radionavigation, both of which can be terrestrial or via a satellite;
- amateur users, both terrestrial and satellite;
- the broadcasting organisations, again both terrestrial and satellite;
- the earth exploration satellite service for surveying and measurement purposes;
- the fixed service, which is sub-divided into terrestrial and satellite, for communications between two fixed points;
- the mobile service, for communications to a movable terminal, which is also sub-divided into terrestrial and satellite;
- maritime uses, similar to aeronautical, for both mobile communications and radionavigation, both of which can be terrestrial or via a satellite;
- meteorological satellites;
- radio astronomy;
- a category for general radars (neither maritime nor aeronautical) which is radiodetermination — this is subdivided into identifying positions ('radiolocation') or navigation purposes ('radionavigation'), both of which are either terrestrial or satellite;
- the space research service, for operations to/from spacecraft.

These principal uses of the radio spectrum, and the sub-categories mentioned are generally referred to as the 'radiocommunication services'.

For most telecommunications operators, nearly all of their applications will fall within the terrestrial and earth-space fixed and mobile services, and possibly the broadcasting services. In some cases they may use the maritime or aeronautical services, but these are normally used by shipping/aviation organisations.

The challenge is that the operators of all of these services have significant demands, to enable them to operate their many and varied applications. It is necessary to balance these demands, within the finite resource of the spectrum. Therefore the spectrum needs to be managed, taking account of the relevant technical issues for each of the varied applications.

Furthermore, because of the strategic capabilities of radio, there are also political and policy considerations, which need to be included. As a consequence, it is necessary to apply a regulatory framework, to ensure that the spectrum is fairly accessible to all users.

As will be discussed later (in section 15.5), although the radio spectrum covers a very wide range (see Table 15.1), particular ranges are best suited for specific applications, based on a combination of physical principles, and the availability of electronic devices at an acceptable cost.

Table 15.1   Radio frequency bands.

Band	Frequency range
Very low frequency (VLF)	3 kHz to 30 kHz
Low frequency (LF)	30 kHz to 300 kHz
Medium frequency (MF)	300 kHz to 3 MHz
High frequency (HF)	3 MHz to 30 MHz
Very high frequency (VHF)	30 MHz to 300 MHz
Ultra high frequency (UHF)	300 MHz to 3 GHz
Super high frequency (SHF)	3 GHz to 30 GHz
Extremely high frequency (EHF)	30 GHz to 300 GHz

Having said that, in practice most of the current spectrum management interest lies in the UHF and SHF bands (300 MHz to 30 GHz), since most of the new technologies and applications being developed are intended for operation in those bands.

15.2.3   Global Spectrum Management — the ITU

The operating conditions (transmitter powers, bandwidths, receiver sensitivities and antenna gains) for the systems operating in the radiocommunication services vary greatly. Hence it is necessary to ensure that they can all share the available radio frequency bands on a fair and equitable basis, particularly since some of them are considered to be, and are referred to as, 'safety-of-life services'.

As a consequence, the International Telecommunications Union [1] (ITU, a member of the United Nations System of Organisations) has developed the Radio Regulations, which is a global international treaty that forms a framework for the use of all the frequency bands from 9 kHz to 3000 GHz.

The Radio Regulations contain a very large table ('Article 5') in which each subdivision of each frequency band is allocated to a particular radiocommunication service (or services) as described in section 15.2.2 above. The upper limit of the allocation table is 275 GHz at present. These allocations are made on either a 'primary' or 'secondary' basis.

- Primary

  Equipment that is operating within a primary service allocation is permitted to operate providing it would not cause harmful interference to any other equipment which is already operating in the band under a primary service allocation.

- Secondary

Equipment that is operating within a secondary service allocation is permitted to operate providing it would not cause harmful interference to any other equipment which is already operating in the band under a secondary service allocation.

Furthermore, equipment operating under a secondary service allocation cannot claim protection from any equipment that may be operated at a later date under a primary service allocation.

To allow geographical flexibility, the allocations in the Radio Regulations can apply to all of the world (a 'global allocation'), or alternatively to just one or two of the three 'regions' of the world:

Region 1 — Europe, Africa and the countries of the CIS;

Region 2 — The Americas;

Region 3 — Asia Pacific.

Furthermore, the allocations in Article 5 of the Radio Regulations are complemented by footnotes, which provide clarifying text or additional regulatory conditions. Footnotes are also often used to identify where one or more country/ies have an additional allocation to a particular service, or they have allocated the band on a different status (e.g. as a primary allocation rather than a secondary allocation).

All of this is intended to provide some harmonisation and structure to the use of the radio frequencies, so as to enable efficient and compatible usage of the bands. Changes to the Radio Regulations are made by the ITU-R World Radio Conferences (WRCs), and normally need to be justified on the basis of good technical arguments. The WRCs are normally held every few years; there was one in 2000 and in 2003, with the following one expected to be held in 2007.

The ITU also has a structure of study groups (and working parties) to undertake the necessary technical studies, which are then used as the basis for discussions and decisions at the WRC (see Table 15.2).

Table 15.2   ITU-R Study Groups.

Study Group	Subject area
SG 1	Spectrum management
SG 3	Radiowave propagation
SG 4	Fixed satellite service
SG 6	Broadcasting services
SG 7	Science services
SG 8	Mobile, radiodetermination, amateur and related satellite services
SG 9	Fixed service

15.2.4   National Spectrum Management

While the Radio Regulations provide a basic framework for the allocation of frequency bands, it does allow flexibility in the choice of systems deployed, and it is still possible for incompatible systems to be operated in the same frequency band. Consequently, the Radio Regulations are not intended to remove the need for licensing of systems, which is done on a national basis.

In the UK, the use of all (non-military) radio transmitting equipment is licensed by 'Ofcam' — the new Office of Communications, although until late 2003 it was done by the Radiocommunications Agency (RA), which was an executive agency of the Government's Department of Trade and Industry (DTI). While Ofcom is free to choose what radio equipment is licensed to operate in any given frequency band, it is, as a signatory to the ITU-R Radio Regulations, obliged to comply with the framework given in the Radio Regulations texts. However, there is a clause in the Radio Regulations that allows for national variations to the allocations in the regulations, providing such equipment does not cause harmful interference to those services which are operating under an allocation in the regulations.

Inevitably Ofcom has to undertake a balancing act, to make spectrum available for a wide range of purposes. In particular:

- there are governmental and national bodies which have radio spectrum requirements, such as the Home Office, Maritime and Coastguard Agency, Civil Aviation Authority, National Air Traffic Service, as well as the Ministry of Defence who manage the licensing of radio spectrum for military purposes in the UK;

- there are various broadcasters, both for radio and television, who obviously require access to the (internationally agreed) frequency bands for broadcasting — additionally, they have requirements for further radio frequencies to enable them to operate 'contribution' links, which they use for collecting audio (and video) material from mobile sources (e.g. a mobile camera) or from temporary locations (e.g. from a news event in a remote location);

- there are telecommunications operators (such as BT) who use radio as part of their network, as an alternative transport medium to cable/wired networks, which may include both national radio links (ranging from a few kilometres to several tens of kilometres), and also satellite links, which are used to provide international connections; radio is also used to provide access connections to customers (either mobile or fixed);

- there are businesses who wish to use radio for private purposes;

- there are numerous 'non-commercial' uses of radio, such as radio-controlled devices, radio astronomy, amateur radio, and the 'industrial, scientific and medical' (ISM) applications (such as microwave ovens) which use radio waves for non-communication purposes (further information on the ISM bands is given in section 15.4).

15.2.5   Regional Spectrum Management

To try to bridge the gap between the global framework of the Radio Regulations, and the national licensing policy, there are a number of regional regulatory bodies, and Europe is no exception. The European Conference of Post and Telecommunications (CEPT) is an organisation of national regulatory administrations, the purpose of which is to harmonise (as far as possible) the national regulations throughout Europe. CEPT has created the Electronic Communications Committee (ECC) which tries to actively harmonise policy on the use and regulation of radio spectrum throughout Europe. Although CEPT consists of 43 member states, it is inevitably influenced by the 15 member states of the European Union, although there is now increasing participation from the Eastern European countries.

One of the main elements of the European regulatory process is the 'European Common Allocation Table' (ECA), which contains those parts of the ITU Radio Regulations which are relevant in Europe, and to which additional relevant information is attached.

This table is augmented by ECC Decisions that identify harmonised usage of the bands across Europe.

The harmonisation of radio spectrum usage and licensing throughout Europe has the benefits of simplifying the co-ordination of radio usage, and maximising the market for new products.

With that in mind, the ECC has a Memorandum of Understanding (MoU) with the European Telecommunications Standards Institute (ETSI), which is the leading body for developing telecommunications product standards in Europe. Because of the inextricable link between radio products and the frequency band in which they operate, that MoU enables standards for radio products to be developed more efficiently.

In addition, as a consequence of the European Commission's new Regulatory Framework for Electronic Communications Networks, the European Commission is now taking a greater interest in the field of spectrum management. To enable this, they have recently created both the Radio Spectrum Committee (RSCom), and the Radio Spectrum Policy Group (RSPG), and it is expected that these will have a greater participation in European spectrum management in the future. The RSCom will assist the Commission by providing advice on radio spectrum matters, which the Commission will use in producing binding technical measures (i.e. the RSCom will form part of the 'committology' process). The RSPG will also advise the Commission on radio spectrum matters, but with a more strategic, longer-term view, and possibly a broader, less technical focus.

15.3   Licensed and Licence-Exempt Operation

As noted above (section 15.2.4), the licensing of radio (transmitting) equipment is a national issue. In the UK the onus is on the Ofcom to ensure that all of the various users of the frequency bands can operate without unacceptable interference (subject to the terms of their use of the band).

15.3.1   Self-managed Licensed Bands

In some cases, each band is identified for a single purpose, and the user is known and can be considered to be accountable for their use of the band. Consequently the whole band is assigned on a licensed basis, to that user. This typically applies where the band is used by a large organisation, which has the capability of ensuring that there is sufficient co-ordination between its links, and hence they are able to manage the band for themselves. Examples of such 'self-managed licensed operators' include the military, other government departments, mobile network operators, and the broadcasters. In such cases, even though the user may have exclusive use of the frequency band in the UK, it is still necessary for Ofcom to apply certain operational restrictions to ensure that they do not cause interference to users in neighbouring bands and/or neighbouring countries.

Although self-managed bands account for a significant proportion of the spectrum, it is only offered to a small number of organisations that have the specialised capability and exclusive access to the band for a given use.

15.3.2   Ofcom-managed Licensed Bands

In some other cases there may be a number of users who wish to use the band (all for the same purpose), in which case it will be necessary to identify frequency channels for each user (i.e. sub-divisions of the band). In this case, each user will be licensed to operate on a particular frequency channel in a particular area, or over a given path (i.e. from 'point A' to 'point B'). In such cases, it is necessary to co-ordinate the different channel allocations that have been given to the different users, to ensure that there is sufficient separation (in terms of both frequency and geographical distance) in order that the users will not cause interference to each other. Ofcom normally undertakes such co-ordination.

In addition, it is sometimes possible to use the band for more than one purpose (i.e. the band is shared by different services), in which case it is often necessary to place additional restrictions on the operation of the equipment in the band. For example, the band might be used for the fixed service, and may also be used by the fixed satellite service from a small number of earth stations. In such a case, the frequencies of the fixed service links would have to be co-ordinated with those of the fixed satellite service links, or alternatively they may be prohibited from operating within a certain distance from the satellite earth stations, to avoid causing interference.

15.3.3   Licence-exempt Bands

In some cases it is possible for a frequency band to be used on an unco-ordinated basis, such that users do not need to plan or co-ordinate the individual radio links. This has the advantage of simplifying the licensing process; however, as a consequence there is no guarantee that they will not experience interference from other users. Such operation is normally applied for mass-market products, where consumers want to be able to buy and operate the devices, without having to worry about licensing and co-ordination. Examples of such devices include cordless telephones, remote controls for car alarms, radio controls for models, etc. The devices are normally (in the UK) referred to as 'licence-exempt'. The term 'licence-exempt' refers to the individual device that is exempt from needing a licence, and is only applicable if devices comply with the relevant regulations.

Since tetherless devices are intended to operate on an unplanned, unco-ordinated basis, licence exemption is the only practicable solution. This is also sensible, since the products will be sold as consumer products for widespread use without any product registration process.

In the UK, to enable equipment to be considered as licence exempt, it has to be exempted by a legislative document called a 'statutory instrument', which has been laid before parliament. Licence-exempt use does not entail payment of a licence fee.

15.4   ISM Bands

Licence-exempt equipment is often operated in what are known as 'ISM bands'. The term refers to the fact that the band has been identified for 'industrial, scientific or medical' purposes, which means that it can be used by equipment which use radiowaves for purposes such as heating, lighting, scanning or measuring, rather than for communications. The ITU-R Radio Regulations state that '... radiocommunication services operating within these bands must accept harmful interference which may be caused by these applications.'

Furthermore, contrary to common belief, the fact that these bands can be used for high-power ISM applications, does not mean that they can also be used for high-power telecommunications applications, or even that the operation of telecommunications equipment in the band is unlimited. These bands can only be used for telecommunications with certain types of approved equipment, and these are normally limited to relatively low power operation.

ITU-R Radio Regulation Footnote 5.150 states that '...the following bands ... 13533-13567 kHz, 26957-27283 kHz, 40.66-40.70 MHz, 902-928 MHz, 2400-2500 MHz, 5725-5875 MHz, 24-24.25 GHz ... are also designated for industrial, scientific and medical (ISM) applications.' For tetherless, licence-exempt communications systems, it is the two bands at 2400-2500 MHz and 5725-5875 MHz which are currently of most interest.

15.5   Choice of Frequency Band for a Radio System

When developing any new radio application, one of the first questions to consider is in which frequency band(s) should it operate? Generally this is a compromise based on the following principles.

- Lower frequency bands

  — cheaper equipment;

  — lower attenuation of the radio signal relative to the distance travelled;

  — radio signals are better at diffracting around corners, and hence there are less areas in a 'radio shadow'.

- Higher frequency bands

  — the bands have higher bandwidths, and hence are capable of providing more capacity;

  — unoccupied (or low-occupancy bands) are easier to find, particularly above 50 GHz.

To put this in context, although (as noted before) the Radio Regulations cover the range from 9 kHz-400 GHz, most of the current spectrum management activity is in the range from 500 MHz-50 GHz.

Within this range, the broadcasting services are at the lower end (to obtain good signal coverage areas), and point-to-point fixed radio links (which can be planned and co-ordinated) are at the higher end of the range.

The identification of a frequency band for a new application is typically a very long process (i.e. taking anything up to 10 years). This is due to the fact that:

- in nearly all cases, new applications are proposed to be operated in bands which are already identified for other purposes, and so it is necessary to undertake technical studies to identify whether the existing and new applications will be able to share the band;

- it may be necessary to make arrangements for the existing user(s) of the band to migrate to another frequency band, or to cease operation;

- it may be necessary to undertake additional sharing studies, for example to identify the necessary guard spaces at the edges of the band, to protect the users of the adjacent bands;

- it may be necessary to develop agreed channel plans for the band (i.e identify the widths and positions of the individual channels within the band).

As a consequence, it is important that for any new application, consideration as to the choice of frequency band is given at a very early stage. This is particularly important, since the outcome of some of those studies could potentially influence the characteristics and design of the radio system. In some cases it is possible to take advantage of the existence of an available band which has already been identified for similar purposes, and so (much of) the necessary work may already have been done.

For tetherless applications, it is typically required to operate over only a short range (i.e. tens or hundreds of metres), and a relatively large bandwidth is required (of the order of MHz or tens of MHz) to provide the necessary data capacity. This would suggest that very high frequencies should be used.

However, since most tetherless devices will be intended to operate as a mass-market, consumer product, the high cost of state-of-the-art devices for the highest frequencies has served to discourage manufacturers away from these bands. Accordingly, tetherless devices are currently aimed at the 1-10 GHz range, and, in particular, the two main areas of interest are around 2.4 GHz and 5 GHz (including the ISM bands).

15.6 The 2.4 GHz Band

The 2.4 GHz band (2400-2483.5 MHz in the UK) is designated in the ITU Radio Regulations as an ISM band, and is therefore occupied by many different types of (non-telecommunications) radio devices, including domestic microwave ovens.

In accordance with European Regulations, the band can be used for telecommunications purposes, subject to various basic parameter limits (including a transmitter power limit of 100 mW 'effective radiated power'). Accordingly, the band can be used by radio local area networks (RLANs) (namely the IEEE802.11, 802.11b and 802.11g standards (see Chapter 3), and other short range communications devices such as Bluetooth (see Chapter 4).

Although the band is over 80 MHz wide, there are concerns that the band might become congested before long. This is due to the fact that:

- the IEEE802.11b RLAN devices (which are nearly all of the RLAN devices currently in use) have 13 channels in Europe (11 in USA and 14 in Japan), but, due to their adjacent channel rejection performance, it is not possible to use more than 3 of those channels at any single location;

- Bluetooth devices operate with frequency hopping across all of the available channels, and therefore they effectively occupy the whole band.

Consequently, while the market for 2.4 GHz tetherless devices is likely to grow for several years, work has also been ongoing to develop other frequency bands for the future.

15.7   The 5 GHz Band

As a consequence of the concerns over the capacity and occupancy of the 2.4 GHz band, 'expansion bands' have been identified at around 5 GHz. The use of the band for tetherless applications was actually anticipated back in the early 1990s, although work on defining its availability is still ongoing. This work was started within Europe, where it was considered (at the time) that the 5 GHz band was being underused, and therefore could also be used for tetherless (RLAN) applications. The bands being considered/developed are:

- 5150-5350 MHz;

- 5470-5725 MHz;

- 5725-5875 MHz.

However, all of these bands are already identified in the ITU-R Radio Regulations for other purposes, including:

- fixed satellite service;

- aeronautical radionavigation service;

- radiolocation service;

- earth exploration satellite service;

- maritime radionavigation service.

The 5725-5875 MHz band is also an ISM band.

15.7.1   Identification of the 5 GHz Band for RLANs

RLANs (and other tetherless devices) are intended to be portable, and therefore they have to be considered as falling within the mobile service.

As Table 15.3 shows, there was no global allocation for the mobile service in this band before the WRC in 2003. Through a footnote (5.447), the Radio Regulations provide an 'additional allocation' for the mobile service in the band 5150-5250 MHz, in 27 countries (including all of the EU countries, except Ireland). Furthermore, following technical studies, CEPT approved in 1999, a European Decision to identify the bands 5150-5350 MHz and 5470-5725 MHz for

HIPERLAN (high performance radio local area networks). However, both of these provisions were specific to a limited number of countries, and did not provide the prominence and long-term security (of access to the spectrum) which was required to ensure the commercial success of these important access technologies.

Table 15.3   Frequency allocations in the 5 GHz band (pre-WRC 2003).

Frequency	Allocations	
5150 MHz		
5250 MHz	Aeronautical radionavigation	Fixed satellite service
	Radiolocation	Earth exploration satellite service
5350 MHz		
5470 MHz	Maritime radionavigation	
5650 MHz	Radiolocation	
5725 MHz		
	Radiolocation (Also identified as an ISM band)	
5875 MHz		

These two provisions (the footnote in the Radio Regulations and the ECC Decision) were intended to enable the implementation of devices complying with the ETSI HIPERLAN standards, which had been in development for a number of years. Following this work in ETSI, the IEEE802.11 committee started work on developing a similar standard for an RLAN in the 5 GHz band; that has now been published as IEEE802.11a.

Following industry lobbying, the US Federal Communications Commission (FCC) identified the bands 5150-5250 MHz and 5725-5825 MHz as 'Unlicensed National Information Infrastructure' (U-NII) bands. The IEEE802.11a standard is designed to operate in those U-NII bands.

Consequently, ETSI and IEEE had produced standards for RLAN devices, even though there was no provision in the Radio Regulations for their operation, other than a footnote which applied in the band 5150-5250 MHz for a limited number of countries. However, this did not preclude the use of these devices, since they can be used at the discretion of the national regulatory authority, providing it has satisfied itself that the devices will not cause harmful interference to the other users of the band (as mentioned in section 15.2.4). For RLAN devices, this was an unsatisfactory situation, since it could lead to fragmented and inconsistent regulation between countries. This would be particularly problematic for consumer products which are intended for operation by users who are likely to be either unaware of, or indifferent to, the varying regulations which apply in each country. For such devices, it is important to harmonise the regulations as much as possible throughout all countries. Harmonised regulations would also create a consistent market for the products, which should enable economic benefits to the advantage of the consumer.

Therefore it was important that the Radio Regulations do include a global allocation in these bands for the mobile service, since it would provide a signal that it is possible to use the band for tetherless devices, which would serve to encourage regulatory authorities to permit their operation. This would tend to lead towards a more harmonised regulation across all countries.

15.7.2   Possible Allocation for RLANs at 5 GHz in the Radio Regulations

As a consequence, a proposal was made to the WRC in 2000, that the next WRC (in 2003) should consider making an allocation in the 5150-5350 MHz and 5470-5725 MHz bands for the mobile service, for the purpose of using the band for RLANs. The 3-year period between the conferences was to provide sufficient time for the difficult sharing studies to be undertaken, to confirm whether, and under what restrictions, the RLAN devices would be able to share the bands, without causing harmful interference to the existing users. Those studies have been complicated by the (understandable) reluctance of the various military organisations to divulge the characteristics of the radar systems ('radiodetermination service', which includes 'radiolocation' and 'radionavigation') which are currently operating in the bands.

The commercial pressures from industry to open up these bands for RLAN devices, have been converted into political pressures to achieve a successful outcome (for all parties). As a consequence, a compromise solution was developed which was accepted by many countries, and formed the basis of a successful solution at the WRC 2003. (It is only at the WRC that the final decision could be made, although preparatory meetings had produced a 'short list' of likely outcomes.) This compromise was close to the solution proposed by Europe; this is not entirely surprising, since that solution had been developed following detailed sharing studies that had been undertaken within Europe during the late 1990s.

The key to solving the 5 GHz sharing problem was the inclusion of a 'smart technology' called dynamic frequency selection (DFS), which would ensure that the RLANs always check the frequency band, both before and during use, to ensure that no radars are occupying the same radio channel. This feature was first specified in the CEPT Decision of 1999, and is now generally believed to be the solution to the band sharing problem. However, it is a technology that has never been used before to ensure protection of one radio service from another, and the efficacy of DFS is still under study. Much depends on the values of the parameters which are specified for the detection and identification of the radar systems. These have been calculated for the specific scenarios which may arise between radars and RLAN devices.

With the specification of DFS, it was possible for an agreement at the WRC 2003 for an allocation to be made in the 5 GHz band, for the operation of RLAN devices.

15.8   Ultra-Wideband Devices

The concept of ultra-wideband (UWB) devices has been widely promoted by many, as a way of operating short-range radio devices (including for tetherless communications), without the 'problems' of conventional regulation (see Chapter 5).

The principle behind UWB is that the devices operate with the same transmit power as any similar communications device, but they operate over an extremely wide bandwidth (relatively speaking), e.g. across the whole of the band from 3.1-10.6 GHz. This results in a very low transmit power spectral density (in mW/MHz). As a consequence, it is claimed that they can operate 'over' many other radio users/ services, since only a tiny ('immeasurable') proportion of the transmit power of the UWB device will fall into the receivers of each of the current users of the bands.

However, this may not necessarily be sufficient to avoid causing interference to the current users, and further studies need to be undertaken. In particular, there are two situations which need to be studied further:

- while each UWB device may introduce negligible interference into each of the bands over which it is operating, the aggregate effect of many UWB devices, could give rise to significant interference levels, particularly if they are as widely adopted as has been suggested (e.g. several in every car for collision detection and avoidance);

- alternatively, it is possible for a UWB device to be sited so that a 'significant proportion' of the transmitted power would fall into the receiver of some other radio system — this could be due to either the short separation distance between the UWB device and the existing user, or because the UWB device is operating within the 'beam' of an existing radio system, and hence the radio would be particularly sensitive to interference from that direction.

At present, the US Federal Communications Commission (FCC) has granted temporary permission for UWB devices to be used for certain specific purposes (including telecommunications), although that is under review. The European Regulators are very interested in the experiences in the USA, but are taking a cautious line, preferring to initiate studies before permitting licensing of such devices.

In Europe, CEPT currently has two project teams addressing spectrum management for UWB devices (including sharing studies), while there is also a Task Group in ETSI looking at developing standards for the devices.

At a global level, there is now a new Task Group (TG1/8) in the ITU-R which will address the questions of regulation, band sharing, etc, for UWB. It is possible that, if this is such a new approach to the use of the radio frequency bands, it may be raised as an issue for further study at a future World Radio Conference — most likely in 2007, although no specific proposal for such work has been developed as yet.

15.9   Open Spectrum — an Alternative to Spectrum Management?

Some visionary people have suggested that the idea of spectrum management is an anachronism, since modern radio devices should not need to be regulated and planned in such a rigid (and time-consuming) manner.

The idea is that a modern radio system should be sufficiently 'intelligent' to be able to scan around and find a frequency band which is currently unoccupied. This would enable radio equipment to use any available channel which is currently unoccupied, and hence maximum use can be made of the radio spectrum. In effect, it is very similar to the DFS mechanism, which was described above (section 15.7) as being the key to enable RLANs to share the 5 GHz band.

In theory, there is no reason why there should not be some 'open spectrum' in the future, which would enable the available frequency bands to be assigned and used more efficiently. However, it depends on one key criterion, namely that there is an understanding of what is meant by 'the band is occupied'.

In practice, before choosing a channel on which to transmit, a radio terminal needs to be able to monitor the frequency band, and identify whether, on each channel, there are any radio signals which are sufficiently strong as to be intelligible to another user.

This would most likely be done by comparing any detected signals against a preset power-spectral-density threshold level.

Unfortunately, considering the vast range of different types of radio systems, used for both communications and non-communications applications, there is a large difference in 'normal receiver signal level' and receiver sensitivities. For example, a simple consumer device (such as an RLAN) would not be able to 'hear' the operation of a satellite downlink, for which a very large dish antenna is normally required to receive an intelligible signal.

As a consequence, to identify whether a channel is available for use, it is necessary to have prior knowledge of the types of system which will be operating in the band. Then it will be necessary for all radio systems operating in the band to be able to detect signals at the lowest power spectral density operating in the band.

Therefore, it may be possible to implement 'open spectrum', but it could not be done equally across all bands (as is currently envisaged by some), because it would require equipment which is capable of operating in similar power spectral density ranges. Consequently, it is most likely that 'open bands' would have to be specified, where each band is identified for a type of application.

Such ideas are starting to be considered, although it is likely to be many years before there is any widespread adoption of such a technique.

15.10   Summary

Spectrum management is a key element in the development of any new radio technology. Due to the multiplicity of existing radio applications, it is necessary for global regulation to be applied over the radio frequency bands, to ensure that all permitted users can operate without causing harmful interference to others, or receiving harmful interference from others.

There is currently a considerable amount of effort ongoing around the world, to address the regulatory requirements to enable tetherless applications to gain access to sufficient suitable spectrum to meet the anticipated demand. This has to be done in the context of the many other demands for radio applications, to enable the available spectrum to be used in the optimum manner.

Therefore, the regulation is essential to ensure that all legitimate users of the radio spectrum can continue to operate reliably. As a consequence, BT maintains a spectrum management capability, both to protect its existing networks and investments, and to provide expertise for the deployment of new technologies, through co-ordination with the national, regional and global radio regulatory bodies.

Reference

1   ITU — http://www.itu.int/

16

MOBILE MULTIMEDIA SERVICES

J A Harmer

16.1   Introduction

Over the last ten years the mobile cellular industry has moved from a niche voice service provider to a mass-market, multi-billion dollar industry. Good quality mobile voice services are now expected at an acceptable price, and international roaming, provided by the GSM mobile standards, is another 'must-have' service.

With aggressive competition and mobile voice usage reaching saturation in many countries, mobile network operators and service providers have seen falling ARPU (average revenue per user) from standard voice services. Right across the mobile industry there is a need to increase revenues from non-voice services. The challenge is to provide appealing applications and services that will drive up usage, and therefore revenue, from data. The contribution of data to mobile operators' revenues has grown steadily over the last two to three years. Many operators currently gain around 15% of their revenues from mobile data, and expect this to increase to 25% by the end of 2004. Some industry analysts such as Yankee Group believe that mobile data will contribute over 30% of revenue by 2007 [1]; the UMTS (Universal Mobile Telecommunications System) forum believes that 50% is achievable by 2006 [2].

Mobile network operators have already invested in the future and mobile data, both through upgrading their networks to offer packet data services over GPRS (general packet radio service), thus laying the foundations for 3G (third generation) mobile services, and also by the purchase of 3G licences for radio spectrum. 3G networks have been launched in the UK, Japan, Italy, Sweden and Austria with more to follow over the coming 12-18 months.

Mobile multimedia has been the goal for technologists and the mobile industry alike, and now many components are coming together to make it a reality, e.g. mobile data networks, devices with colour screens, multimedia content. This chapter takes a look at the present status and future prospects for mobile multimedia, from

SMS-based services through to 3G services, complemented by wireless LAN hot-spots. The chapter also considers the importance of appropriate business models, value chains and billing, along with some potential solutions and developments.

16.2   Mobile Multimedia Vision

We have all seen videos and films where pocket-sized devices are used in the home, outdoors, abroad, on trains and taxis to videoconference, send pictures, make reservations and payments, and be entertained. In these scenarios, the services and applications are the sole focus — the network and its capabilities are hidden from the user and taken for granted. The appeal of mobile communication is not in doubt, enabled by:

- personalised devices and services — the phonebook on the mobile itself is essential for many users, while ringtones, screensavers and games are used to customise telephones to individual preferences, and, in future, the services and applications used will also be personalised;

- immediacy — the mobile telephone has revolutionised our ability to communicate without regard to time of day, location or even country;

- localisation — handover and international roaming means that we can be contactable virtually anywhere, and, although some exist now, the range of location-based services will increase over the next 5 years [3];

- ubiquity — as many countries enjoy 99% population coverage for GSM and GPRS services, users have come to expect good coverage as the norm, and so, for new 3G operators, service continuity across 3G and 2G networks is a very important factor in meeting user expectations;

- ease of use — while the explosive growth of SMS shows that people are prepared to use services that are valuable even if the user interface is difficult, ease of use is important on small mobile devices; colour screens and icon-based menus add to the usability of mobiles, and will also encourage users to try new data services and downloads.

Compared to other devices, many people have a special 'relationship' with their mobile. Indeed, some suppliers are starting to take different approaches to selling mobile telephones. For example, '3' has adopted a 'beauty salon' type feel to their stores, while Chinese manufacturer DBtel is planning to open outlets in the UK modelled on jewellery stores, offering high-end jewel-encrusted mobiles which are popular in the Far East [4].

Although initial hype for 3G has dampened industry expectations, there is still a vision for mobile multimedia fuelled by the extraordinary growth of SMS, the realis-ation of MMS (mobile multimedia messaging service), and the launch of 3G networks.

3G has always been considered an 'evolution' and not a 'revolution' and the development of mobile multimedia should be viewed in the same way. The multimedia roadmap starts with the here and now, for example ringtones delivered over SMS. MMS adds the capability for picture messaging, and GPRS enables faster access to mobile portals and information services. 3G will offer greater capability for video and multi-tasking of applications. Although the topic of this chapter is multimedia, it is important to note that multimedia applications are the richest of a broad set of applications that include information, commerce, messaging, e-mail, personal information management (PIM) and, of course, voice, and the great majority of these services will be available on GPRS networks as well as 3G.

Potential revenues from mobile data and multimedia could be large. The analyst specialists, Strategy Analytics, expect mobile multimedia revenues to grow to $30bn by 2007 [5]; the UMTS forum predicts worldwide demand for 3G services to generate $320bn by 2010, of which the majority ($233bn) is from new data services [2].

Studies by the UMTS forum support the 'evolution' view — it is expected that the adoption of 3G will be slow in the early years, and used by less than 30% of the global subscriber base even by 2010.

For mobile network operators and service providers it is essential to start offering and marketing new multimedia services now. To date, mobile networks have offered users a new way of accessing existing applications (voice and messaging) and in doing so have created a new user base of mobile customers. Personalised multimedia gives the opportunity to offer new services to an existing customer base. This is essential if ARPUs are to rise in the future. There will certainly be challenges in encouraging customers to use the new services and to quickly upgrade to new multimedia terminals (why have the latest PC if you're quite happy with your old 486 one...?).

Many of the opportunities for mobile multimedia will arise when mobile operators and service providers work more closely with businesses, content providers and device manufacturers, for example:

- marketing and product awareness, through fun interactive games and downloads;

- a channel to market as part of an enterprise's multimedia call centre strategy;

- a channel to market for the media industry — building on the success of ringtones to include audio-track streaming and download, video clip download, pop group trivia and 'mobile merchandise', sports highlights, and cartoons;

- vertically integrated devices and applications (e.g. cameras/camera phones, surveillance);

- location-based multimedia, travel news, maps and pictures, picture messages sent to subscribers within a location (e.g. university campus).

16.3   The Roadmap for Mobile Multimedia

As previously discussed, mobile multimedia is not for 3G alone. In fact, important building blocks have already been laid on existing 2G and early 3G networks. This section reviews the devices, standards and applications of mobile multimedia and how they form a roadmap for mobile multimedia that culminates with 3G.

16.3.1   SMS Messaging — R U RDY 4 IT?

Although mobile operators' data revenues are growing, SMS — or 'texting' — is still the most dominant form of mobile data. Most industry analysts believe that SMS traffic will continue to grow and will be the mainstay of mobile data revenues for several years to come. This is illustrated in Yankee Group's forecasts [1] in Fig 16.1.

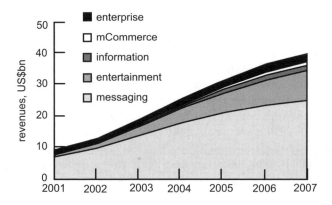

Fig 16.1   Messaging forecasts (Yankee Group [1]).

Despite high use of SMS in the consumer segment, it is important to note that today this is nearly all 'person-to-person' messaging. A major challenge of mobile multimedia is to increase the use of 'machine-to-person' applications such as multimedia downloads and games. SMS is already playing an important role in the growth of machine-to-person applications through the delivery of and payment for ringtones. SMS is used to collect revenue for value added services though premium rate SMS charges. Other popular SMS applications include voting, competitions and text-based games. The Mobile Data Association [6] has been tracking SMS usage since 1998 and has reported on consistent growth year on year. Daily SMS traffic in the UK is now 58.5 million (compared with 38 million in 2001). The MDA forecasts 20 billion messages for this year.

16.3.2   Mobile Multimedia Messaging Service (MMS)

SMS is very limited in terms of the content that can be carried (text only) and the size of message (160 characters, although some telephones enable message concatenation). MMS extends the richness of messaging by adding pictures, captions, sounds, music and moving images.

16.3.2.1   MMS and How It Works

MMS enables multimedia messages to be sent to other MMS telephone users, or to e-mail users. Messages can be composed on an MMS mobile telephone (for example incorporating a picture taken with an integral camera) or on a Web site. Messages are sent and received using either circuit-switched data or over GPRS using WAP. Use of the GPRS data service by MMS applications is very attractive to mobile operators as a means of generating higher data revenues. Incoming messages to MMS telephones are received as 'push' alerts notifying the location of the multimedia message which can then be retrieved. Depending on the configuration of the telephone, retrieval can be automatic or by the user accepting the message. An example MMS architecture is shown in Fig 16.2. MMS standards are set by 3GPP (3rd Generation Partnership Programme) and the OMA (Open Mobile Alliance). To improve interoperability, conformance documents have been agreed between vendors to detail the minimum specifications required for interoperability. Nevertheless, the standardisation of some aspects of the MMS architecture has been slow, for example the interfaces between mobile multimedia service centres (MMSCs).

16.3.2.2   MMS Devices

MMS-enabled mobile telephones have colour screens, greater capacity for message storage, polyphonic ring tones, and integral or accessory cameras. Some MMS telephones, without cameras, offer 'receive only'. Although the network infrastructure — MMSCs — were available in 2001, telephones were not available until April and June 2002 when Ericsson's T68i and Nokia's 7650 respectively were launched. A wider variety of MMS and picture-phones is now available with dozens of MMS capable handsets on the market now from all the major handset vendors. Handset availability and price will continue to be an important factor in determining the speed of take-up. MMS is only likely to truly take off when a 'critical mass' of handsets exists — between 15-20% penetration. People want to send their pictures to others! There is evidence of this phenomenon in Japan, where there are over 10 million camera telephone subscribers and picture messaging is extremely popular. There are estimated to be around 750 000 MMS capable phones in the UK compared with 50 million SMS enabled [7] but attractive prices and wider choice will accelerate uptake.

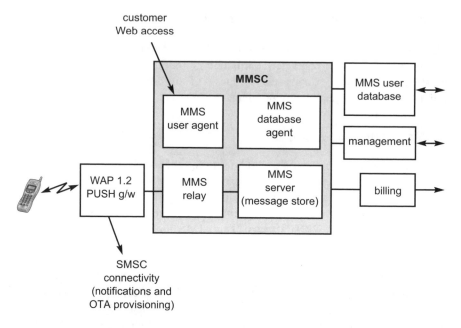

Fig 16.2   MMS architecture.

16.3.2.3   *MMS Applications*

The initial focus for MMS has been person-to-person picture messaging. However, the ultimate goal for MMS must be to stimulate use of application-to-person messaging. BT Exact has been involved in MMS application development and deployment through its Rocking Frog personalisation and profiling project (see Fig 16.3). A suite of applications has been developed which uses MMS to good effect. Examples are:

- 'now playing' — information about current radio tracks or TV broadcasts, linked to picture galleries about artists;

- dedication service — linked to a radio station play list, users can dedicate a track with a personalised MMS video or image and greeting;

- programme guide — sends personalised information about programmes and features that are of interest, along with reminders to watch or listen.

Another example of launched MMS services is Vodafone's 'Live!' service which encouraged 380 000 user registrations across its European footprint during October and November 2002. By March 2003 this had grown to 1 million. Services include picture messaging, instant messaging and games download. In Japan J-phone's 'Sha-mail' service enables millions of customers to send photos using integral cameras using a proprietary MMS-like service (although J-Phone plans to migrate to

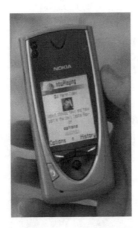

Fig 16.3   Rocking Frog MMS application.

the MMS standard). Following the popularity of Web logs 'blogs', a mobile equivalent — coined 'photoblogs' — is developing as a means of capturing and storing on the Internet pictures taken with MMS phones.

As the number of MMS users grows, it is easy to see how person-to-person messaging could substitute for 'traditional' postcards and birthday cards, enabling personalised greetings, pictures and sounds to be sent. For the consumer market, application-to-person messaging will be driven by the availability of suitable and fun content, such as picture galleries, ringtones, and games.

The business market for MMS will be more challenging. Mobile marketing and advertising are natural opportunities for businesses to make the most of technologies and devices that their customers will be using. Business-to-employee applications could also include 'push' images for surveillance and security use, delivery of multimedia stock, share and performance data, sending pictures of incidents or equipment for expert advice.

In November 2003 there were 115 operators worldwide offering MMS services to their customers. Of these 67 are in Eastern and Western Europe [8]. Interoperability and interconnection has been a problem initially for mobile operators. There was a parallel in the early days of SMS, when it was only possible to send messages to devices on the same network. But today's users are accustomed to full SMS interoperability and expect the same for MMS. Difficulties arise for both technical and commercial reasons.

- Between mobile devices

  Different handsets format pictures differently and vary in terms of screen size and colour representation. It is the task of the MMSC to convert images to be displayed on different devices. There is a risk, however, that the user experience will be poor, or messages reduced to the lowest common denominator — which is SMS!

- Between MMSCs

  In order to get MMSCs into the market, many vendors' solutions pre-date the 3GPP standards for interconnection. For this reason interconnection between MMSCs is proprietary and can be difficult.

- Between networks

  Interconnect agreements are starting to take place, and the GSM association has issued "MMS Interworking Guidelines", Permanent Reference Document IR.52. As with SMS, termination charges between operators are likely.

- Billing

  There have been problems integrating new MMSCs within existing billing and rating systems.

These problems are in the process of being resolved and mobile operators in the UK and other European countries have announced interoperability.

16.3.3   Downloads, Games and Multimedia Streaming over GPRS

There are now around 190 GPRS networks launched worldwide — with another 30 under way and a further 25 planned [8]. Network operators are also enabling international roaming for their GPRS data subscribers. There is a wide variety of handsets available offering colour screens, multimedia messaging services and downloadable games and applications.

Now that both GPRS networks and terminals are available, businesses are starting to tackle some of the dilemmas that they face in the launch of mobile solutions into their businesses: 'What is the return on investment?' 'What devices should be used?' 'Which users in the organisation should have them?' 'Which applications should be mobilised?' 'Who can help?' Pilots and trials are proving an effective way of overcoming many of these barriers and BT Exact has developed bespoke solutions for customers including travel and retail companies. Typically these solutions provide access to existing corporate e-mail or data systems and are not multimedia! However, GPRS is an important solution for enabling businesses to access critical information on the move.

Consumers, on the other hand, want to access content and information and the fact that the GPRS network is used is not of importance. In this context the customer is buying 'games download' or 'video clips' rather than the technology used to deliver the content (which in most cases will be WAP over GPRS). However, the increasing availability of GPRS-enabled telephones and the improved user experience, when compared with WAP over circuit switched data (CSD), is gradually having an impact on the use of mobile data. According to the Mobile Data Association [9] which monitors monthly WAP usage, there are well over 25 million data-enabled telephones in the UK. WAP usage is growing, with a daily average of 30 million WAP page impressions in September 2003, compared with 11 million in September 2002.

The latest multimedia mobile devices, combined with the GPRS mobile data network, offer new potential for applications and services, including applications downloaded to the device. However, there are also challenges that ultimately affect the user's end-to-end multimedia experience:

- terminal compatibility with downloadable applications — as not all mobile devices support the same standards, or have proprietary extensions, not all devices can run all applications;

- network bandwidth and delay — when compared with wired networks, cellular networks offer limited data capacity (around 40 kbit/s), and only 'best-effort' delay, which can result in long download times and poor performance.

The mobile industry continues to seek new opportunities to generate revenues from data applications. Giving users the ability to download applications will generate revenues, but this is potentially a double-edged sword — once downloaded the user can use the application repeatedly, in most cases without generating additional network traffic. In addition to the technical issues described in this section, the business model for downloads must also be understood. Users will be able to receive applications from a number of sources, e.g. from network operator/ service provider sites, application portals, third party sites and also peer-to-peer — transferring applications directly between devices. For this reason, a commercial framework is needed that includes authentication, security, billing and payment, and digital rights management to control the use and reuse of content.

16.3.3.1   *Mobile Device Application Download and Execution Standards*

Just as there are a number of operating systems for mobile devices (Symbian, Palm, Microsoft), there are also different solutions for application run-time environments. The two most prominent are Java2 micro edition, popular for European and Japanese devices, and BREW which, although later to market, is enjoying success in the USA and Korea. Overall the number of telephones with run-time environments supporting downloadable applications is growing quickly.

Java2 Micro Edition (J2ME)

Sun's J2ME is a compact wireless version of the Java programming and run-time environment, designed for devices with limited memory capacity and power. Elements of the standard directly relevant to mobile devices are:

- connected limited device configuration (CLDC), which includes a small footprint (186k) Java virtual machine (KVM), a Java run-time environment which has been designed to fit within the constraints of mass-market, low-price, mobile devices;

- mobile information device profile (MIDP), which provides application functionality such as connectivity, user interface, and storage on the device — application programming interfaces (APIs) are provided to enable application developers to use the MIDP functionality (MIDP2.0 includes extensions for audio, gaming, security and improved user interface support);

- wireless messaging API (WMA), which enables applications to use a messaging service for communication, for example (but not exclusively) SMS;

- mobile media API (MMAPI), which enables applications to carry out video or audio playback or recording — specified media types are to be specified and supported.

A major problem with current implementations of J2ME is that of incompatibilities. MIDP1.0 has been launched into the market but has many proprietary elements (around 50 million Java-enabled mobile devices have been deployed worldwide) [10]. This makes it difficult to economically develop applications for large numbers of different devices. MIDP2.0 is close to standardisation to alleviate these problems and could give a 'kick-start' to the market for downloaded applications.

An important step forward for J2ME in the mobile industry is an activity known as 'JTWI' (Java Technology for the Wireless Industry). An international expert group comprised of mobile vendors, mobile operators and software vendors is working on a specification (called JSR185) that identifies all the mandatory technologies that must be included in devices in order for them to be JTWI compliant. Mandatory technologies include CLDC 1.0, MIDP2.0, WMA and MMAPI. Through standardising mobile telephone functionality and APIs and tools for programmers, it will be easier to develop and launch mobile applications on a wide range of telephones, smartphones and PDAs and on a wide range of wireless networks. A roadmap for JSR185 was issued in January 2003 and compliant devices are likely to be available in early 2004.

Qualcomm's Binary Run-time Environment for Wireless (BREW)

Like J2ME, BREW enables applications to be developed that can be downloaded and executed on mobile devices, without the need for developers to have access to or knowledge of the mobile telephone chipset and operating system. BREW was developed by Qualcomm [11] and provides an application software interface layer for devices using Qualcomm's chipset. For this reason BREW is found on CDMA networks and has been very successful in Korea and the USA. BREW provides a client-side development environment based on C and C++. However, BREW extensions enable other environments (including J2ME) and browsers to run on the BREW platform as an application. The BREW2.0 client enables these to be delivered and provisioned automatically over the air. Applications can be recalled and upgraded by mobile operators, over the air, without user intervention.

Unlike J2ME, BREW includes a distribution and billing system as well as a development environment, so applications can be brought to market quickly. An example in the USA is Verizon's 'get it now' application download service which has spurred 8.5m application downloads in the six months since the service has been launched; 3.2m enabled telephones have been deployed.

16.3.3.2 Games, Downloads and Multimedia Streaming — Who's Doing It?

Games and Downloads

Many industry analysts are bullish about the future for mobile gaming. Types of games range from those embedded in telephones (such as the ever popular 'snake'), to WAP-based games and multiplayer games with a common server. However, downloaded games are currently of high interest to mobile operators and service providers, where the game is purchased, downloaded and played off-line.

Mobile gaming is most prominent in the Asia-Pacific region where advanced handsets have been available for longer. According to Ovum [12] it took 18 months for 50% of i-mode users to have Java-enabled handsets. In South Korea, games downloads are the second most popular application (character downloads are most popular).

Closer to home, all UK mobile operators are offering downloadable games. For example, mmO$_2$ has launched a 'games arcade' [13] featuring over 100 games, each compatible with selected Nokia or Siemens multimedia telephones. Vodafone Live! includes a games arcade for Java-compatible devices including the Nokia range and Sharp GX10.

BT has developed avatar messaging to complement video and picture messaging. Avatar messaging allows a text message to be spoken to the recipient using an avatar of the sender, or a selected avatar which could be a celebrity, cartoon character, etc. An advantage of avatar messaging is that the amount of data sent over the mobile network is significantly less than for mobile video-streaming. The solution could be used for customer care and user information and instructions, in addition to fun messaging applications. The roadmap for BT's 'Erica' platform (described later in section 16.5.1) also includes provision for games download.

Multimedia Streaming

MMS-capable telephones on the market today offer picture messaging and video messaging — where a video is captured on the camera-phone and sent to a recipient. True video-streaming from network to mobile telephone is only just starting to appear. A major challenge is delivering good enough quality over the restricted bandwidth of today's mobile networks.

BT Exact has developed a solution for mobile streaming, called Fastnets (Fast Start, Network Friendly Streaming — see Fig 16.4), that is highly optimised for the challenges of mobile networks (see Chapter 18 for more details). Initially developed for PDA clients such as the Ipaq and xda, the player has been ported to mobile multimedia telephones such as the Nokia 7650.

Fastnets provides a number of features that makes this solution so successful for mobile use, some of which are described in more detail below.

Fig 16.4   Fastnets for PDA and mobile telephone.

- Lightweight player

    At around 75 kb, the player is small enough to be downloaded to the device over the air via GPRS, or via Bluetooth, an important consideration for future upgrades and enhancements.

- Fast start

    The player starts to play out the video-stream with minimum waiting for the user. For mobile people who 'want it now' this is an important feature; users who have to wait too long for buffering to happen will not come back again to use video-streaming services, nor — as importantly — pay for them.

- Network optimised

    Through optimal compression and error correction and adapting to network conditions such as congestion, great quality voice and video can be achieved over today's GPRS networks.

- Standards based

  Fastnets is compatible with 3GPP standards.

GPRS is a shared medium. It should be noted that under extreme scenarios, such as accessing GPRS for video-streaming at a football stadium, the performance of this and many other applications suffers. This is due to the priority given to voice services, the lack of available GPRS data bandwidth or quality of service. The deployment of 3G networks addresses some of these issues as it is able to support greater data rates, higher capacity of users and dedicated data channels for applications such as video.

16.4   Multimedia and 3G

The auction of licences for 3G radio spectrum created a vast amount of interest — and hype — in the capabilities of 3G and when new multimedia services would be available. 110 3G licences have been awarded globally, but to date few networks have been launched:

- mmO$_2$ launched its Manx Telecom showcase on the Isle of Man in October 2001 with around 150 single mode 3G handsets;

- NTT DoCoMo's 3G launch in Japan in October 2001 has now been extended to cover 69% of the Japanese population;

- Sonera has launched multimedia applications under the banner of 3G but over a GPRS network;

- Hutchison's '3' service was launched in the UK (on 3 March 2003) and has also been launched in Italy, Austria, Sweden, Australia and Hong Kong;

- Mobilkom launched its 3G service in Austria in April 2003.

With around 2-3 times the data rate per user compared with GPRS, 3G networks provide the capability and capacity to offer richer mobile multimedia to users. True two-way videoconferencing is offered over a circuit-switched bearer delivering 64 kbit/s — similar to ISDN today. However, for packet data services including downloads, browsing and for network-to-telephone streaming (e.g. video clips), higher capacity is available — in theory up to 384 kbit/s. 3G handsets available today are limited to 64 kbit/s uplink (reasons for this include power consumption and heat dissipation). This gives an asymmetric service with higher data rates downlink. However, it is unlikely that the full 384 kbit/s data rates will be achieved. Compared with GPRS, 3G can support more concurrent users at 64 kbit/s. A practical rule of thumb for 3G data rates is 64 kbit/s uplink with 64-144 kbit/s downlink.

Each 3G launch has helped to move the technology and industry a big step forward — mmO$_2$ for the network and radio access, FOMA for terminals, and '3' for 3G/2G interworking.

16.4.1   UK — Manx Telecom Applications Showcase

The first 3G network with a strong focus on applications and multimedia was Manx Telecom's 3G showcase on the Isle of Man. Manx Telecom is part of the mmO$_2$ group. BT Exact was a showcase partner, delivering voice and data network design, interworking, subjective testing, security, applications evaluation and integration, and end-to-end testing. Understanding the performance of multimedia applications over a 3G network was an important part of the applications selection process. The combination of bandwidth, delay and jitter determines the end-user experience. For some applications it would be better not to launch at all rather than risk poor performance. Network simulation was used (even before the 3G network was available) to test applications under varying bandwidth, round-trip delay, jitter, packet loss and breaks and pauses in transmission. Using both network simulation and the characteristics of the live 3G network, applications were selected on the basis of their performance. The showcase currently exhibits a broad range of applications including video-streaming, conferencing, information services and games (Fig 16.5).

Fig 16.5   3G applications — from information to games.

16.4.2   Japan — from i-mode to 'FOMA'

16.4.2.1   i-mode — 2G

The Japanese market has demonstrated a leading role in mobile multimedia through the popularity of NTT DoCoMo's 'i-mode' applications. i-mode was launched in 1999 and since then has attracted over 36 million subscribers (over 80% of NTT DoCoMo's cellular subscriber base) [14, 15]. The company has also managed to grow data ARPU to around 20% and reduce customer churn to less than 2%.

i-mode offers a huge array of applications, through a combination of 'official' i-mode sites (of which there are about 3000) and 'unofficial' sites which exceed 53 000 in number. Official sites are required to comply with strict guidelines to ensure quality of applications. The most popular applications are e-mail, screen savers, character downloads, banking, games and ringtones. The i-mode business model is based on revenue sharing with content providers and application developers. Users pay a subscription fee for most i-mode services, and also pay per packet for airtime when accessing the services. Further revenues are derived from transactions and advertising. In 2001 the i-appli service was launched on DoCoMo's 2G network, enabling downloads to Java-enabled telephones. Camera-phones were launched in June 2002 and 10 million have been sold so far and are used for the 'i-shot' service.

i-mode has been very successful, but there is increasingly strong competition in Japan. For example, J-phone launched its very popular Sha-mail photo-messaging service in late 2000, far in advance of i-mode's launch in 2002 and KDDI has launched telephones with integrated GPS satellite positioning to enable location-based services. Despite these challenges, i-mode remains strong as a result of NTT DoCoMo's power over suppliers, in specifying the system end to end (including terminals which are NTT DoCoMo branded), and having an established business model.

16.4.2.2   Freedom of Multimedia Access — 3G

In Japan, NTT DoCoMo's 3G network was launched in October 2001 and now provides coverage for 90% of the population. As on the Isle of Man, the network provides a packet-based service offering 64 kbit/s uplink and up to 384 kbit/s downlink. There is also a 64 kbit/s circuit-switched data service. 3G services are offered under the name FOMA (freedom of multimedia access); multimedia applications include:

- digital video and picture messaging — sending video images and still pictures via the telephone;

- live videophone which can interwork with PCs on the fixed network;

- 'i-motion' applications are provided for compatible FOMA terminals, including news, sports highlights, music videos, film previews — the video services are based on MPEG-4 and are multiplexed using the ASF mobile profile format, with a maximum file size of 100 kb;

- i-appli services enable applications to be downloaded to the telephone, including video games, maps, weather reports, and stock applications — there are currently four devices available that are compatible with download services;

- 'M-stage visual' requires a specific telephone (the FOMASH2101V) and offers access to film content over a 64 kbit/s circuit-switched link — around 80 channels of content are available from a range of broadcasters, examples being Hollywood channel, Cartoon Network, sports content, pasta recipes, and fashion.

Initially the take-up of 3G services in Japan had been slower than expected, especially given the previous success and high usage of i-mode services. Part of the problem was low penetration of 3G handsets — many of the services (such as videoconferencing) are only available between FOMA devices. In recent months there has been an upturn in usage, stimulated by a number of factors — NTT DoCoMo has invested heavily in handset development, with the latest devices having far improved battery life, and prices having been capped to make 3G telephones more affordable. There were around 420 000 FOMA users in April 2003 (with 138 400 new additions contributed during March) [16]. This growth is very encouraging for the mobile industry.

16.4.3   Europe — '3'

Unlike FOMA which launched with single-mode (3G only) telephones, Hutchison 3G (using the brand '3') in the UK had to contend with the challenges of interworking between 3G and 2G networks to enable their customers to use the mmO_2 2G network when outside 3G coverage. A new entrant to the mobile market-place, '3' offers a selection of handsets with video calling, video messaging, video download, Java games and multi-tasking capability.

Services include video calling (two-way video), news and finance, TV and cinema listings, group talk, UK business listings, maps and directions, 'find my nearest' using postcode, and exclusive mobile Barclaycard premiership football video. Like NTT DoCoMo, '3' is planning to offer video-calling interworking with fixed PCs.

16.4.4   WLAN Multimedia Services

The use of wireless LAN has been stimulated by both availability (WLAN is now factory fitted to many laptops and PDA devices) and falling prices. The scene is now set for adoption in the office and home for private networking. But public access is also a reality; with the USA currently leading this market, public wireless LAN provides 'hot-spots' for people to access when they are 'on the pause'. While there has been considerable debate whether WLAN is friend or foe to GPRS and 3G the widely held view is that these technologies are complementary, offering different benefits for the user. In fact, attempts in the USA to roll out WLAN to emulate cellular networks has resulted in some spectacular failures.

In the UK BT launched its Openzone [17] product in August 2002, offering Internet access via WLAN hot-spots in hotels, coffee shops and other public areas.

There are now 400 live hotspots with a continuing ambitious plan to significantly increase the number of sites.

At present, WLAN is used for connectivity, replacing wired connections and few new applications are being developed. However, the combination of high data rates and coverage of hot-spots, offices and campuses offers new potential for delivering multimedia applications. As a single example, opportunities may exist in the health sector to electronically store patient records, photographs, x-rays, and charts so that medical professionals can access them from anywhere in the hospital — without the need for large numbers of fixed workstations. The key here is that the wireless potential gives rise to new applications and databases that include multimedia, which is not the case today. Other applications could include closed-circuit television and surveillance, presence and conferencing, and intranet TV. It is also possible to use WLAN access points to provide location-based services within a campus or building. A further development of WLAN applications is the integration of voice and data over a common network. BT Exact has extended its voice over IP (VoIP) trials to include VoIP over WLAN and Bluetooth.

Public WLAN (P-WLAN) is now becoming established, particularly in the USA where there are many P-WLAN providers and aggregators. Europe is now following suit with launches in the UK, Germany and Scandinavia.

16.5 Mobile Multimedia Challenges

Mobile multimedia is already a reality, but, as is the case for most new technologies, there are some challenges to be overcome in the early phases of launch.

16.5.1 Creating and Enabling Mobile Multimedia Services

As illustrated by i-mode and the media-rich applications offered on the new 3G networks, delivery of applications and content is critical to the future success of mobile multimedia. However, a challenge for network operators and service providers is to quickly evaluate, launch and deliver applications, set up billing and payment capabilities, and collect revenue — at the lowest cost.

The best mobile applications are well integrated with the network and use the special features of the network, for example the user's location. However, mobile networks have components from many vendors, and do not have the same APIs for application developers. The goal of 'write once, use on any phone or device' is a dream! Reality is that applications and content need to be adapted to the capabilities of different devices.

BT's 'Erica' project has developed a platform (Fig 16.6) to demonstrate the value of open interfaces to enable application development, hiding the complexities of underlying networks. Exemplar applications currently include a fleet-tracking application and mobile video-streaming. Further enhancements are planned to integrate J2ME and Web Services to expose APIs that are easy for application developers to work with.

In a similar way, developers will be able to access intelligent network capabilities in the mobile network such as location gateways, billing and call set-up. Through integrating SIP (the session initiation protocol), the Erica platform will also be able to demonstrate the development of 'all over IP' services required for future releases of 3G networks. Furthermore, voice interaction will be provided on the platform to ensure that applications are available to the very widest base of fixed and mobile devices.

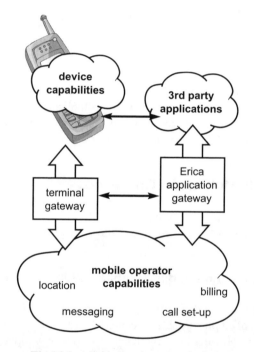

Fig 16.6   Erica development platform.

16.5.2   Interoperability Challenges

While multimedia, downloads and games offer new opportunities, there are some obstacles and barriers to be overcome.

- Device compatibility

    As can be seen with mobile operator games arcades, it is important to check the compatibility of the game with the device. Some games are compatible with one device only.

- Device configuration

  Over the air (OTA) programming is offered by mobile operators to enable users to automatically configure their mobile telephone and GPRS settings by receiving an SMS. But once again there are vendor-dependent elements to OTA programming which mean that not all telephones can be configured in this way. Web site help pages providing configuration information and help are essential to avoid very high impact on call centres.

- Lack of multi-channel, pervasive game play

  The ideal gaming experience would be one that enables participation through a variety of media, for example broadband when in the home and wireless when on the move. Pervasive gaming is a new concept that involves multiple people playing in multiple locations over multiple media, over a long period of time. Users can be alerted to developments via telephone, SMS, e-mail, etc. In time, this approach can create on-line communities of interest.

- Usability challenges

  The enhancement of mobile terminals with multimedia capabilities and programming environments, such as J2ME and BREW, should lead to a massive expansion in the range and variety of mobile applications. The poor usability of WAP services generated a lack of interest in the services offered, especially when coupled with a charging mechanism based on connection time.

The device format for many MMS users will not change significantly from that used for early WAP services, apart from the possible addition of a small joystick for menu navigation and slightly larger, colour screens with better resolution. Many devices include MMS as an additional messaging feature alongside SMS or e-mail. The danger is that users consider MMS as a simple replacement to SMS — offering little more than is possible with e-mail.

Usability extends through the user interface, to how users, particularly those who just want to use a service, can be guided through a service such as MMS without the need to understand the technology beneath.

A customised user interface is at the heart of the Vodafone live! offering, having technologies available to support a range of devices with a consistent set of options — one of these is the Trigenix user interface development environment from 3Glab.

Vodafone is not the only company developing new user-friendly suites of mobile data services, as mentioned earlier there is an industry-wide set of guidelines that ensures these services share the same capabilities. The M-Services guidelines were announced by the GSM Association in June 2001 and updated in February 2002.

They are not intended as a standard, but more a framework for a set of key technologies (for example, MMS and Java), a common look-and-feel to a user interface, and widespread interoperability within which operators are able to develop their own services. Vodafone Live! is really their interpretation of M-Services under the Vodafone brand.

Another solution that does not require the devices to be made physically larger is to add voice control and audio feedback to the traditional mobile telephone interface to make it multimodal. Using speech recognition technology, navigation through an application becomes easy with voice commands, while data entry is made hugely simpler. Audio feedback can be used to supplement what can be shown on the limited screen acreage.

Speech recognition, however, is computationally expensive and the more complex recognition tasks are beyond the capabilities of current MMS devices. Distributed processing solutions are therefore required to deliver the full range of multimodal enabled applications, with the demanding speech recognition tasks carried out on network-based servers.

Multimodal interfaces are explored in more detail in Chapter 17.

16.6   Business Models, Billing and Pricing for Multimedia

The established popularity of ringtones and growing popularity of games downloads show that people are happy to spend around £1-2.50 per item for relatively small-scale multimedia components. The launch of 3G services raises further questions regarding how much people will be willing to pay. What is the ceiling for mass-market multimedia pricing? Are these substitutional revenues? How much mobile multimedia do you get for, say, the equivalent price of a newspaper and a chocolate bar each day?

As previously described, the i-mode business model has three tiers:

- subscription for services;

- content payment;

- conveyance charge for traffic carried over the network.

These three tiers currently support a 'pay per item' model. However, this model may not suit all types of content or all types of user. Mobile operators and service providers will have to consider different and more flexible ways of billing. When users have to pay separately for content and conveyance, it is difficult to appreciate the value for money, and the value of content is not necessarily related to the volume of data. People are already familiar with paying for 'bundles' of services: pay TV is a good example. Customers using mobile applications and multimedia will certainly benefit from different pricing structures including: 'as much as you can eat' for a fixed fee, multiple uses for a given fee, bundled packages, location-based charges, 'buy one get one free', etc. Mobile operators and service providers will need to develop sophisticated mediation, rating and billing systems in order to do this. However, benefits will be reaped in terms of knowing much more about their customers and their preferences — a great 'kick-start' to personalised applications and services.

Through payment for multimedia services, mobile operators and service providers are seeking to tap into new revenue streams by being a channel to market for content providers. Some providers are offering exclusive content to their customers. It is tempting for providers to go down a 'walled-garden' path that prevents access to third party content and applications. However, the SMS experience has shown that it is important for mobile users to be able to share and communicate across network boundaries. Through offering both approved and non-approved sites, the i-mode experience in Japan has shown that these two modes can co-exist and earn revenues for all parties. Content providers are in a strong position since they own the multimedia that users will want to buy. A key strength for mobile operators and service providers is to enable their customers to purchase content quickly and easily, and in a managed way. BT's 'click and buy' service is an example of such payment schemes that could encompass the mobile domain, to provide a purchasing service that is independent of any network access.

Personalisation and profiling will also be a way in which mobile operators and service providers can make the experience better and faster for mobile data users. Content owners, spurred on by the success of ringtones and interest in mobile games, are starting to realise the opportunities offered by mobile multimedia and need to work with mobile operators and service providers to offer new services.

16.7   Summary

Mobile multimedia is already a reality, and is being used by millions of users on existing 2G networks as well as new users on 3G networks, primarily for entertainment purposes. Many European eyes still look to Japan as a case study where picture messaging has been extremely popular and where mobile data is making a significant contribution to ARPU and reducing churn. In addition, now that handset quality has improved, 3G seems to be starting an upturn. Whether in the Far East or closer to home, mobile telephones with colour screens and GPRS data access are key enablers for mobile multimedia and will encourage customers to experiment and try new services.

Compared with Japan and Korea, Europe is just starting out on the mobile multimedia journey, but a roadmap is already set which develops from MMS. As with all new technologies there are challenges to be overcome in the first 2-3 years. There are challenges for interoperability and compatible standards to enable a good end-to-end experience for all users — irrespective of their chosen application, network, or device. The industry is working together to deliver interoperability. But in the long term the winners will be the providers who can offer content and services that are personalised, consistently available (and via different access, e.g 2G, 3G, WLAN) , localised (be it content or bundled billing and pricing options), ubiquitous (despite roaming, sending cross-network, accessing from WLAN) and easy to use, coupled with an appealing and well understood charging model.

This is just the beginning. RU UP 4 IT? — imagine the sounds, animations and pictures.

References

1   Yankee Group Mobile Briefing to BT (February 2003).

2   UMTS Forum White Paper No 1 (August 2002).

3   Ralph, D. and Searby, S. (Eds): '*Location and Personalisation: Delivering Online and Mobility Services*', The Institution of Electrical Engineers, London (2004).

4   '*Mobile*', Weekly newsletter (May 2003).

5   Strategy Analytics insight (December 2002).

6   Mobile Data Association — http://www.mda-mobiledata.org/resource/hot_topics.asp

7   The Guardian, Thursday (3 July 2003).

8   GSM Association — http://www.gsmworld.com

9   Mobile Data Association — http://www.text.it/wap/default.asp?int PageId=480

10   Java — http://www.java.sun.com/products/jtwi/

11   BREW — http://www.qualcomm.com/brew/

12   Ovum Report: 'Wireless Games', (August 2002).

13   mmO$_2$ — http://www.o2.co.uk/games/

14   Gartner report: '*NTT DoCoMo: i-mode wireless Internet services*', (July 2002).

15   i-mode — http://www.nttdocomo.com/home.html

16   3G— http://www.3gnews.com/

17   Openzone — http://www.bt.com/openzone/

17

MULTIMODALITY — THE FUTURE OF THE WIRELESS USER INTERFACE

S P A Ringland and F J Scahill

17.1 The Importance of the User Interface

With revenues from voice calls projected to fall, the mobile industry has a huge challenge ahead of it to drive revenue growth from mobile data applications. Multimedia messaging applications will clearly be the first major step in this direction — however, they will not be the whole answer.

The mobile environment presents some unique challenges for data applications not seen in the traditional PC world, centring around the user interface. In the PC arena, mass-market adoption began with the move from command line interfaces to windowing interfaces. This enabled PCs to be used by a wider range of people and generated the computer industry as we know it today. However, we cannot expect to see the same explosion of applications (particularly Web applications) on mobile devices unless there are radical improvements in the mobile user interface. By comparison with current mobile devices, the PC has a couple of huge advantages in its user interface:

- it has a large screen that can display a lot of information;

- it has a full size keyboard and mouse on a stationary desk through which information can be entered with great ease.

History has shown that technologies can stand or fall by the quality of the user experience and how it matches up to user expectations. Take the examples of WAP and interactive TV, both promising access to the Internet without a PC. In reality the range of applications was limited by their poor user interfaces and both failed to live up to their marketing campaigns.

For mobile data applications to succeed, they must overcome the challenges of the mobile environment. Existing user interfaces are simply not up to the job and until they improve, mobile data applications will be restricted to niche markets for very motivated user groups. Fortunately a new breed of user interface is starting to appear which may just be in time to meet this need. Combining the power of speech recognition with the traditional graphical user interface, multimodal interfaces offer the best of both worlds. Data entry by voice is simple, while the screen is ideal for graphical output and audio can be used to supplement what can be presented on the small screens of mobile terminals.

In this chapter we look at what constitutes a multimodal interface and the benefits it brings. We examine the range of architectures that can be used to deploy these interfaces in the mobile environment. We explore the issues with content authoring and the directions current standards efforts are taking, finishing up with a brief look at the current state of the market.

17.2   What is Multimodality?

17.2.1   Unimodal versus Multimodal Interfaces

In everyday life people typically prefer face-to-face meetings over telephone calls because the multi-channel interaction provides a much richer interaction. More information can be exchanged in a limited time, often with different types of information transferred by different channels. The core business of a meeting may be conducted by voice, but facial expressions carry important information on moods and attitudes.

In man/machine interaction, if we look at the user interfaces currently available, we see almost exclusively unimodal interfaces, where there is only a single channel for data output (e.g. screen) and only a single channel for data entry (e.g. keypad). A much richer user interface can be created if multiple channels (or modes) are used for input and output. For example, screen output can be supplemented with audio, while data entry through a stylus might be augmented with spoken data entry using speech recognition technology. More exotic options may also be possible, like gaze tracking or gesture recognition (e.g. for sign language). For the rest of this chapter, we will assume that we are dealing with only two modes, specifically a voice interface combined with the traditional graphical interface.

17.2.2   Sequential Multimodal

While some service providers currently claim to provide 'multimodal' access to e-mail, in reality what they generally mean is that you can either access the mail through a graphical user interface (GUI) on a wireless personal digital assistant (PDA), or you can telephone an interactive voice response (IVR) system and have your e-mails read to you. There is no link between the modes and you must choose one mode and complete your transaction in that mode.

The most rudimentary form of interaction that can genuinely claim the description of multimodal is referred to as 'sequential multimodal'. In this style of interaction, the user will alternate between modes within the course of a transaction. BT Exact has developed a prototype WAP and voice-based stock quotation application in which the user starts the application as a traditional WAP application, and then switches to voice to specify the stock in which they are interested. They hear the headline price information read to them, and then receive confirmation of the stock price together with supplementary trend information through the WAP graphical interface.

At any one time, the user is interacting either through one mode or the other but not both. With the advent of multimedia messaging, this style of interaction is likely to become popular with voice being used to control the delivery of multimedia content.

17.2.3   Simultaneous Multimodal

'Simultaneous multimodal' is altogether more interesting. Here the user can interact through more than one mode at once. The different modes can be used to perform different aspects of the task or can be combined in a single action. For example, in filling out a hotel booking form, a user could specify the required dates by voice while at the same time selecting the room type on-screen through a drop-down list. This is known as 'unco-ordinated', as the two modes are essentially being used independently. By contrast, in the 'co-ordinated' case the two modes provide different pieces of information needed to specify a single action. For example, in a route-finder application the user could say 'Show me the quickest route from here to here', while indicating two locations on an on-screen map using a stylus. No action can be taken on the basis of input from a single mode, but by combining the inputs and taking account of their relative timing, the full request can be understood.

This combination of data from different modes is known as semantic fusion. In the mapping example above, the job of merging the two streams of input into a single command is relatively straightforward. However, it is not always that simple. The inputs from the two modes might be contradictory rather than complementary so the system must be capable of conflict resolution. These conflicting responses can arise from two sources:

- the user actually made two conflicting inputs (e.g. said 'yes' and clicked on 'no');

- the system misunderstood one of the inputs (this is particularly relevant with voice interfaces which are prone to speech recognition errors).

There are a variety of strategies for dealing with conflicting inputs:

- accept most reliable modality;
- accept latest input;
- accept first input;
- accept neither and confirm.

17.2.4   Advantages over Unimodal

The benefit of using multimodal interfaces lies in the fact that different modes are fundamentally better suited to entering and delivering different kinds of information. If you imagine a telephone-based (i.e. voice only) theatre booking application, it is very easy to indicate which show you want to see: '... I'd like 3 seats for next Tuesday's performance of Cats ...', but how do you find out which seats are still available? The system is not going to read out an exhaustive list of which seats are free. In a multimodal version of the same application, you could still indicate which performance you wish to attend by voice, but then the application would show you a plan of the theatre indicating the available seating. Using a mouse or stylus you would select the seats you want and the transaction would continue.

The mobile environment presents a number of unique challenges for user interfaces not encountered in the desktop PC. Multimodal interfaces can meet these challenges and offer greatly enhanced usability.

- Device size

  The pressure from users to keep devices small and light has unavoidable implications for the user interface. Screens must be kept small, limiting the amount of information they can display. Similarly keypads must be kept small or omitted, making entry of text awkward. In applications that need this (e.g. for city names in time-table applications), the level of difficulty can be sufficient to put the users off entirely. Multimodal applications score here by making use of the most appropriate mode for each part of the interaction, and allow the user to choose whichever is most convenient for them at any given moment.

- Environment

  Unlike the desktop user environment, which is fairly static, the mobile user environment is potentially a continually changing one. The user may wish to change their style of interaction at any time due to changing circumstances. For example, if they have entered a high ambient noise area then speech recognition performance will become more error prone. A good multimodal interface should recognise this and allow the emphasis of the interaction to move towards the visual interfaces. Alternatively, they may have entered a railway carriage and may want privacy, so may wish to bias the interaction towards the graphical interface.

- Multi-tasking

  While desktop PC users are generally focused on the single task of interacting with the PC, this is often not the case with mobile users who may be performing another task at the same time (e.g. carrying out stock control, walking). Exactly what they are doing at the time may influence how they wish to interact. If they need hands free operation, a standard GUI is unusable, but the combination of voice input with audio and visual output might be ideal. Similarly use of a wireless PDA application can be changed from a two-handed operation to one-handed by the addition of voice input.

- Inclusivity

  A well-designed multimodal interface gives control of the mode of interaction to the user. Different individuals will have different preferences. Someone hard of hearing may wish to use a screen as their primary means of receiving information, while someone with sight impairment might prefer audio output. Preferences might be down to cultural factors or simply personal whim. There are many reasons why a user will choose to interact in one way rather than another. Some are due to the practicalities of using the interface but some are due to softer psychological effects that arise from personality or cultural differences. Multimodal interfaces are inherently more inclusive of the user population than any single unimodal interface.

- Confirmation and error correction

  Particularly when compared with voice-only applications, multimodal interfaces offer great advantages for confirmation and error correction. Spoken confirmation dialogues can be avoided by use of on-screen confirmation. Error-correction dialogues, the bane of voice applications, can similarly be side-stepped by the use of the graphical interface to correct speech recognition errors.

The overall message is that there are a host of reasons why people will want to interact with applications using different modes of interaction at different times. The key to success is allowing the user the flexibility to interact in the manner that they choose at any given time. For the user the multimodal interface offers only benefits, while the application provider benefits from an increased user base. The only downside comes by way of increased complexity in authoring and delivery of the application. In summary there are more advantages than disadvantages:

- pros:

  — users can select the most convenient mode or combination of modes to use for any given circumstance;

— errors in one mode can be corrected using another;

— can interact through multiple devices;

— can switch between devices through the course of a transaction;

- con:

— increased complexity in application authoring.

17.3   Multimodal Architectures

In implementing multimodality in the mobile environment, the limits of the wireless networks and the terminal devices place significant constraints on how the application may be deployed. Although many mobile terminals are able to handle visual-browsing-type applications and even some limited Java applications, the available memory and CPU power is still very limited. Unfortunately, speech recognition is both memory and CPU intensive and even the highest powered PDAs are currently unable to perform the most complex recognition tasks, though they may be able to cope with the simpler tasks.

A second issue with adding the voice modality is that the voice interface requires many more application level constraints than a visual application. Dialogue flow and speech recogniser vocabulary constraints can be both dynamic and large in size. The bandwidth and latency supplied by the mobile network can mean that these constraints cannot be downloaded within a reasonable time for the voice interface to be usable. For example, a speech recognition grammar might be as large as several megabytes, so passing this over a 10-20 kbit/s general packet radio service (GPRS) link would cause unacceptable application delay, even assuming the mobile terminal was capable of handling it.

Multimodal architectures therefore fall into two categories:

- thick client — appropriate for high-powered devices or applications with fairly static user interfaces, e.g. scheduling, messaging applications;

- thin client — appropriate for low-powered devices or applications with a high proportion of dynamically changing voice constraints, e.g. transactional applications.

17.3.1   Thick-Client Solutions

The thick-client implementation (see Fig 17.1) is appropriate for use with high-powered mobile terminals for applications that do not require large recognition grammars (or for applications which do not change and can be downloaded once only, but are used repeatedly). This architecture utilises speech recognition resident on the mobile device.

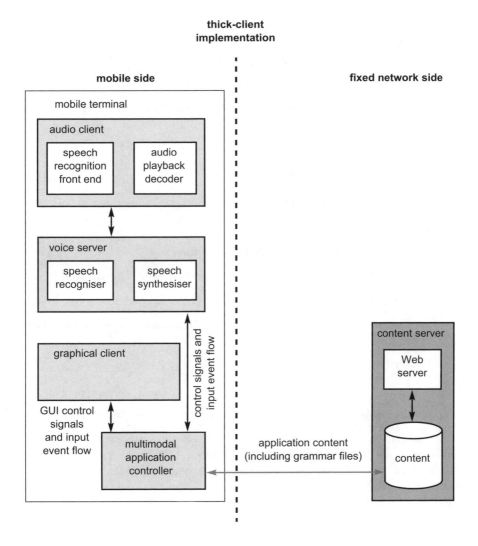

Fig 17.1   Thick-client implementation.

There are both advantages and disadvantages to this approach:

- pros:

  — keeps network traffic low for simple applications;

  — latency low once application is downloaded;

  — speech recognition can be speaker dependent;

- cons:

  — application download may take some time, as multimodal application, and visual and voice content must all be passed over the mobile link;

  — only suitable for high-powered mobile terminals (i.e. wireless PDAs);

  — not suitable for applications requiring large recognition grammars (e.g. name and address look-up).

17.3.2   Thin-Client Solutions

The thin-client implementation (see Fig 17.2) is generally more flexible:

- pros:

  — can deliver multimodality to low-power terminals;

  — can support solutions that use multiple devices to deliver the solution;

  — does not require multimodal application to be passed over the mobile link;

  — does not require voice content (other than the actual audio) to be passed over the mobile link;

  — latency low for application start-up;

- con:

  — requires audio and visual data to be passed over the mobile link.

A variant on the thin-client solution moves the multimodal application interaction processing on to the device. All the speech recognition is still performed in the network and typically the speech recognition constraints would be downloaded by the network voice server directly from the content server, thereby avoiding the transport of large amounts of data over the mobile link. The advantages and disadvantages to this are:

- pro:

  — can deliver multimodality to intermediate power terminals;

- cons:

  — requires audio and visual data to be passed over the mobile link;

  — requires the multimodal application to be passed over the mobile link;

  — difficult to support multiple device solutions.

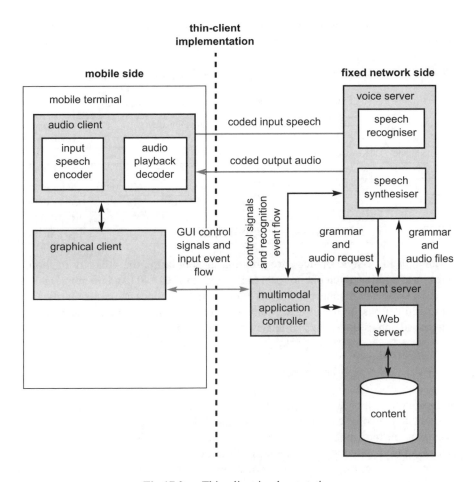

Fig 17.2   Thin-client implementation.

A further variant on the thin-client solution is the multiple-client solution, here the graphical client and the audio client may be on multiple devices, e.g. a PDA and a telephone. The multimodal interface is then delivered by combining these devices.

17.3.3   Real-Life Architectures

In real life, the mobile market is characterised by a wide range of devices with varying network connectivity, and a variety of thick- and thin-client solutions are possible. From the situation a few years ago, where the only common mobile terminals were simple mobile telephones offering only voice calls and SMS, the variety of wireless enabled terminals has expanded significantly. You can now purchase anything from a basic telephone to a general-purpose wireless personal digital assistant (PDA) like the mmO$_2$ XDA.

Table 17.1 summarises the main range of mobile terminals and form factors.

Table 17.1   A summary of available devices.

Device	Form factor	Functionality	Network bandwidth
2G voice-phone	Keypad + small screen	Voice + SMS	N/A
2.5G browser-phone	Keypad + small screen	Voice + SMS + WML	20-40 kbit/s
2.5G smartphone	Keypad + colour screen	Voice + WML/XHTML, J2ME, MMS	20-40 kbit/s
2.5G PDA	¼ VGA touchscreen	Voice + WML/XHTML, J2SE, SMS	20-40 kbit/s
3G smartphone	Keypad + colour screen	Voice + WML, XHTML, J2ME, MMS	64-384 kbit/s
WLAN PDA	¼ VGA touchscreen	XHTML, J2SE	2-11 Mbit/s

Although PDA-style devices are growing in popularity, the majority of mobile devices are based on a telephone form factor with limited CPU and memory, and for the foreseeable future one can expect multimodal solutions to be primarily thin-client architectures. In the next sections we show how thin-client architectures can be deployed across the variety of legacy and current mobile devices.

17.3.3.1   2G Voice Telephone

The most common mobile user is the 2G basic voice telephone user. The user has a circuit-switched voice connection and a simple SMS text-messaging display. For the basic user, current mobile applications and services consist of either interactive voice response applications or interactive SMS applications.

Interactive voice applications provide a very usable and convenient interface for entering data, but their complexity is limited by peoples' short-term memory which is easily overloaded. A simple multimodal system can be produced by enhancing the interactive voice application with an SMS output mode. This supplements the audio interface reducing the load on short-term memory.

If we take a common voice application providing share price information, we can enhance this by providing a summary of additional share-related information via SMS to the user. The key time-critical, current-price information is delivered by audio while supplementary information, such as trends, volumes and prices of other shares in the same sector, are displayed as an SMS.

SMS as an output modality has a number of limitations:

- it is a variable-delay output and there is typically no quality of service guarantee that the SMS will be delivered within a reasonable latency — however, the experience of the UK network is that SMS can typically be delivered within 5-10 sec and so SMS as a complementary output modality can be acceptable;

- it can only display simple text — with the arrival of MMS, we now have a means to provide more useful graphical feedback and can imagine MMS messages containing maps, product images, etc.

An alternative to SMS, for the display mode, is to combine the basic telephone with another visual device. This is most likely to take the form of a public display screen. The voice telephone now becomes a remote control device providing the user with a much more powerful interface than touchscreens. This leads to interesting applications in the space of interactive billboards which we discuss later (see section 17.5.1.1).

17.3.3.2 Browser Telephones

WAP in Europe and iMode in Japan brought in a new generation of mobile telephones capable of a limited form of Web browsing. The user interfaces on the telephones were largely very similar to the previous generation, but with slightly larger screens, typically capable of displaying up to half a dozen lines of text and, in the case of iMode, capable of supporting colour graphics.

For data entry, the situation is the same as for SMS —information, including items like user name and password, needs to be entered via the numeric keypad. This user interface hurdle has proved sufficiently large and the consumer's motivation to persevere sufficiently small that very few WAP-based commerce sites have been created. The user interface has fundamentally limited the viable application space for browser telephones.

However, WAP, and in particular WAP over GPRS, offers a number of features which enable more usable multimodal interfaces:

- GPRS sessions are 'always on' and can be automatically suspended and resumed in the event of incoming calls to the handset;

- the WAP2.0 specification includes a number of features such as WAP 'push' and wireless telephony application interface (WTAI) which allows outbound voice calls to be initiated from a WML application.

Given these features it is possible to produce multimodal applications which combine a GPRS WAP session with a circuit-switched voice call to deliver sequential multimodal applications. If we look at our earlier example of an enhanced stock quote, we can now produce a WAP application which offers as an option the ability to switch to voice control for data entry. By clicking on a 'use voice' option on the WML page we can either make the device place an outbound call to a voice service or get a voice service to call the WAP device. Once the voice service has collected the data, then the service can terminate and the WAP session will be re-established. An updated WML page containing the query results can then be pushed to the device.

Of course it is not as simple as that:

- WAP browser support for WAP push and WTAI varies from one device to another;

- GPRS implementation of session suspend and resume varies between devices;

- for devices without WTAI support, one needs to use a call back from the network — call set-up time on mobile networks can be of the order of 5-10 secs, and this delay in switching to the voice mode can wipe out any efficiency gains achieved from using voice.

Given these issues, one needs to be careful in implementing a sequential multimodal interface for enhancing WAP if one is actually to deliver an enhanced interface for the user.

17.3.3.3   2.5G and 3G Smartphone

The latest consumer trend for mobile devices is for smartphones which include MMS clients and Java support. Again the same basic limitation of data entry through a numeric keypad is a hurdle to interactive applications.

Fortunately the addition of Java and MMS also enables more complex multimodal architectures to be delivered on these telephones. So now, in addition to the messaging enhanced voice interface of the 2G and sequential multimodal interfaces on a browser telephone, the smartphone can additionally support simultaneous multimodal interfaces, where voice control is freely mixed with interaction via a Java application.

The support for J2ME on these smartphones enables the download of the thin client required for the architecture shown earlier in Fig 17.2. Combining this with the low-bit-rate speech codecs, required by the MMS client, allows us to run simultaneous distributed speech recognition over the relatively low bandwidth (10 kbit/s upstream and 20 kbit/s downstream) network connections provided by GPRS.

With the launch in April 2003 of the UK's first public 3G network we are also starting to see some 3G devices. Essentially 3G devices are little different in functionality from the current generation of 2.5G smartphone devices. However, they do provide two features that make multimodal applications development easier:

- higher bandwidth connection and more cost-effective support of IP networks;

- class A device support, allowing simultaneous IP network access and voice telephone call.

Both of these features mean that it is easier to deliver simultaneous multimodal interfaces. The higher bandwidth allows us to use higher speech coding rates which improve recognition performance. Class A devices essentially provide a guaranteed

QoS for the voice channel which can ensure latency between voice commands and the corresponding effects on the visual application are minimised.

17.3.3.4   WLAN-enabled PDA

Wireless PDAs offer the best user interface of the mobile terminals. With their large touch-sensitive screens and soft keyboards or handwriting recognition, they provide a usable interface for entering data. Usability issues still remain for the mobile user. Operation requires the use of two hands, one to hold the device and one to enter information and control it, while data entry is typically impossible when moving due to the difficulty of hitting small key pads.

From an architecture perspective the wireless LAN-enabled PDA provides an easy environment in which to implement thin-client solutions. The high-bandwidth low-latency network connection allows good synchronisation of the modes while also supporting high-bandwidth audio for speech recognition performance. The general-purpose programming environment allows the easy development and deployment of the clients.

17.4   Content Authoring and Standards

There is currently no broadly accepted approach to creating multimodal applications. In an ideal world, an application writer would be able to create an application independently of which devices and modes of interaction would ultimately be used to access it. However, we are currently a long way from the ideal of a standardised mode-independent authoring language. So for the time being, application writers will have to decide in advance which modes their application will support, and specify in detail how the user can interact in each mode and specify how the modes will interact with each other.

17.4.1   Unified versus Collaborating Applications

For creating multimodal applications, there is a spectrum of approaches currently being adopted. At one extreme, is the 'unified application' approach and at the other is that of 'collaborating applications'.

In the unified application approach, the author writes a single core controller which manages the flow of the interaction with the user. The speech recogniser is essentially considered as just another input device like a keyboard, passing events to the core controller. The author has to write the controller software in such a way that it caters for all possible combinations of input and output across the different modes. Such a system was developed by BT in the mid-1990s and is described in Wyard et al [1]. Today, an implementation might be written in Java using the Java Swing API to manage graphical input and output, and the Java Speech API for voice interaction. This kind of approach offers the author great control over the fine detail of how the modes interact with each other. However, this control comes at a cost in the level of effort required to define the application and in the difficulty of testing that it has been implemented correctly. The unified approach is well suited to adding voice control to fixed applications such as word processing or device control, where the application is not changing rapidly and can justify the level of effort needed to implement the solution.

At the other end of the spectrum, the collaborating applications approach starts from the standpoint of having two separate applications, each implementing one mode of the interaction. The separate applications are kept synchronised with each other via a co-ordinator. To convert these from being two independent applications, the content author then has to define how the two are to be synchronised in a manner that the co-ordinator can interpret. Figure 17.3 shows the basic differences between the approaches.

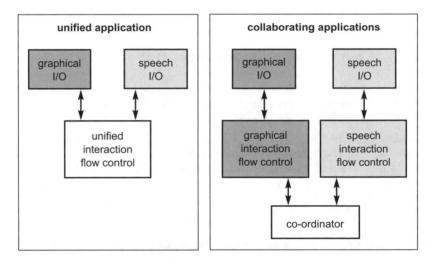

Fig 17.3   Unified versus collaborating applications.

The collaborating application approach benefits in that the individual applications can be written and tested inde-pendently in familiar unimodal application development environments, prior to being integrated via the co-ordinator. It also makes it easy to deliver solutions where the different modes are delivered on separate devices or as different styles of application, e.g. the voice interface might be written in VoiceXML while the GUI is written as a Java application, or the GUI could be written in HTML in combination with a legacy voice application.

The collaborating application approach is also flexible in not tying the multimodal interaction to a specific visual interface. For example, you might want to provide a voice plus GUI multimodal application which delivers the graphical interface either on an HTML browser or on a WAP browser. The approach allows the content author to write the voice side of the interface once and use it with both of the graphical implementations.

This approach works well for Web-based applications. The graphical interface can be written in familiar HTML and delivered through a standard Web browser while the voice application might be written in VoiceXML and delivered through a VoiceXML browser.

In practice, the two main approaches being put forward for standardisation of Web-based multimodal applications fall in between the two extremes.

17.4.2   W3C Standardisation Activities

The World Wide Web Consortium (W3C) [2] is the leading body for the standardisation of application description languages. The W3C has a number of working groups looking at the issues around the user interface. The Voice Browser working group has been active for a number of years in defining mark-up languages for interactive voice response applications. This WG [3] has produced a number of specifications which aid the definition of portable voice applications, primary among these are the VoiceXML2.0 candidate recommendation, which standardises the dialogue description, and the Speech Recognition Grammar Specification, which standardises the format of speech recognition grammars. Between them these specifications allow the portability of voice applications but importantly move voice application development into the same Web application framework as WML/XHTML visual applications. This opens the door for portable multimodal applications.

In January 2002 W3C announced the formation of the Multimodal Interaction working group which would work in conjunction with the Voice Browser WG to specify mark-up languages for multimodal applications. This Multimodal WG is currently determining the requirements of a multimodal application framework and reviewing proposals for multimodal mark-up languages.

There are currently two competing public proposals for a multimodal mark-up language:

- speech application language tags (SALT) [4] — a Microsoft-led industry group proposing a set of extension tags to allow the control of voice-processing engines from within XHTML;

- XHTML+VoiceXML (X+V) — a Motorola/IBM/Opera industry grouping which proposes a scheme for mixing XHTML and VoiceXML mark-up within the same document.

It is early days yet and there are many issues to be resolved before either of these would be accepted.

- SALT

  SALT essentially provides a small set of XML tags which allow the invocation of speech recognition and audio prompting through asynchronous methods together with a binding mechanism to allow the speech recognition results to bind to the Document Object Model (DOM). The SALT specification reuses some of the W3C specifications, most notably the speech recognition grammar format, but it does not use the VoiceXML dialogue specification.

  Dialogue flow control is achieved by scripting within the document. This use of low-level primitives and scripting has the advantage of providing very fine control over the voice interface and allows the implementation of arbitrary voice dialogues. The disadvantage is that content pages are typically larger and may prove to be more difficult to debug and maintain than an equivalent VoiceXML document. For it to be successful, SALT is likely to be dependent on the emergence of libraries of reusable components.

- HTML+VoiceXML

  The X+V proposal allows the embedding of VoiceXML *<form>* elements within an XHTML document. Like SALT it relies on a shared document object model between the voice and visual mark-up to enable the merging of user inputs between modes. The VoiceXML forms instances are invoked by XML event bindings within the XHTML document. The reuse of VoiceXML for dialogue flow-control reduces the flexibility of the application designer, but, conversely, makes applications easier to debug. The form elements can be in separate documents, referenced by URI, which potentially leads to smaller documents and better reuse.

The fundamental difference between SALT and X+V lies in the fact that X+V reuses VoiceXML elements for the dialogue control and maintains a declarative approach to the interface definition, whereas SALT is more a scripting approach.

Both X+V and SALT are first attempts to standardise multimodal mark-up languages, and a number of aspects of multimodal user interfaces are not currently addressed by these proposals. Most important of these for the mobile environment is the current lack of support for non-mark-up languages such as Java.

Regardless of the final choice of standard mark-up, experience suggests that the barrier to the appearance of multimodal applications is not the time taken by application developers to learn a mark-up language but rather the time taken to learn good interface design. Users can cope with badly designed visual content much more easily than badly designed voice content. The potential of multimodal interfaces to confuse the user is high; good design skills are essential if the advantages multimodal offers are to be exploited.

17.4.3   Other Standards Activities

Although the bulk of the effort into multimodal standards is currently happening within the W3C Voice Browser and Multimodal Interface working groups, it is worth noting a number of other standards activities that will complement and aid the delivery of multimodal applications.

The IETF [5] Speech Service Control Protocol working group is developing a control protocol to allow distributed control of network servers providing speech recognition, synthesis and verification. The speech SC protocol would be an enabler for thin clients to be able to invoke speech functionality over a mobile IP network.

The ETSI [6] Speech Transmission Quality sub-group, Aurora, is developing compact speech parameterisations to allow robust speech recognition in distributed environments such as mobile. This attempt to standardise on a speech parameterisation is difficult since for most speech recognition systems parameterisation is proprietary and so there are potential technical and commercial barriers to acceptance by the speech industry of a distributed speech recognition standard.

17.5   The Mobile Multimodal Market

Why is now a good time to be thinking about multimodal services? The mobile Internet market is characterised by two recent trends:

- a huge investment in wireless packet data networks and the associated need to provide transactional services which will generate return on this investment;

- increasing complexity and diversity of mobile terminals.

At the same time, speech recognition technology and standards, such as VoiceXML, are maturing. In the USA and Europe a number of voice portal services vie with mobile data portals, e.g. Tellme, AOLbyPhone, and AT&T Wireless. The consumer and business markets are likely to exhibit significant differences in device and networks as well as the more obvious applications (see Table 17.2).

Table 17.2   Consumer and business market differences.

	Consumer	Business
Applications	Messaging, games/entertainment, information	Messaging, sales force automation, logistics, field force automation, information
Network	GPRS	GPRS/WLAN
Device	Single integrated device, e.g. smartphone	Multiple devices, e.g. Wireless PDA + telephone

It is also worth noting that mobile content is virtually always a subset of existing on-line content. Multimodal interfaces offer the opportunity to make this subset closer to the original desktop application in functionality; however, they must do this within the constraints of the user group's device form factor and network connectivity.

17.5.1   Multimodality in Action

This section describes two examples of current multimodal application that lie at the two extremes of mobile devices.

17.5.1.1   *Interactive Billboards*

The BT Exact voice-controlled shop window (see Fig 17.4) is an example of how multimodality interfaces can be delivered to the most basic mobile device. The application allows a user to remotely control a shop-window-mounted visual display using voice commands issued via their mobile telephone. By simply dialling the telephone number on the screen, customers can ask for information on products and services, see them displayed on the screen complete with audio commentaries, and have information and special promotions sent by SMS to their telephone, personalised to their individual requests. The system works by linking an HTML visual application displayed in the shop window with a VoiceXML application hosted on a voice gateway within the mobile operator's network.

Fig 17.4   The voice-controlled shop window.

The application offers a number of advantages over traditional touchscreen interfaces:

- the display does not have to be aligned in any particular way with the window;
- user identity can be captured through CLI, providing important marketing information;
- the user can be transferred to a human agent at any time to close deals;

- SMS can be used to enhance the user experience with product information and discounts.

The display can, of course, be any form of public display terminal and does not necessarily need to be in a shop window. Indeed it can be used to create a virtual shop presence in locations where a real shop window may be too expensive or impractical.

17.5.1.2   *Voice-enabled WLAN PDA*

At the other extreme of the mobile UI spectrum is the wireless LAN-enabled Pocket PC. In this application the visual interface and voice interface are integrated on the same device and are seamlessly combined to provide a simultaneous, co-ordinated, multimodal interface.

The visual interface is provided by the pocket IE HTML browser running an HTML application downloaded over the wireless LAN network. Voice control is provided through a voice over IP (VoIP) audio client running on the PDA and communicating via VoIP to a VoiceXML gateway in the network. The user is able to interact with an enterprise Web application using any combination of voice and GUI interactions.

This solution is ideal for corporate users within a corporate LAN. Applications that can particularly benefit include those with a requirement for hands-free or eyes-free operation such as warehousing and technical field engineers. Other applications include in-vehicle WLAN deployments for emergency services.

17.5.2   Commercial Multimodal Products

The first commercial systems and software are just beginning to appear. The majority of these solutions are in trial developments and we can expect to start seeing commercial services exhibiting some form of multimodal user interface during 2004.

Current commercial systems fall into one of two categories.

17.5.2.1   *Client/Server Solutions*

Multimodal client/server platforms are available from a number of companies. These solutions tend to support a range of types of multimodal interface from simple SMS-enhanced voice applications through to simultaneous multimodal on wireless-enabled PDAs.

17.5.2.2   Terminal Resident Solutions

These solutions are almost exclusively targeted at the Microsoft Pocket PC PDAs. The available power and memory means that command and control type voice functionality is possible. The most common application to be voice enabled is PocketOutlook. These solutions are typically sold for around $50. The jury is out on whether individuals will purchase such software or whether, like desktop PC voice recognition, it will be consigned to niche user-domain markets.

17.6   Summary

Mobile devices themselves are fundamentally limited in size and form factor by the need to be mobile. Voice control offers the means to circumvent these form-factor limitations and deliver usable service to the mobile market. Such multimodal interfaces are one step along the way to creating more natural interactions between people and applications.

As mobile applications and networks become more intelligent, we can expect the UI to become even more responsive to user needs at any one time, adjusting the combination of modalities in use in response to location, device availability, network availability and user preferences.

As networked computing becomes more ubiquitous, we can also expect multimodal interfaces to be delivered through more varied combinations of devices and utilise more varied modes of interaction which may include gaze tracking, gesture recognition and electronic ink.

Already available today are multimodal solutions that can enhance the user experience for any mobile user from those with a basic telephone to those with the most powerful PDAs. Take-up of these solutions for mobile applications and services is essential if mobile data applications are to deliver a return on the huge investment in 2.5G and 3G networks.

References

1   Wyard, P. et al.: '*Spoken language systems — beyond prompt and response*', BT Technol J, **14**(1), pp 187-205 (January 1996).

2   World Wide Web Consortium — http://www.w3.org/

3   Voice Browser Working Group — http://www.w3.org/Voice/

4   SALT Forum — http://www.saltforum.org/

5   IETF — http://www.ietf.org/

6   ETSI — http://portal.etsi.org/

18

MOBILE VIDEO-STREAMING

M D Walker, M Nilsson, T Jebb and R Turnbull

18.1 Introduction

Mobile video has been lauded as one of the key service innovations that justified the heavy investment in 3G licences across Europe some three years ago. While 3G undoubtedly has the capacity to provide compelling video-based applications, existing 2G-based technologies such as general packet radio service (GPRS) can be used to deliver mobile video services today.

This chapter considers the state of the art for mobile video streaming, discusses the international standards that are applicable for mobile streaming, and describes a mobile video-streaming system developed by BT, which enables high-quality video-streaming using both 3G and GPRS networks.

18.2 Technical Details for Mobile Streaming

18.2.1 IP Streaming Principles for Mobile

Real-time video applications require media packets to arrive in a timely manner — excessively delayed packets are useless and are treated as lost. This causes the synchronisation between encoder and decoder to be broken, and errors to propagate in the rendered video. The technical implication for IP networks is that the data representing the compressed audio and video cannot be considered as ordinary data files — transfer methods such as ftp are not appropriate. The actual transfer methods for streamed media are discussed later.

By way of example, consider an audio/video clip of duration one minute, that has been compressed at an average data rate of 20 kbit/s. To stream this clip to a user, not only must the network provide an average throughput of 20 kbit/s, but the instantaneous throughput of the link must not fall below an average of 20 kbit/s, for the duration of the streaming. However, IP is non-deterministic — without quality of service (QoS) guarantees, there is no way of knowing for sure what the future bandwidth characteristics of the network link will be.

This problem is exacerbated for delivery to a mobile device — mobile IP networks, particularly GPRS, provide very low and variable data throughput. The device itself may not have sufficient processing power to employ complex error concealment and recovery schemes. This can lead to a poor experience when viewing streamed media that has not been designed for mobile networks.

18.2.2   Common Techniques to Improve Performance

Typically, streaming systems employ a number of well-known techniques to provide resilience to network congestion. These include the insertion of regular key-frames within the data-stream to provide re-synchronisation points should timeliness be lost, at the cost of reduced compression efficiency. A second common technique is to employ buffering at the player — delay is traded for packet loss and jitter: provided the average throughput requirements of the video-stream match the average available bandwidth, the receiver buffer size can be dimensioned to accommodate the expected jitter, and allow for retransmissions. In fact, buffering can also be used to artificially increase the perceived throughput of the network link, by forcing the player to buffer a large proportion of the entire clip before playback starts, but at the cost of high start-up delay.

18.2.3   Compression

Compression is a method for reducing the amount of data required to represent digital media. In the same way that a utility such as WinZip^{TM} is popular for compressing documents, so there are techniques to compress audio and video data. The higher the degree of compression, the lower are the storage and transmission requirements for audio/video clips. Compression is achieved using an encoder/decoder pair, termed a 'codec'.

18.2.3.1   Video Compression

This section gives an overview of standard video compression techniques. Standards-based video codecs fall into the category of block-based interframe prediction algorithms. Such systems are designed to reduce the data rate of video by exploiting spatial and temporal redundancies between pictures, and by removing features of the video that are less perceptible by the human visual system (HVS).

A generic block-based interframe prediction video coder has the following components:

- motion estimator — previously encoded pictures are searched for similarities to the current picture (motion tracking);
- inter-frame loop — code differences between the previous picture and motion-compensated previous pictures;
- transform — picture pixel differences are transformed to frequency coefficients;
- quantiser — frequency coefficients are quantised, with the least significant bits being discarded;
- entropy coding — quantised frequency coefficients are entropy coded.

The combination of motion tracking and inter-frame coding removes temporal redundancy. The frequency transform puts the resulting pixel differences into a better format for the removal of spatial redundancy, which is achieved by the combination of quantisation (a lossy process) and entropy coding (the lossless representation of the result of quantisation — similar in concept to Morse code in that short codes are used for frequent symbols and long codes for less frequent symbols). The encoder also replicates some decoder functionality to allow the encoder to track the operation of the decoder, and produce 'previous' pictures for future motion tracking.

For a fuller description of standards-based video encoding, see Ghanbari [1].

A generic inter-frame video encoder is presented in Fig 18.1; the diagram also shows which components are fixed by video codec standards, and which components are open to allow implementor value-add. An example of value-add is given by noting that a valid bitstream can be generated by only coding pictures in the 'intra' mode (i.e. with no reference to previous pictures). However, this will typically require five times as many bits to achieve the same quality as inter-frame coding with motion compensation.

Figure 18.2 shows the components of a video decoder. While the majority of picture-quality gains can be made at the encoder, the implementation at the decoder can include intelligent error masking, and post-processing to increase the quality of the decoded pictures.

Consequently, a vendor's implementation of a video standard may be good, bad or indifferent, depending on the nuances of their coding decisions and pre-/post-processing options.

Video Compression Standards

International standards on video compression are essential for interoperability. There are two well-known standardisation bodies that have developed video compression standards: the International Telecommunication Union (ITU-T), which has developed the H.26x series of Recommendations, and the International Standards Organisation (ISO), whose Moving Picture Experts Group (MPEG) has

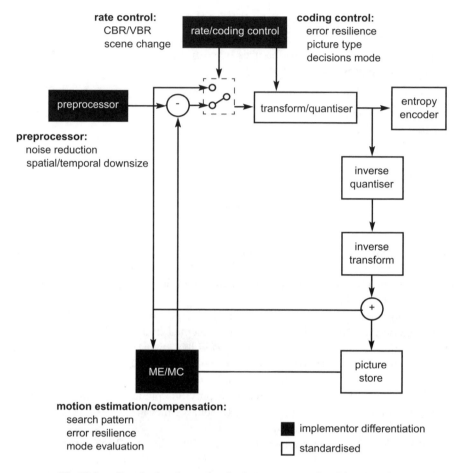

Fig 18.1   Standard and non-standard components of a video encoder.

developed the MPEG series of standards. The ITU-T has produced the following Recommendations:

- H.261 (also called 'P times 64'), the first digital video coding standard with practical success;

- H.262 (common text with MPEG-2), for high-quality videoconferencing;

- H.263 (and its updates: 'H.263+' and 'H.263++') for videoconferencing and streaming.

MPEG has produced the following standards:

- MPEG-1 (ISO/IEC 11172-2) for video CD applications;

- MPEG-2 (ISO/IEC 13818-2) — used widely in DVD, broadcast, satellite, and cable DTV;

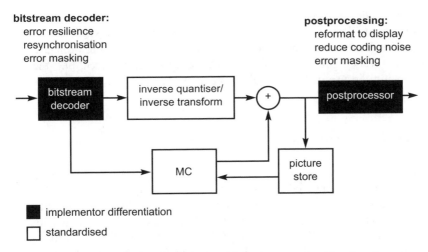

bitstream decoder:
error resilience
resynchronisation
error masking

postprocessing:
reformat to display
reduce coding noise
error masking

■ implementor differentiation

□ standardised

Fig 18.2   Standard and non-standard components of a video decoder.

- MPEG-4 Visual (ISO/IEC 14496-2), for streaming applications (including compatibility with the H.263 baseline profile).

For mobile streaming applications, it is normally recommended that H.263/ MPEG-4 are used. The similarity between the two coding schemes is acknowledged on the MPEG Web site [2]:

'Although the schedules and requirements for MPEG-4 video and ITU-T's video coding efforts are different, their requirements do overlap in many ways, for example both efforts seek at least to define a standard that can efficiently code natural video at bit rates in the range of 24 Kbit/s to 64 Kbit/s. Because of this overlap in requirements, many participants in MPEG-4 Video also participate in the ITU-T's low bit-rate coding (LBC) effort. Through these common members and through formal liaison statements the MPEG-4 Video and ITU-T frequently share their results with each other in hopes of improving their standards and achieving a high degree of interoperability between their standards'.

H.263 and MPEG-4 for 3GPP Streaming

3GPP specifies H.263 and MPEG-4 may be used for video streaming, with compatibility between the two achieved at baseline. While this baseline compatibility is mandatory for 3GPP compliance, the baseline implementations only provide basic functionality. Both standards can be enhanced by use of different profiles/annexes. The implementor chooses which optional extras to add. For example, with MPEG-4 streaming, the scalability modes are popular as they enable mechanisms to provide an adjustment of the quality of video received by a user both efficiently and gracefully, depending on the perceived performance of the access network. It should be noted that the scalability mode on its own does not provide efficient and graceful degradation to changes in network performance — the application vendor must still measure the network performance, and then make decisions on the use of the scalability modes to adjust the quality of video being transmitted.

H.264 Video Compression — The Future of Video Compression

The ITU-T and ISO, working as the Joint Video Team (JVT) and ISO/IEC 14496-10, have recently approved a new video codec that is being heralded as a key enabling technology for mainstream use of video applications. The standard is known as both H.264 and MPEG-4 part 10. Many results for the new codec show that it requires about half the data rate of H.263 and MPEG-4, and as little as a third of the data rate for MPEG-2, to achieve the same picture quality [3, 4]. The H.264 design is architecturally equivalent to previous codecs as detailed above; however, by applying a 'back-to-basics' principle, the JVT has managed to refine each component to achieve this remarkable performance. The major differences between H.264 and H.263/MPEG-4 are described below.

- Motion Compensation

  The motion compensation process differs from that of prior standard designs in its support of a wider variety of block shapes and sizes and in its fractional-sample interpolation filtering. Further enhancements are achieved by motion displacement accuracy of ¼ of a pixel[1] — meaning that motion is much more accurately tracked. In addition, multiple previously encoded and transmitted pictures are available for motion prediction.

- Transform

  H.264 uses a 4×4 transform, in contrast with the 8×8 shape typically found in prior designs. The inverse transform is specified in terms of precise integer operations, rather than as a rounding tolerance relative to a real-valued ideal transform as found in most prior designs. In some cases the 4×4 shape is extended by applying an additional 2×2 or 4×4 transform to the DC coefficients of the first-stage transform.

- Entropy coding

  Entropy coding is performed either by using context-adaptive variable length coding (CAVLC), or by use of context-based adaptive binary arithmetic coding (CABAC).

---

[1] Pixel: a picture element (also known as 'pel').

- Loop filter

  A strong deblocking filter is used within the prediction loop.

The standardisation process was completed in mid-2003. As reported in IT Week in April 2003 [5]:

'... trial implementations and system tests have begun in earnest. Perhaps the most impressive demonstration was given by BT's research labs when it showed an H.264 decoder running on a Nokia 7650 mobile phone. BT's implementation uses Quarter Common Intermediate Format (QCIF), which offers a resolution of 176×144 pixels, and can decode at about 14 frames per second. The video bit rate was about 20 kbit/s, which can be handled by either GPRS or HSCSD mobile data technology ...'

While the 3rd Generation Partnership Programme (3GPP) [6] standardisation body specifies H.263 and MPEG-4 for video compression, it is currently considering additional video codecs, and is expected to recommend H.264/MPEG-4 part 10.

18.2.3.2   Audio Compression Standards

3GPP specifies two codecs for the coding of speech and audio for a transparent end-to-end packet-switched streaming service. The codecs achieve compression by making assumptions about the signals such that they can exploit the redundancies and irrelevancies pertinent to speech or music.

Speech codecs are explicitly based on a model of speech production. The encoder uses a mathematical model of the human vocal system and by analysing the input signal, produces parameters describing the activity of the glottis and vocal tract. These parameters are transmitted to the decoder, where they are used to re-synthesise the original speech.

This method is very effective at coding speech sig-nals and can produce toll quality at bit rates as low as 4.8 kbit/s (e.g. AMR mode 0). 3GPP has specified that the AMR decoder shall be supported for narrowband speech [7] and the AMR wideband [8] operating at 16 kHz sampling for wideband speech.

The algorithms used for speech codecs rely on assumptions about the source signal. They are therefore not suitable for the effective coding of non-speech signals such as music. For this reason 3GPP has specified the use of the MPEG-4 AAC codec [9], which has been developed by MPEG. Like previous MPEG audio codecs such as the popular MP3 codec, it uses a perceptual noise shaping scheme, which uses the signal-masking properties of the human ear in order to achieve compression. AAC supports sample rates between 8 kHz and 96 kHz and any number of channels between 1 and 48, although 3GPP have specified an upper limit of 48 kHz and either mono or stereo operation only.

18.3   Streaming Techniques

Current IP networks are not well suited to the streaming of multimedia content as they exhibit packet loss, delay, jitter and variable throughput, all of which can detract from the end-user's enjoyment. Although quality of service schemes can reduce the severity of these effects, they are generally not widely deployed.

As IP networks generally have no connection admission control they can become congested causing a sustained drop in throughput. If transmission continues at the same rate, the result is high levels of packet loss and 'unfairness' to other network users — it is usual, and 'polite', to reduce transmission rate during congestion.

This section considers, in more detail, the international standards applicable to mobile streaming, and some well-known techniques for ameliorating the effects of network throughput variation and packet loss.

18.3.1   Video-Streaming — Congestion Control

This section examines some of the popular methods of providing video-streaming over general IP networks. It should be noted that, while a good implementation of a video codec can provide value-add in terms of better picture quality for the same bit rate, the video packets must still be transmitted over a network to a receiver. It is this transmission that can differentiate a good implementation from a bad implementation.

A generic video-streaming architecture is shown in Fig 18.3.

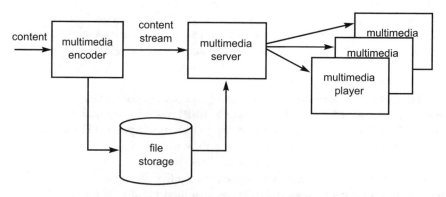

Fig 18.3   Generic end-to-end video transmission.

The role of the multimedia server is to provide the player with multimedia data that has been transmitted across the network. Most server/client applications operate by employing an end-to-end congestion control mechanism, coupled with a buffering strategy at the player. This enables the player to supplement data received from the network with data that has been built up in this buffer. This provides better utilisation of the network, and better performance. However, it is not enough for the congestion control algorithm to simply measure the available network resources — the compression stage must include mechanisms to transmit the video at varying data rates, without compromising the requirement for timeliness of delivery. A streaming system may need to be able to reduce its transmission rate requirements, which necessitates reducing the compressed rate of the media to be streamed. We use the term 'elastic' for media encoded in such a way.

The use of real-time encoding would allow the bit rate to be varied dynamically, but is not practical due to the high processing power requirements which prevent such a solution from scaling to support many users. So to achieve a system that is scalable, media must be compressed without knowledge of the state of the network at transmission time, but, as soon as they are compressed, they become inelastic, and need to be transmitted at their encoded bit rates.

Layered encoding [10] is a well-known technique for creating elastic video sources, but such techniques are inefficient in terms of compression [11], and significantly increase the processing power required at the client which is undesirable for streaming to limited capability mobile devices.

Transcoding [12] is another technique for creating elastic video sources, which can be designed to have much lower computational complexity than video encoding. However, the computational complexity is not negligible, and so would have an impact on the scalability of the architecture.

18.3.2   Streaming Protocols

The transport mechanism for streaming media is not adequately catered for [13] by either the transmission control protocol (TCP) [14], or the user datagram protocol (UDP) [15]. While it is clear that TCP is the protocol of choice for file-transfer type applications such as e-mail, and that UDP is the favoured transport protocol for conversational applications, there is no clear choice for streamed media. The standards bodies do not give many clues either — both UDP and TCP are supported by 3GPP. However, most 3GPP applications use UDP for streaming — even though research has shown that TCP may provide acceptable quality, at a lower application complexity [13].

18.3.2.1   UDP Streaming

UDP is often the favoured transport protocol for delivery of audio and video data over IP networks. However, as UDP provides no guarantee of delivery, a video streaming application needs to be provided with extra information to enable the detection of lost, delayed and unordered packet delivery. UDP is usually supplemented with framing and feedback information, such as that provided by the real-time transport protocol (RTP) and the associated RTP control protocol (RTCP) [16].

RTP provides framing and timing structure to audio and video streams and allows inter-stream synchronisation. RTP provides additional information, such as payload type identification, sequence numbering and delivery monitoring, which can be used by the streaming application to detect, and act upon, transmission errors. It should be noted that RTP itself does not provide any mechanism to ensure timely delivery or provide other QoS guarantees, but relies on lower-layer services to do so. RTCP provides a mechanism for providing feedback to the source.

18.3.2.2   TCP Streaming

While UDP is often the transport protocol of choice for many streaming applications, TCP can be preferred because of its relative implementation simplicity. For example, although TCP can offer no timeliness guarantees, there is now no requirement for extra framing information such as RTP and RTCP — with TCP, packets are guaranteed to arrive eventually. With TCP streaming, the application need only ameliorate the effects of delay, which is usually through the use of novel buffering techniques.

There is still much research comparing UDP to TCP for streaming applications [13] — but, with other schemes (see section 18.3.2.3) still in their infancy, the mix of solutions is likely to continue for the forseeable future.

18.3.2.3   TCP-Friendly Streaming

While TCP and UDP can both provide some benefits to streaming technologies, each has drawbacks. There has recently been effort in the research community to define a 'half-way' solution — termed 'TCP-friendly' [17]; this congestion control mechanism is designed to be used for non-TCP applications. The goal of this research is to develop a congestion control scheme that is 'friendly' to other TCP traffic (i.e. a TCP-friendly source will fairly share network resource with TCP traffic) yet is more efficient and responsive than TCP.

Although a number of research papers have been published showing the benefits, this scheme is still relatively immature, and is not widely adopted.

18.4   Fastnets

This section provides an overview of the Fastnets[2] system, developed by BT. Fastnets is a multimedia streaming system designed to overcome the characteristics

---

[2] Fastnets: FAST-start, NETwork-friendly Streaming — http://mobile.f-nets.com

of IP networks, and in particular mobile IP networks, to provide users with continuous multimedia of consistent quality with minimal start-up delay. H.264 video compression is used to achieve acceptable video quality over GPRS to mobile devices, such as the Nokia 7650.

18.4.1   Introduction

Our underlying philosophy when designing the Fastnets system was to develop technology that would enhance the end-user experience of streamed mobile video. In particular, Fastnets was designed to achieve 'Fast-start' streaming (reducing the delay normally associated with buffering) and to provide the best compression efficiency, without losing the ability to adapt to changes in network throughput. While the Fastnets system is an end-to-end streaming architecture, each technology component has been designed to 'plug-in' to an existing, standards-based streaming system, to potentially provide 'turbo-charging' of the existing video-streaming system.

18.4.2   Key Technology Components

Fastnets has a number of key technology components that each provide enhanced streaming, and together enable Fastnets to achieve high-quality streaming over 2.5G networks such as GPRS:

- uses H.264 for compression (also runs with H.263/MPEG-4);
- features fast-start, adaptive video — starts in 2 sec for 3G networks, and 5 sec for GPRS;
- uses CBR and VBR streaming (VBR provides up to 30% better compression);
- uses both TCP, and UDP operating a TCP-friendly transmission algorithm;
- uses bit-rate switching for adaptation — efficient and effective;
- fully featured with 'trick-modes' such as fast-forward, random access;
- includes in-band bandwidth detection for optimal streaming.

18.4.3   System Architecture

The key attributes of the overall system are:

- decoupling of the transmission rate from the media encoding rate;
- varying the transmission rate in a network-friendly manner;
- building up a buffer of data at the client without incurring a start-up delay;

- smoothing short-term variations in network throughput by use of client buffering;

- adjusting long-term average bandwidth requirements to match the available resources in the network by switching between streams encoded at different bit rates.

The Fastnets architecture, shown in Fig 18.4, consists of a Fastnets encoder for encoding live or stored multimedia content for either onward transmission via a Fastnets server, or for storing into a Fastnets file, for subsequent transmission using a Fastnets server.

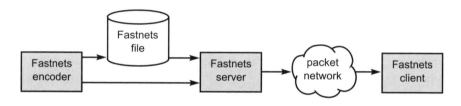

Fig 18.4   The Fastnets architecture.

18.4.3.1   Fastnets Encoder

The Fastnets encoder encodes multimedia content into an elastic representation. Audio is encoded using AMR encoding at 4.8 kbit/s, and hence is inelastic, but as it requires a lower bit rate than video, the combined encoding can be considered to be elastic if the video is encoded in an elastic fashion.

Video is encoded using the H.264/MPEG-4 Part 10 standard into a hierarchy of independent video streams, with means to link between them.

The video hierarchy includes an 'intra stream', to allow random access, and one or more 'play streams', for ordinary viewing, each being encoded at a different bit rate, thus making the hierarchy elastic. The hierarchy also contains switching streams, which allow switching from the intra stream to play streams, and between play streams.

A hierarchy with two play streams and one intra stream is illustrated in Fig 18.5. The purpose of the hierarchy is to allow the server to transmit play streams to a client to achieve an optimal balance between building up a buffer of received data to provide resilience to network degradation, and providing video at the highest rate that the network instantaneously supports.

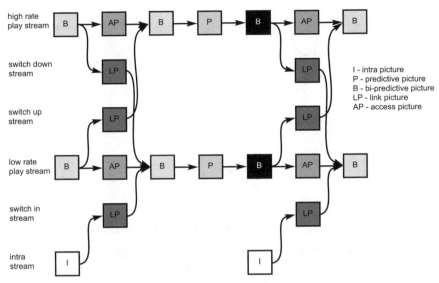

Fig 18.5   An example video hierarchy.

The intra stream consists of a series of intra pictures for providing random access and recovery from severe error conditions. The play streams consist of predictive and bi-predictive pictures, and periodic access pictures. The switching streams consist of a series of linking pictures.

Access and linking pictures are encoded as SP pictures, as defined in the H.264/ MPEG-4 Part 10 standard. The concept underlying the encoding of SP pictures is shown in Fig 18.6. The source picture and the motion-compensated prediction are independently quantised with the same quantiser and transformed, before being subtracted and entropy coded. The reconstructed picture is simply a quantised version of the source, independent of the motion-compensated prediction. Hence identical reconstructions can be obtained even when different reference pictures are used, and hence mismatch-free switching is achieved. The motion compensated prediction is relevant, as it reduces the entropy of the signal to be variable-length encoded, and hence reduces the number of bits produced by encoding a picture.

18.4.3.2   Fastnets Server and Client

The Fastnets server receives encoded multimedia content either from the Fastnets encoder or from a Fastnets file, and serves this content to one or more clients. The server scales to support many clients accessing many pieces of content

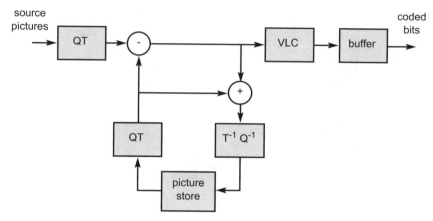

Fig 18.6   Concept diagram for encoding SP pictures.

independently as it performs little processing, just selecting packets for onward transmission. No encoding or transcoding of media is performed in the server.

The server works in the same way for both live streams from the encoder, and for pre-encoded streams from file. This chapter therefore focuses on the details of serving live content. Figure 18.7 shows the main features of the server and client.

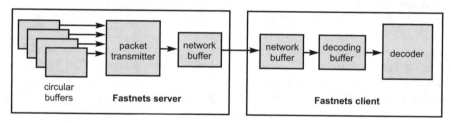

Fig 18.7   The Fastnets server and client.

The server consists of a number of circular buffers, one for each intra, play and switching stream for each piece of live content. For each client there is one instance of packet transmitter which determines when and from which buffer the next packet is read, and sends it to the client. The transmission ensures that all packets are received, and that the transmission rate is controlled to be 'network friendly'. Fastnets currently uses TCP, but there is also a variant that is more suited to multimedia streaming, by using a different transmission-rate-adaptation algorithm that allows a higher average throughput to be achieved.

18.4.3.3   Start-Up

When a client connects to the server, the server locates an appropriate intra picture from the circular buffer, and sends this to the client. It then sends the linking picture to switch to the play stream with the lowest bit rate, and continues to serve from that play stream.

The transmission of packets to the client is an independent process, with the rate of transmission depending on the state of the network. The intention is that initially the transmission rate is greater than the bit rate of the play stream with the lowest bit rate. This allows the client to start decoding and presenting media to the user immediately, while also building up data in its decoding buffer.

18.4.3.4   Client Buffering

The decoding buffer at the client reduces the impact of network jitter, packet loss and variable throughput on the quality of media presented to the user. By decoupling the transmission rate from the media encoding rate, the client's decoding buffer can be filled when network conditions are favourable, providing resilience for less favourable times.

The accumulation of tens of seconds of data in the decoding buffer allows jitter of the same magnitude to be masked from the user and provides time for the retransmission of lost packets before they are needed for decoding, thus allowing recovery from most instances of packet loss, including short periods of severe loss where the network throughput drops to zero, without affecting decoded media quality.

18.4.3.5   Streaming

The server attempts to send packets as quickly as possible. Initially a number of packets are sent back-to-back regardless of the network capacity, as they are simply building up in the network buffer. When the network buffer becomes full, the rate at which packets can be sent to the network buffer matches the rate of transmission over the network.

As the transmission is reliable, the server knows how much data has been received by the client — this is all of the data that has been written to the network buffer less what is still in it, which is calculated assuming it is full.

By assuming that the first transmitted data was decoded as soon as it was received, the server knows, by dead-reckoning, how much data has been decoded at any later time, and so knows how much data is in the client's decoding buffer.

When the amount of data buffered in the client's decoding buffer reaches a threshold, say 30 sec, the server restricts the transmission rate to maintain this level of fullness.

The network throughput is estimated by counting bytes that have been sent and dividing by the time over which the measurement is made. This technique works satisfactorily provided the server is attempting to stream as quickly as possible. But when the amount of data in the decoding buffer exceeds a threshold, the server restricts its transmission rate to maintain a constant buffer fill. In this case, to make an estimate of the network throughput, the server will periodically ignore the client decoding-buffer-fullness threshold, and stream at full rate for a given period of time. So no feedback is required from client to server, beyond that needed by TCP, for the server to know how much data has been received by the client and the instantaneous network throughput.

18.4.3.6   *Play Stream Switching Decisions*

The server initially streams the play stream with the lowest bit rate, to allow the client to decode and present media to the user immediately while also building up the level of the decoding buffer to provide resilience to network impairments. If the network has sufficient capacity to support transmission of a higher rate, the server switches at an appropriate moment to streaming a higher rate play stream.

The server makes use of its knowledge of the amount of data buffered in the client's decoding buffer and the instantaneous network throughput to select the play stream to transmit to the client.

We have considered two criteria for determining when to switch to a higher rate stream. Firstly, the client should have sufficient data in its decoding buffer to be able to continue decoding and presenting media for a specified period of time. Secondly, the network throughput that has been achieved in the recent past should be sufficient to sustain streaming of the higher rate play stream indefinitely for live streaming, and until the end of the file when streaming from file, taking into account the data buffered at the client and the amount of data in the file still to be transmitted.

The aim is to avoid frequent switching as this can be more annoying to the user than constant quality at the lower rate.

Also to achieve this aim, the switching-down decision includes hysteresis relative to the switching-up decision. The Fastnets system considers just one criterion — switching down to the next lower bit rate play stream when the client no longer has sufficient data in its decoding buffer to be able to continue decoding and presenting media for a specified period of time. In the case of a configuration with three or more play streams, and the currently streamed play stream being the third or even higher rate play stream, this strategy does not result in an immediate drop to the bottom of the hierarchy. As access pictures only occur periodically, and the decoding buffer fullness would normally recover after an initial switch down, a second switch down would not usually be necessary.

18.4.3.7   *Streaming of Live Content*

In the case of live streaming, it is not possible to build up data in the client's decoding buffer if the first data is sent to the client with minimal delay from the time it was captured. Hence, the first data that is sent to the client is 'old' data, that is, data representing events that took place some time before the client connected. Then, as the decoding buffer fills, the most recent data in it becomes more and more recent, while the media presented to the user remains at a constant delay from the actual time of its occurrence.

The system is only acceptable for the streaming of live events where users are prepared to view the media with a constant delay of tens of seconds — users will generally accept such delays for the majority of uni-directional media streamed over IP networks.

18.4.3.8   Streaming from File

Streaming from file is simpler than live streaming. There is no need for circular buffers since data can be read from file as and when needed. The server, however, uses the same techniques to fill up the decoding buffer at the client and to switch between play streams. Trick modes, such as fast forward, fast reverse and random access are achieved using the intra stream.

Live streamed content can be made available for streaming at a later time by writing 'old' data in the circular buffers to file just before being overwritten. In addition to allowing users to view the content later, this functionality also allows a client to pause the presented media for an indefinite period of time, and continue streaming afterwards. It would also allow the user to fast forward after such a pause to catch up with the live stream.

18.4.3.9   Packetisation

We have implemented a proprietary packetisation derived from TPKT, with the header specifying the length of the packet, an RTP timestamp, and a stream identifier to allow audio and video to be multiplexed into a single TCP connection. This ensures synchronisation of audio and video transmission. If separate TCP connections were used, it is possible that they would respond slightly differently to network characteristics and achieve different throughputs, which would result eventually in vastly different amounts of data in the decoding buffers. Although these differences could be managed, the issue is totally avoided by using a single TCP connection.

18.4.4   Fastnets System Performance

A demonstration of the Fastnets system is described in this section.

A desktop Pentium PC is used to run the Fastnets encoder and server. The client is a Nokia 7650 mobile telephone, shown in Fig 18.8. In a typical configuration, to a telephone that supports four time-slots in the downlink, two switching streams are used, with bit rates of 10 kbit/s and 20 kbit/s.

Fig 18.8   Fastnets on the Nokia 7650.

The system performs as expected. Transmission starts with the intra stream and then switches to the 10 kbit/s play stream, where it stays for some time, accumulating data in the client as a result of actually transmitting faster than 10 kbit/s. Then when sufficient data has been accumulated, and the short-term average receiving rate is more than 20 kbit/s, it switches to the higher rate play stream.

At times during a lengthy session, occasional switches back to the lower rate play stream occur as a result of reduced network throughput. And very rarely, media presentation is interrupted because of a significant period during which the network could deliver no data to the client.

The overall effect during most sessions is that the user can view continuous media presentation, with occasional changes in quality, but no distortions of the type usually associated with bit errors and packet loss. Only very rarely are complete pauses in media presentation observed as a result of severe network impairments and loss of throughput.

18.5   Summary

This chapter has discussed a number of well-known and standardised techniques for audio-/video-streaming. It has also described the Fastnets multimedia streaming architecture for mobile IP networks. The system makes use of a hierarchy of independent video streams encoded according to ITU-T Recommendation H.264 and a bit-rate adaptive transmission technique to match the transmitted multimedia data to the instantaneous network capability. The system allows users to access multimedia content over GPRS networks, and in the near future 3G networks, using devices such as the Nokia 7650, and to experience continuous multimedia of consistent quality with minimal start-up delay.

References

1 Ghanbari, M.: '*Video coding — an introduction to standard codecs*', IEE Telecommunications (1999).

2 MPEG — http://mpeg.telecomitalialab.com/

3 Topiwala, P., Sullivan, G., Joch, A. and Kossentini, F.: '*Overview and performance evaluation of the Draft ITU-T H.26L video coding standard*', Proc SPIE Appl Dig Im Proc (August 2001).

4 http://standards.pictel.com/ftp/q6-site/0109_San/VCEG-N18.doc

5 IT Week: '*Codec frees video telephony*', — http://www.itweek.co.uk/Analysis/1139947

6 3rd Generation Partnership Programme — http://www.3gpp.org/

7 3GPP TS 26.071: '*Mandatory speech codec speech processing functions, AMR Speech Codec; general description*' — http://www.3gpp.org/

8 3GPP TS 26.171: '*AMR speech codec, wideband; general description*' — http://www.3gpp.org/

9 ISO/IEC 14496-3 (1999): '*Information technology — coding of audio-9 objects — Part 3: audio*' — http://www.iso.org/

10 Ghanbari, M.: '*Two-layer coding of video signals for vbr networks*', IEEE Journal on Selected Areas in Communications, **7**(5), pp 771-781 (June 1989).

11 Walker, M. and Nilsson, M.: '*A study of the efficiency of layered video coding using H.263*', The Ninth International Packet Video Workshop, New York (April 1999).

12 Kasai, H., Nilsson, M., Jebb, T., Whybray, M. and Tominaga, H.: '*The development of a multimedia transcoding system for mobile access to video conferencing system*', IEICE Trans Communications special issue on Mobile Multimedia Communications, **E00-B**(1) (January 2002).

13 Krasic, C., Li, K. and Walpole, J.: '*The case for streaming multimedia with TCP*', 8th International Workshop on Interactive Distributed Multimedia Systems (iDMS 2001), Lancaster, UK (September 2001) — http://www.cs.ubc.ca/~krasic/publications/krasic-idms2001.pdf

14 Transmission control protocol — http://www.ietf.org/rfc/rfc0793.txt?number=793

15 User datagram protocol — http://www.ietf.org/rfc/rfc0768.txt?number=768

16 Real-time transfer protocol (RTP) and RTP control protocol (RTCP) — http://www.ietf.org/rfc/rfc1889.txt?number=1889

17 TCP friendly unicast rate-based flow control — http://www.psc.edu/networking/papers/tcp_friendly.html

19

A SOCIAL HISTORY OF THE MOBILE TELEPHONE WITH A VIEW OF ITS FUTURE

H Lacohée, N Wakeford and I Pearson

19.1   Introduction

The history of the mobile telephone is as much about social and political developments as it is about the emergence of new technologies, standards and systems. As Agar has pointed out, the mobile telephone has been:

'... a way of rebuilding economies in eastern Europe, an instrument of unification in western Europe, a fashion statement in Finland or Japan, a mundane means of communication in the USA ... an agent of political change in the Philippines ...' [1]

Even within one culture, such as that of the UK, the mobile telephone may have multiple meanings. For example, it may be linked to youth culture through texting, business activities via data services, or motherhood as it allows the notion of shifting roles between work and home. Therefore any history of the mobile telephone must take on board the links between technical features and social relations, between functionality and cultural norms.

The growing penetration of mobile telephony and mobile communications in the UK suggests that none of us can remain immune to the social and cultural consequences in our everyday lives. Non-ownership of a mobile telephone has become an identity as important as ownership. While it took the domestic telephone approximately thirty years to migrate from an instrument found most often in the hallway of the home in the 1960s, to its ubiquitous position today in the living room, kitchen, and bedroom, the mobile telephone found its way into our pockets in less than half that time. Prior to 1985, no one in Britain had a mobile telephone, now most people own, or have access to one; in 2002 the World Telecommunications

Development Report stated that every sixth person in the world had a mobile telephone [2, 3]. The number of mobile subscribers around the world is likely to reach 1.4 billion this year, far more than the number of land lines (1.1 billion). Over the last decade in particular, mobile telephone use has escalated dramatically and for many people, the ability to communicate while on the move is now seen as essential to business, commerce, individual lifestyles and everyday social interaction.

A social history of the mobile telephone is not just a history of a shifting concept of mobility. In fact, the linking of mobile telephones with mobility may be premature; young people, for example may use text to communicate across very small distances, even across the room. Even if mobility is the key social concept, it is cross-cut by cultural behaviours and beliefs about intimacy, the role of public space, the changing place of women in the labour market, customisation of commodities, to name just a few. The mobile telephone has a global history in the sense that it has been developed or stalled by national politics as much as engineering challenges, exemplified by the different ways in which third generation (3G) licences were sold in the UK, France, Germany, Sweden and the USA [1]. Meanwhile there have been vast societal changes in terms of production and consumption, largely embedded in cross-national processes of globalisation. Political influences on design have been accompanied by huge social changes, such as the development of travel and the increasing car culture during the period of the mobile's early development.

19.2   Invention and Adoption

In the UK the first land mobile services were introduced in the 1940s and commercial mobile telephony began in the USA in 1947 when AT&T began operating a 'highway service' offering a radio-telephone service between New York and Boston.

In the mid 1950s the first telephone-equipped cars took to the road in Stockholm, the first users being a doctor-on-call and a bank-on-wheels. The apparatus consisted of a receiver, a transmitter and a logic unit mounted in the boot of the car, with the dial and handset fixed to a board hanging over the back of the front seat. With all the functions of an ordinary telephone, the car telephone was powered by the car battery. Rumour has it that the equipment devoured so much power that it was only possible to make two calls — the second one to ask the garage to send a breakdown truck to tow the car with its flat battery!

These first car-phones were too heavy, cumbersome, and expensive to use for more than a handful of subscribers and it was not until the mid 1960s that new equipment using transistors were brought on to the market. Weighing a lot less and drawing less power, mobile telephones now left plenty of room in the boot but were still the size of a large briefcase and still required a car to move them around (Fig 19.1).

Fig 19.1   Early car phone.

In the USA, in 1977, the Federal Communications Commission (FCC) authorised AT&T Bell Laboratories to install the first cellular telephone system. AT&T constructed and operated a prototype cellular analogue system and a year later, public trials of the new system were started. The FCC were convinced that cellular radio was practical but the sheer size of America presented problems that were not to be encountered elsewhere. Hence, although the first working examples of cellular telephony emerged in America, the Nordic countries of Denmark, Norway and Sweden and also Finland were soon to overtake their lead. In 1969 the Nordic Mobile Telephone group (NMT) had been set up to develop a cellular telephone system and by 1981, in Sweden, there were 20 000 mobile telephone users — higher than anywhere else in Europe. Cell-phones became standard kit for truckers and construction workers and by 1987 some were being sold for private use. Spain, Austria, the Netherlands and Belgium were quick to follow suit and order NMT services while Germany, France, Italy and Britain decided to design their own systems.

In 1979, the first commercial cellular telephone system began operation in Tokyo and by the mid-1980s there was a significant expansion of services offered to the general public that rapidly attracted large numbers of subscribers wherever services were available. In the UK, two companies were granted operating licences; Telecom Securicor Cellular Radio Limited (Cellnet) and Vodaphone. In January 1985, both companies launched national networks based on analogue technology and customers were able to avail themselves of the service using a mobile telephone the size of a brick (Fig 19.2).

Fig 19.2 An early analogue mobile 'brick'.

In the late 1980s there was a move to develop standards for a second generation (2G) of mobile telecommunications and digital technology; the global system for mobile telecommunications (GSM) was introduced throughout Europe in order to provide a seamless service for subscribers. Analogue technology was phased out in the UK in 2001 and digital technology (GSM) is now the operating system for 340 networks in 137 countries. Although Europe is the dominant user of GSM, it has also been accepted in other areas such as the Asia Pacific region.

Digital networks and an increase in the number of service providers to the market in the early 1990s served to further increase the number of subscribers and consumer popularity rose immensely. In 2000, 50% of the UK population owned a mobile and in 2001 almost 50% of British children aged between seven and sixteen had one.

Today at least 65% of households in the UK have access to a mobile and there are approximately 47 million mobile telephones in the UK [4]. It is estimated that during the next five years the percentage of calls made from mobile telephones will increase by 25% and user numbers are expected to rise to fifty million [4].

Huge advances in technology have undoubtedly played a role in the rapid and unprecedented take-up and widespread availability of mobile communication. Agar [1] maintains that in the 1990s technical trends, particularly miniaturisation and improvements in battery technology for example, triggered our mobile world. Once batteries became powerful and portable, mobile telephones could became small and light enough to carry around. There was a leap from car-phones to hand-phones and new, lighter, more portable designs proved attractive to a new customer base. It was at this point that the mobile telephone began to move from a business tool to the pervasive communications device we see today.

Much like the landline telephone, in its early inception the mobile telephone was an elitist device mainly used for business by middle and upper class males [5]. In the UK, in 1965, an exclusive (and expensive) service called System 1 was launched in West London that was used primarily by the chauffeurs of diplomats and company chairmen [1]. By 1967, use had trickled down to 14 000 privileged and wealthy users of the System 4 mobile telephone and, by 1981, the Post Office Act meant that the telephone business left the hands of the Post Office and was renamed British Telecommunications. In the 1980s the marketing of mobiles was aimed at business people and it was not until the early 1990s that mobile ownership among the general public began to take off.

Since then, just as the landline was quickly adopted for more sociable purposes [6], the mobile telephone has quickly been integrated into more and more aspects of our daily lives. By the late 1990s, across the world an economic split developed between those who paid for mobile telephones on a monthly contract basis and others, mainly the young and the poor who used pre-paid services. Pre-pay customers tended to use text messaging rather than voice calls because it was cheaper, and economic differences in use are still in evidence today. Text messaging is far less popular in the more affluent USA than in poorer countries like the Philippines which is the 'texting' capital of the world, although this has as much to do with network functionality as simple economics [1].

19.3   Developing Communication on the Move

Despite the rapid technological advances since the 1940s, the technology itself did not lead to an inevitable mobile revolution. Agar [1] suggests that organisational change may yield a clue as to the rapid contemporary adoption patterns:

'... there has been a correlation, a sympathetic alignment, between the mobile phone and the horizontal social networks that have grown the last few decades in comparison with older, more hierarchical, more centralised modes of organisation ...'

He goes on to argue that the mobile telephone activated, and was activated by, new forms of social network just as the mainframe computer of the 1950s was tied to a centralised, hierarchical, bureaucratic organisation. Central to contemporary social networks is the linking of communication to mobility. Geser has suggested that, seen within an evolutionary perspective, our ability to communicate has been shaped by two highly consistent physical constraints — geographical proximity (in order to initiate and maintain interactive relations), and a stable dwelling place (necessary for the development of more complex forms of communication and co-operation) [7]. We live in an increasingly complex world and geographical dispersion means we often need to maintain a significant proportion of our social network across distances where opportunities for face-to-face meetings are intermittent. Telephony, and in particular mobile telephony, is a key contributor to enabling us to keep track of, and successfully participate in, a set of relationships in a complex, social world. Much of the imagery and talk of the mobile telephone (particularly in early advertisements) reinforce the claim that it serves to free us from the need to be bound to a specific location (see Fig 19.3). Hence much of the significance and value of the mobile lies in apparently empowering human beings to engage in the fundamental activity of communication, free of the constraints of physical proximity and location specificity. Yet reading off social practices from advertisements is at best a risky activity. The fantasy of the mobile telephone, as with personal digital assistants, relies on consistently masking the 'hidden work of mobility' [8].

Fig 19.3   The 'freedom' of the mobile telephone.

The mobile telephone also facilitates communication that might not otherwise take place at all and hence it is the perfect tool for increased levels of social grooming, i.e. letting someone know that you are thinking about them [9]. Text messaging (SMS) is an excellent example — messages are often low in informational value but high in terms of social grooming. Our desire to keep in constant touch is perhaps never more in evidence than in our use of text messaging, and interaction need not be within a close social network. Text messaging surged in the week that war broke out recently in Iraq [10]. The BBC World Service was inundated with thousands of text messages from mobile telephone users across the globe wanting to express their views about the war. Nigel Chapman, deputy director of the BBC World Service said:

'... suddenly text messaging appears to have moved on from personal communication to personal statement. People have strong views about the war and are using the technology they use every day to tell us. New technologies are giving us a level of interaction with our audiences that we have never seen before ...'

Text messaging was an accidental success that took the mobile industry by surprise; there was very little promotion or mention of SMS by network operators until after it had taken off. SMS was a user triumph, particularly among young adolescents. Text messaging from the tiny keypad of the mobile telephone was and is a cumbersome exercise, but, paradoxically, because entry barriers to learning to use the service are high, or at least, higher than making a voice call, adolescents saw this as an advantage in that it enables them to exclude adults [11]. Allied to exclusivity is the fact that young people evolved a new alphabet around text messaging that makes messages virtually unintelligible to outsiders. The cost of sending a text message is also pertinent to this group in that it costs less to send a text than to make a call, and, where finite funds and the need to maintain a wide social circle are at odds, this is of consequence. It should also be noted here that text messaging is one of the few services in consumer history that has grown rapidly without a corresponding decrease in pricing. The price of SMS has remained steady and is likely to do so until networks can deal with the inevitable increase in message volume that would accompany reduced prices. Costed per character it is currently one of the most expensive ways for a user to communicate text across digital networks, suggesting that a strictly economic analysis of texting would not capture the rich social norms of the activity.

In many social situations, text messages are far less intrusive than a phone call and, while they have some of the advantages of e-mail (the recipient can choose when and whether to respond), they are far more accessible. As a result there is a very low threshold for sending messages and while the informational content is often minimal, it is highly valued as social grooming. Plant [12] also points out that texting is particularly popular with individuals and cultures who have a tendency to be reserved, because the necessary brevity of messages provides the opportunity to be direct, informal, and even cheeky.

'She' [13] reports that this is highly valued by teenagers in Bangkok and Thailand because it avoids the necessity to voice feelings and thoughts and ice can be broken without the risk of embarrassment.

19.4   The Emergence of Mobile Telephone Behaviour

The belief of technology companies is that successful technologies owe a large part of their success to the fact that they fulfil or enhance an existing human need, or fit well into an already well-established social context. This serves to shape the way that technology will be used. Nevertheless the uses for new technologies may be quite unexpected. For example, video was developed primarily as a tool for business but has been widely adopted as a child-minding device to entertain children while carers are busy with other tasks. As Geser [7] points out, users gradually change their habits and learn to use new technology in a variety of ways across an increasing range of situations, but it is very common for them to be unable to predict their future usage patterns accurately. For example, few would have predicted that the mobile telephone might potentially be used as an instrument for divorce but it is interesting to note that in 2001, senior Islamic figures in Singapore ruled that Muslim men cannot divorce their wives by sending a text message saying 'talaq' (Muslim men are able to divorce their wives by saying the word 'talaq' — 'I divorce you' — three times [14]). While technological innovations in general have been the focus of a wealth of research, telephony, and more specifically mobile telephony, is only just beginning to be studied in any depth. And yet, as noted by several researchers [7, 15], in a global context mobile telephony is used by a far broader stratum of the population than PCs and the Internet. As with many technologies, usage expands over time as the technology is adopted into an increasing spectrum of circumstances. For example, women who typically might have said that they needed a mobile telephone for emergencies when travelling alone at night quickly found that it served as an ideal instrument for remote mothering [16].

Given the capacity of the mobile telephone to retain significant social relationships over distance it is not surprising to find that in the UK, when children leave home for the first time to go to university, parents often supply a mobile telephone as part of the essential 'leaving home' kit. As Geser [15] describes, this helps to cushion the child from the potentially traumatic experience of living in a foreign environment by enabling them to remain tightly connected to loved ones at home; hence he describes this function of the mobile telephone as 'a pacifier for adults.'

Mobile telephones are also used to fill up 'void' spaces in time, for example when travelling, or waiting at a bus stop. Plant [12] reports that in Japan, where arriving in good time for an appointment is a very important part of social etiquette, many Japanese use their mobile telephone to while away the time they have gained by being early. As Plant points out, this means that other means of killing time such as reading books and newspapers are losing out to the preferred 'keeping in touch.' In the UK, anyone travelling on a train will be familiar with the cacophony of different voices announcing 'I'm on the train', or the rush to mobile communication if there is a delay in their journey. Indeed, the infringement on the peace and quiet of other passengers has led to the creation of 'quiet carriages' where the use of mobile telephones and personal stereos are banned. Just as netiquette was touted as a rules-based way of governing communication on the Internet, there is a development of what might be called 'm-etiquette' for mobile telephone use. Infringement of these norms has social consequences. The temptation to talk on the telephone while driving long distances is so prevalent that legislation has been introduced in many countries to curtail this practice.

Much of mobile voice call behaviour has been explained in terms of the potentially unstable or unknown sense of place between communicators. When making a call to a friend on a landline we usually know something of their environment, have knowledge of other family members who may be present, and the kind of activities that they may be engaged in at the time of the call [17]. However, in making a call to a mobile telephone this information is absent — hence the question 'where are you?' enables a communicative context to be established. As Chihara [18] points out, responses to this question may of course be intentionally false, but, nonetheless, a shared context of supposed location is still created. An increasing emphasis on the consumption of place is evident in the work of technologists seeking to develop location-based services. These both reflect and reinforce the relationship between actual geographical locations and mobile telephone behaviour.

Far from place becoming less important, these developments suggest that location will become accentuated.

19.5   The Mobile Telephone in a 'Risk Society'

Widespread use of a relatively new technology inevitably raises questions concerning health and mobile telephone safety has been the subject of much public debate. Conflicting reports about possible health risks of mobiles appeared in the late 1990s and are centred on emissions of radiofrequency (RF) radiation from the handset and from base-stations that receive and transmit the signals. Public concerns about possible ill effects from mobile telephones, base-stations and transmitters led to a UK Government decision in 1999 to establish an independent expert group to examine any possible adverse effects. This group used input from a wide variety of sources across the UK and abroad to produce a report in 2000 known as the Stewart Committee Report [19] and points out that:

'... the balance of evidence does not suggest mobile phone technologies put the health of the general population of the UK at risk ...' [20]

Although this report would suggest little evidence to support the idea that carrying a mobile telephone in a trouser or jacket pocket is a health risk, in September 2002 jean manufacturer Levi Strauss launched a new line of trousers which it says protects the wearer against any radiation emitted by a handset.

Although Levi say that they are not implying in any way that mobile telephones are dangerous, the company decided to launch the new range after extensive market research showed that fashion conscious consumers were also health conscious. The trousers, are fitted with pockets which have an 'anti-radiation' lining and are expected to go on sale early in 2004. The World Health Organisation has suggested that there is no need for this lining and this is supported by the mobile telephony industry who warned that the lining could stop mobile handsets from working properly. The new product has also provoked criticism suggesting that the company is playing on consumer fears. A spokesperson for Levi Strauss said that it is merely responding to consumer desire. Call volume would seem to suggest that people, fashion conscious or otherwise, do not take any supposed health risks very seriously at the point of use. However, the mobile has become part of a culture of technological risk, in which everyday practices are conducted in the context of competing scientific claims about harm and danger.

While mobile telephones have received much negative attention in the press there are also numerous reports in newspapers world-wide describing how mobiles have saved lives. According to the Cellular Telecommunications and Internet Association, over 140 000 calls are placed to the emergency services from mobile telephones every day. The prevalence of mobile telephones means that in an emergency situation the chance of someone witnessing the event having a mobile and being able to contact the emergency services is high. Examples of mobiles being used in this context are too numerous to mention but certainly they are reported across the globe — from being trapped in the rubble of an earthquake in El Salvador [21] to a car crash victim in British Columbia [22], to saving the life of a British explorer trapped near the North Pole [23]. Recently, the first case of a text message being used to locate a casualty in the UK was also reported [24].

A culture of public reporting that goes further than accidents and emergencies has also emerged; many people ring in to radio stations to report traffic hazards, congestion, and bad weather conditions. This has led to claims that mobile telephones are making a significant contribution to social capital by providing a means for people to become more active citizens by engaging in small acts of social responsibility [25].

19.6   The Object as Fashion Accessory

The mobile handset has become a widely recognised consumer artefact. Particularly notable in this regard is the way in which faceplates have become circulated independently of the handset in order to enable constant customisation, for example to match a colour scheme, or to show allegiance to a football club, or a popular icon. Many advertisements in the UK for new mobile telephone upgrades play on the public's understanding of technological novelty. The idea that the mobile is on constant show and is therefore a fashion accessory has fed into an advertising rhetoric of continual upgrading to avoid being shamed.

Recently advertising has particularly focused on the youth market, both in terms of handsets and services. Although it is not obvious how this might have a positive impact on the health of the nation's teenagers, there is some evidence to suggest that a decline in teenage smoking [26] is correlated with mobile telephone ownership. There was a sharp decline in smoking among boys and girls aged 15 in the late 1990s during which time mobile telephone ownership sharply increased. The British Medical Journal [27] reported that the fall in youth smoking and the rise in ownership of mobiles among adolescents are related because the functions that smoking offers to teenagers are similar to those offered by owning a mobile telephone. Mobiles consume teenagers' available cash, particularly topping up pay-as-you-go cards. It is also argued that the mobile is an effective competitor to cigarettes in the market for products that offer teenagers adult style and adult aspiration because the marketing of mobiles is rooted in promoting self-image and identity. As ownership increases, mobiles become essential for membership of peer groups that organise their social life on the move — hence the need to own a mobile provides vigorous competition for money that might otherwise be spent on cigarettes.

The appeal of mobile telephones to teenagers goes far beyond fashion and text messaging; more than anything it provides them with a private, personal piece of technology that enables them to exclude adults from their communication circles and practices.

An important issue in teenagers' lives is the process of emancipation from parents and in some respects, owning a mobile can be seen as a rite of passage. Fortunati [28] argues that as a result, the family social system is weakened because children are now managing their own communications networks and Plant [12] notes that mobile telephone use among children and teenagers in Japan enables many to lead lives that are totally opaque to their parents.

One of the dangers in approaching the mobile telephone as an artefact of mass consumption is that we can assume that the buying of the new device is when 'meaning-making' happens. However, the social relations around consumption extend out from this point of purchase, both before the sale and afterwards. This would suggest that more attention needs to be paid to research on the circulation of telephones, for example as they are bought, and sold as 'second-hand', or are passed through friendship or family networks. Furthermore it is clear through the circulation of faceplates that the mobile may not always be consumed as a whole object, but needs to be considered as an artefact which may be disaggregated. This is also evident in the way SIM cards may circulate independently of their original handsets or owners. The flip side of the conventional talk of consumption is the role of mobile telephones in supporting all kinds of illegal activities — understandably hard to research, but significant to their social history.

19.7   Global Contrasts

One of the dangers in presenting descriptions of behaviour and consumption in a European context is that the pattern of adoption shows vast global variation, as the quotation from Agar [1] indicated at the beginning of this chapter. As noted by Townsend [15], one major impact of the mobile telephone is its capacity to include partly illiterate mass populations in less developed countries.

Mobile telephones are undoubtedly changing how, when, where, how often, to whom, and about what we communicate, but if changes in western society have been remarkable, elsewhere in the world they are even more dramatic. For example, in Britain we have long enjoyed an efficient network of public and private access to telecommunications but many other countries, particularly poorer countries, have not. Traditional state telecommunications monopolies in the former Soviet Union, for example, made people wait for months or even years to have a telephone connected and vast regions were not connected at all. Mobile telephones, and, in particular, pay-as-you-go schemes, have enabled millions of people world-wide to bypass landline-based systems that are too expensive, poorly maintained, difficult to access, or even non-existent. For example only 3% of Africans have a mobile telephone but they account for 53% of all telecommunications subscribers on the continent. In Gabon there are 37 000 land lines but mobile subscriptions exceed 250 000, and China has more mobile subscribers than the USA and Canada put together at 145 million.

In rural India, mobile telephone demand is growing rapidly, and its usage flourishes where fixed-line service is non-existent. In Kerala, for example, fishermen use mobile telephones to compare the prices they might receive for their catch at different ports. One fisherman claimed that his profit on each eight-day fishing run in his trawler had doubled because he was now able to use his telephone to compare prices at Cochin with those at Quilon, a port 85 miles away [29]. Similarly, Plant [12] describes a Somali trader in Dubai who exports small electrical goods who says that the mobile is his livelihood — 'no mobile, no business'. In Europe browsing Web pages via a mobile phone has had a very luke warm reception but among market traders in Senegal it is proving very popular. A project called Manobi [24] run by French and Senegalese entrepreneurs gathers information about the prices of food and goods being sold in Dakar and uploads them to its central database. Farmers can use their mobiles to dial in to the server via WAP and find out what the prices are at different markets so that they can get the best return for their produce. Many of the farmers using the system are illiterate but they are familiar with using calculators and treat the mobile telephone in a similar fashion. Even though the project is in its early stages it is already having an effect on the way the farmers grow crops and now many of them are only producing crops for particular markets and will only bring produce to a market where they know they will get a good price.

19.8 Social Synopsis

The mobile telephone appears to be a global artefact, attracting attention from technologists and policy-makers, as well as consumers. In this chapter we have suggested that the multiple dimensions of social influence stem not just from the technological developments of smaller batteries and compatible standards, but from the ways in which these become integrated into social changes which are already under way and are supported by the new kinds of communicative forms which the mobile telephone permits. A measure of the mobile as a cultural icon in the UK is the role it plays in contemporary conceptual art, such as the wall of mobile telephones constructed by Thomson and Craighead [30]. In thinking about cultural importance it is crucial to look beyond traditional technical functionality of, for example, the SIM card, the network and the keypad. The handset as a symbol of status and fashion does not necessarily require connection to a network.

This chapter necessarily provides only snapshots in social history. Social research on mobile telephones has only become a sustained endeavour in the last 4-5 years, and many topics, particularly in a non-Western context, remain underexamined. Among the most interesting of these is mobile use by specific local subcultures which are outside the marketers' gaze. It is likely that if a fuller social history were written we would discover even more competing accounts of the emerging role of the mobile, and richer stories about the ways in which it has been resisted or rejected.

Currently much research has been funded by industry and this has led to a strong emphasis linking telephones to consumption, and a forward-looking stress on new business offerings such as mobile picture messaging. Yet looking back to the earlier developments, particularly through previous conflicts about standards or unanticipated functions, is equally important in assessing the on-going impact not just on individual behaviour, but on the social structure of contemporary societies.

The final section looks at what might happen over the next ten years.

19.9 A View of the Next Decade of Mobile Communications

The coming decade will see rapid convergence of bio-technology, materials science (including nanotechnology), artificial intelligence, robotics, computing and tele-communications. This convergence will create whole new classes of device that we can scarcely imagine today. BT is already discussing with other companies the potential for electronic devices printed on to or even into our skin. These would enable video displays and tattoos, smart make-up and perfume, medical monitors, computer interfaces and communications. It will eventually be possible to print a telephone into your wrist! We can expect technology of all kinds to become cheaper and more ubiquitous, resulting by 2010 in a smart environment bristling with sensors, processing and communication, with displays everywhere, and millimetre accurate positioning.

It is reasonable to assume that a number of high-speed wireless networks will be available that do not directly charge users for communication. This will force communications companies to move further up the value chain and charge for added value, that is value that we can add to people's lives or businesses. Fortunately, the explosion of new technologies resulting from convergence will provide us with many opportunities to do just that.

There are strong hints that we will soon have wirelessly networked credit card sized polymer displays for just a few pounds. These could be used for hundreds of new applications.

Apart from being a light, portable TV, they would have extensive marketing capability, perhaps giving us the video cornflake packets illustrated in the film *Minority Report*. Devices such as that would also give us smart tickets that can guide you through an airport terminal and keep you up to date with flight progress, while acting as an advertising medium to show you special offers in the shops as you walk past.

They could also be used as compact notebooks, diaries, and mail terminals. They will be used for clothing and body adornment, portable games and entertainment. They are very likely to be used for social and tribal display, assisting networking in nightclubs, and marking tribal allegiance on the street.

It is the social uses of telecommunications that have traditionally been overlooked in favour of one-to-one communication. Most people exist as part of several social groups, and communications and relationships between members of these groups are a potentially large, mostly untapped, revenue source. People can already send text messages to groups of friends, but we must go much further. We should expect instant voice messaging, coherence utilities and tribal maintenance. A quick glance at a display when we are shopping should tell us if a close friend has also come to town and is within a short distance. Adding value to our lives by improving and managing social interaction opportunities will replace much of the revenue lost by the dropping of call charges.

Mobility itself will not contribute any extra volume to devices in the far future, so we should not expect mobility to have any effect on form factor. At presumed low cost, wireless connectivity will be taken for granted in any electronic gadget. Design will become increasingly important as the technology is commoditised.

We may expect that as devices become less bulky, there will be less need for integration of functionality. Furthermore, ubiquitous high-speed wireless (probably over LANs in the long term) will ensure that remote functionality is available even via modest gadgets. So people may carry a range of different devices with them for different purposes, without too much concern for their built-in function, and of course these will interwork well. Some possible devices of the future are shown in Fig 19.4.

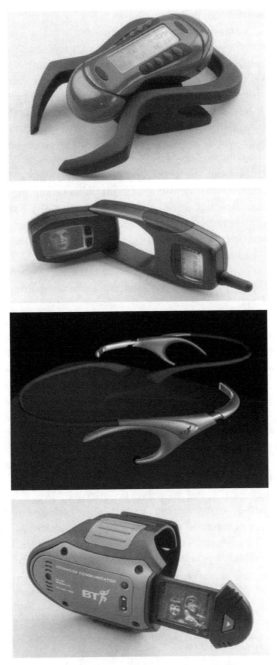

Fig 19.4 Examples of some possible future devices.

Increased connectedness makes networks more powerful. In the future, most people would be constantly accessible (when they want to be) everywhere, and 'always on.' This makes it much easier to make strong social networks. Pressure groups and ideological communities in particular will benefit, able to instantly co-ordinate the behaviour of large groups of supporters, and canvass new support. Network communities are already making good use of text messaging, but future networks will link not only the people, but all of their IT resources, seamlessly. Messaging would be instant and virtually free, so current barriers would be very much reduced. Large global pressure groups could therefore wield their increased muscle much more easily and effectively.

But having ubiquitous access to the Internet will make the Web more useful too. Today, it is still painfully difficult to access the Web on the move at any speed — only a few locations offer high-speed access, and this usually has to be bought first. As free access everywhere becomes routine, we will make much more frequent use of the Web, accelerating the virtuous circle linking the number and value of users and the quality and usefulness of applications. This of course will feed back into other areas of development. It will be seen that free access actually increases the revenue for the network providers and communication companies, who will adapt to new business models.

An area that is often overlooked is the creation of applications by users themselves. Text messaging may have been created by engineers, but its take-up was still surprising to most (a text-messaging-phone invention in 1991 was dismissed as irrelevant by senior BT managers — why would anyone want to send text when they can talk to someone?). However, we can expect that artificial intelligence and generally better software tools will enable almost anyone to create new applications with the technologies freely available. This would cause a huge increase in the range of applications, most of which would not come directly from engineers in IT companies. If we are to make the most of the potential opportunities, we must have platforms on which customers can easily do this, and more importantly, try to guess the areas of life that our customers are likely to address, and the things they are trying to solve or improve. We might be able to do something even better, even earlier. This role is perhaps more appropriate to psychologists than engineers. Furthermore, if the domain of social interaction services proves to be as lucrative as hoped, engineering might have a much more direct social focus, which might help to attract more women engineers.

References

1   Agar, J.: '*Constant Touch: A Global History of the Mobile Phone*', Icon Books, Cambridge (2003).

2   ITU: '*World Telecommunications Development Report: reinventing telecoms*', (2002).

3   ITU: '*Trends in Telecommunication Reform: effective regulation*', (2002).

4   Mobile Operators Association — http://www.mobilemastinfo.com/information/history.htm

5   Roos, J. P.: '*Sociology of cellular telephone: the Nordic model*', Telecommunications Policy, **17**(6) (August 1993).

6   Fischer, C.: '*America Calling: A Social History of the Telephone to 1940*', University of CA Press (1992).

7   Geser, H.: '*Towards a sociological theory of the mobile phone*', (2002) — http://www.socio.ch/mobile/t_geser1.htm

8   Churchill, E. and Wakeford, N.: '*Collaborative Work on the Move*', in Brown, B., Green, N. and Harper, R. (Eds): 'Wireless World', Springer Verlag (2001).

9   Haddon, L.: '*The Social Consequences of Mobile Telephony*', Framing, Oslo (2000).

10 Guardian unlimited (28 March 2003) — http://www.guardian.co.uk/

11 Rautiainen, P.: '*Mobile communication of children and teenagers*', Case, Finland, Tampere (1997—2000) — http://www.telenor.no/

12 Plant, S.: '*On the mobile: the effects of mobile telephones on social and individual life*', (2000) — http:/www.motorola.com/mot/documents/0,1028,333,00.pdf

13 'She' magazine — http://www.she.co.uk/

14 BBC World Service report (2001) — http://www.bbc.co.uk/worldservice/

15 Townsend, A. M.: '*Life in real-time city: mobile telephones and urban metabolism*', Journal of Urban Technology, **7**(2), pp 85-104 (2000).

16 Ling, R.: '*Traditional and fixed and mobile telephony for social networking among Norwegian parents*', in Elstrom, L. (Ed): '*Human Factors in Telecommunications*', 17th International Symposium, pp 209-256 (1999).

17 Lacohée, H. and Anderson, B.: '*Interacting with the telephone*', in Kraut, R. and Monk, A. (Eds): '*Home Use of Information and Communications Technology*', Special Issue of the International Journal of Human-Computer Studies, **54**(5), pp 665-699 (May 2001).

18 Chihara, M.: '*Lying on the go*', Boston Phoenix (16 March 2000) — http://www.bostonphoenix.com/

19 Stewart Committee Report — http://www.iegmp.org.uk/

20 Sir William Stewart, Chairman IEGMP (April 2002) — http://www.iegmp.org.uk/report/announcement.htm

21   Wireless News Factor — http://www.wirelessnewsfactor.com/perl/story/7311.html

22   Wired — http://www.wired.com/news/wireless/0,1382,52363,00. html

23   vnunet — http://www.vnunet.com/News/1131965

24   BBC — http://news.bbc.co.uk/

25   Chapman, S.: *'Lifesavers and cellular samaritians: emergency use of cellular (mobile) phones in Australia'* (1998) — http://www.amta.org.au/files/issues/pdfs/emergency.pdf

26   Office for National Statistics: *'Drug use, smoking and drinking among teenagers in 1999'*, London, ONS (2000).

27   Charlton, A. and Bates, C.: *'Decline in teenage smoking with rise in mobile phone ownership: hypothesis'*, British Medical Journal, **321**, p 1155 (2000).

28   Fortunati, L.: *'The mobile phone: new social categories and relations'*, University of Trieste (2000) — http://www.telenor.no/prosjekter/Fremtidens_Brukere/seminarer/ mobilpresentasjoner/

29   New York Times (4 August 2001) — http://www.nytimes.com/

30   Thomson and Craighead — http://www.thomson-craighead.net

ACRONYMNS

1G	first generation
2G	second generation
3G	third generation
3GPP	3rd Generation Partnership Programme
4G	fourth generation
AAA	authentication, authorisation and accounting
ACK	acknowledgement
ADSL	asymmetric digital subscriber line
AES	advanced encryption standard
AIFS	arbitration inter-frame space
AKA	authentication and key agreement
ANP	anchor point
ANWR	access network wireless router
AODV	*ad hoc* on-demand vector
AP	access point
AP	adaptation path
AP	application provider
APE	*ad hoc* protocol evaluation
API	application programming interface
ARPU	average revenue per user
ARQ	automatic repeat request
ASIC	application specific integrated circuit
ASK	amplitude-shift keying
BAN	BRAIN access network
BAR	BRAIN access router

BC	bearer control
BCMP	BRAIN candidate mobility protocol
BGP	border gateway protocol
BMG	BRAIN mobility gateway
BPF	bandpass filter
BPSK	binary phase shift keying
BQB	Bluetooth Qualification Body
BRAIN	Broadband Radio Access for IP-based Networks
BRAN	broadband radio access network
BRENTA	BRAIN end terminal architecture
BREW	Binary Run-time Environment for Wireless (Qualcomm)
BSS	basic service set
BSS	business support system
BTS	base-station transceiver system
CA	collision avoidance
CABAC	context-based adaptive binary arithmetic coding
CAR	committed access rate
CAVLC	context-adaptive variable length coding
CC	call control
CCC	conference call control
CD	collision detect
CDK	complementary code keying
CIS	Commonwealth of Independent States
CMOS	complementary metal oxide semiconductor
CEPT	European Conference of Post and Telecommunications
CLDC	connected limited device configuration
CP	content provider
CS/CCA	carrier sense/clear channel assessment
CSD	circuit-switched data
CSMA	carrier sense multiple access
CTS	clear to send
DAB	digital audio broadcasting

DARPA	Defense Advanced Research Project Agency (USA)
DCF	distributed co-ordination function
DCS	dynamic channel selection
DECT	digital enhanced cordless telecommunications
DFS	dynamic frequency selection
DHCP	dynamic host configuration protocol
DIFS	DCF inter-frame space
DMAP	DECT multimedia access profile
DOCSIS	Data Over Cable Service Interface Specification
DoD	Department of Defense (USA)
DOM	Document Object Model
DoS	denial of service
DPRS	DECT packet radio service
DSDV	destination-sequenced distance-vector
DSL	digital subscriber line
DSM-CC	digital storage media command and control
DSR	dynamic source routing
DSSS	direct sequence spread spectrum
DTH	direct to home
DTI	Department of Trade and Industry
DVB	Digital Video Broadcast Forum
DVB-T	terrestrial digital video broadcasting
E2ENP	end-to-end negotiation protocol
EAM	electro-absorption modulator
EAP	extensible authentication protocol
EAPOH	EAP over HIPERLAN
EAPOL	EAP over LAN
ECA	European Common Allocation (Table)
ECC	Electronic Communications Committee
EDCF	enhanced distributed co-ordination function
EHF	extremely high frequency
ESI	enhanced service interface

ETSI	European Telecommunications Standards Institute
FCC	Federal Communications Commission (USA)
FEC	forward error correction
FOMA	freedom of multimedia access
FTTH	fibre to the home
FTP	file transfer protocol
GAP	generic access profile
GGSN	gateway GPRS support node
GLC	gateway location centre
GloMo	global mobile information systems
GPRS	general packet radio service
GPS	global positioning system
GSM	global system for mobile telecommunications
GUI	graphical user interface
HF	high frequency
HIPERLAN	high performance radio local area network
HiSWAN	high speed wireless access network
HLR	home location register
HVS	human visual system
IAPP	inter access point protocol
IDL	interface definition language
IETF	Internet Engineering Task Force
IFS	inter-frame space
IIOP	Internet interoperability protocol
IMSC	instant messaging service centre
IN	intelligent network
INAP	IN application part
InP	indium phosphide
IP	Internet protocol
IP$_2$W	IP to wireless (interface)
IPSec	IP security
ISA	industry standard architecture

ISM	industrial, scientific and medical
ISO	International Standards Organisation
ISP	Internet service provider
IST	information society technologies
ITU-T	International Telecommunication Union
IVR	interactive voice response
IWF	interworking function
JAIN	Java APIs for intelligent networks
J2ME	Java2 micro edition
JCC	Java call control
JTAPI	Java telephony application programming interface
JTWI	Java technology for the wireless industry
LAN	local area network
LBC	low bit-rate coding
LDAP	lightweight directory access protocol
LF	low frequency
LoS	line-of-sight
MAC	medium access control
MANET	mobile ad hoc network
MEMS	micro-electrical mechanical system
MF	medium frequency
MF-TDMA	multiple frequency time division multiple access
MIDP	mobile information device profile
MIND	Mobile IP-based Network Developments
M-ISP	mobile Internet service provider
MMAC	multimedia mobile access communications
MMAPI	mobile media API
MMCC	multimedia call control
MMR	MIND mobile router
MMS	mobile multimedia messaging service
MMSC	multimedia messaging service centre
MN	mobile node

MoU	memorandum of understanding
MPCC	multiparty call control
MPEG	moving picture experts group
MQW	multi-quantum well
MVNO	mobile virtual network operator
NAT	network address translation
NGS	next generation switch
NMT	Nordic Mobile Telephone (group)
NO	network operator
NPV	net present value
NTDR	near-term digital radio
NTIA	National Telecommunications and Information Administration
OFDM	orthogonal frequency division multiplexing
OMA	Open Mobile Alliance
OOK	on-off keying
OSA	open service architecture
OSS	operational support systems
OTA	over the air
PAM	presence and availability management
PAM	pulse amplitude modulation
PAN	personal area network
PAWLAN	public access wireless LAN
PCF	point co-ordination function
PCI	peripheral component interconnect
PCMCIA	Personal Computer Memory Card International Association
PCP	portal context provider
PDA	personal digital assistant
PDH	plesiochronous digital hierarchy
PEP	performance enhancing proxy
PHB	per hop behaviour
PIM	personal information management
PIN	personal identity number

PN	pseudo-random noise
PoP	point-of-presence
PP	portal provider
PPM	pulse position modulation
PRNET	packet radio network
PSD	power spectral density
PSK	phase shift keying
PWA	personal wireless assistant
P-WLAN	public WLAN
QAM	quadrature amplitude modulation
QCIF	quarter common intermediate format
QoS	quality of service
QPL	qualified product list
QPSK	quadrature phase shift keying
RA	Radiocommunications Agency
RCS	return channel by satellite
RF	radio frequency
RLAN	radio local area network
RoF	RF-on-fibre
RP	resource proxy
RRM	radio resource management
RSCom	Radio Spectrum Committee
RSPG	Radio Spectrum Policy Group
RTCP	RTP control protocol
RTP	real-time transport protocol
RTS	request to send
RTT	round-trip time
SALT	speech application language tag
SB3G	systems beyond 3G
SC	service control
SCP	service control point
SCS	service capability server

SDH	synchronous digital hierarchy
SDP	session description protocol
SDPng	next generation SDP
SDR	software defined radio
SGSN	serving GPRS support node
SHF	super high frequency
SIG	Special Interest Group
SIM	subscriber identity module
SINR	signal to interference plus noise ratio
SIP	session initiation protocol
SLA	service-level agreement
SMS	short messaging service
SMSC	short message service centre
SOAP	simple object access protocol
SP	service provider
SURAN	survivable adaptive radio network
TCAP	transactional capabilities application part
TCP	transmission control protocol
TDD	time division duplex
TDM	time division multiplexing
TDMA	time division multiple access
TINA-C	Telecommunication Information Networking Architecture Consortium
TM	transverse magnetic
TORA	temporally ordered routing algorithm
TPC	transmission power control
UDDI	universal description, discovery and integration
UDP	user datagram protocol
UHF	ultra high frequency
UML	unified modelling language
UMTS	Universal Mobile Telecommunications System
U-NII	Unlicensed National Information Infrastructure

UTRAN	UMTS terrestrial radio access network
UWB	ultra-wideband
VHF	very high frequency
VIE	visual information engineering
VLF	very low frequency
VLR	visitor location register
VoIP	voice over IP
VPN	virtual private network
VRL	virtual radio link
W3C	World Wide Web Consortium
WAP	wireless application protocol
W-CDMA	wideband code division multiple access
WDM	wavelength division multiplexing
WEP	wired equivalent privacy
WiFi	wireless fidelity
WIG	Wireless Interworking Group
WISP	wireless Internet service provider
WLAN	wireless local area network
WMA	wireless messaging API
WRAN	wireless radio access network
WRC	World Radio Conference
WSDL	Web service description language
WTAI	wireless telephony application interface
X+V	XHTML+VoiceXML
ZRP	zone routing protocol

INDEX